26th Annual Meeting of the North American Membrane Society (NAMS 2016)

Bellevue, Washington, USA
21-25 May 2016

ISBN: 978-1-5108-5115-3

Printed from e-media with permission by:

Curran Associates, Inc.
57 Morehouse Lane
Red Hook, NY 12571

Some format issues inherent in the e-media version may also appear in this print version.

Printed by Curran Associates, Inc. (2017)

For permission requests, please contact North American Membrane Society (NAMS) at the address below.

North American Membrane Society (NAMS)
University of Arkansas
Ralph E. Martin Department of Chemical Engineering
3202 Bell Engineering Center
Fayetteville, AR 72701-1201
USA

Phone: +1 479 575-7926
Fax: +1 479 575-7926

names@uark.edu

Additional copies of this publication are available from:

Curran Associates, Inc.
57 Morehouse Lane
Red Hook, NY 12571 USA
Phone: 845-758-0400
Fax: 845-758-2633
Email: curran@proceedings.com
Web: www.proceedings.com

Page:

GENERAL SUBMISSIONS
BIOMIMETIC MEMBRANES

1 **Biomimetic and Bioinspired Membranes: Opportunities, Challenges and New Directions**

Manish Kumar[1], *Yuexiao Shen*[2], *Patrick O. Saboe*[3], *and Tingwei Ren*[4]
(1)Department of Chemical Engineering, Pennsylvania State University, University Park, PA, (2)Department of Chemical Engineering, The Pennsylvania State University, University Park, PA, (3)Department of Chemical Engineering, Pennsylvania State Univerity, University Park, PA, (4)Chemical Engineering, the Pennsylvania State University, State College, PA

2 **Functional DNA As Versatile Building Blocks for Biomimicking Membranes.**

Thomas Schafer
Institute for Polymer Materials (Polymat), University of the Basque Country, San Sebastián, Spain

Mangrove Membranes.

Mariusz Grzelakowski
Applied Biomimetic Inc., Cincinnati, OH

Mangrove Membranes

Mariusz Grzelakowski[†]

Applied Biomimetic Inc. 2180 E Galbraith Road 45237 Cincinnati Ohio, USA

(Email: mg@appliedbiomimetic.com, telephone: +1 513 558 6090)

Porins based membranes have been considered a promising filtration technology and have been intensively studied over the last few years[1]. Most common approach is to embed aquaporin, using lipid or block copolymer (BCP) vesicles, into membrane forming polymers on the porous substrates[2]. This approach has shown its limitations on number of occasions. The commercial membranes shown up to date failed to utilize selectivity and permeability of aquaporin or other porins in the separation layer. We have chosen an alternative approach to formation of active layer, where block copolymer vesicles serve as a vehicle for protein delivery and a coting material at the same time. Cross-linked block copolymer vesicles were successfully used to form separation layers. Membranes built this way were characterized by significantly increased water flux if polymer vesicles were reconstituted with aquaporins.

Here we will discuss the improvements to the original proof of concept[3], novel filtration membranes and scale up activities for production of membranes utilizing the functions of membrane proteins.

References

1. M. Kumar, M. Grzelakowski, J. Zilles, M. Clark, W. Meier. Highly permeable polymeric membranes based on the incorporation of the functional water channel protein Aquaporin Z. *Proceedings of the National Academy of Sciences of the United States of America* (2007), 104(52), 20719-24.
2. Desalination by biomimetic aquaporin membranes: Review of status and prospects C.Y. Tang, Y. Zhao, R. Wang , C. Hélix-Nielsen, A.G. Fane DES-11385; Desalination 2012
3. M. Grzelakowski, M. F. Cherenet, Y.Shen and M. Kumar. A framework for accurate evaluation of the promise of aquaporin based biomimetic membranes. Journal of Membrane Science Applied Biomimetic Inc. and Department of Chemical Engineering, The Pennsylvania State University, University Park, PA 16802

Acknowledgement

Applied Biomimetic A/S is part of Elsweb, a Danish high-tech venture portfolio company owned by Jørgen M. Clausen

Aquaporin-Based Biomimetic Reverse Osmosis Membranes: Stability and Long Term Performance.

***Saren Qi**[1] and Rong Wang[2]*
(1)Singapore Membrane Technology Centre and School of Civil and Environmental Engineering, Nanyang Technological University, Singapore, Singapore, (2)Singapore Membrane Technology Centre, Nanyang Technological University

This study investigated the stability of the aquaporin-based biomimetic (ABM) membrane and its long-term reverse osmosis (RO) performance. A series of experiments showed that the ABM membranes exhibited good chemical stability after soaking in three commonly used chemical cleaning agents, namely ethylenediamine tetraacetic acid (EDTA), sodium hydroxide (NaOH) and citric acid, respectively. In addition, it was found that both water permeability (flux per unit applied pressure) and solute rejection of ABM membranes are sensitive to the temperature and applied pressure. In the long-term RO performance evaluation, a real RO feed water from water reclamation process was used as feed; this is the first reported application of real RO feed water to ABM membranes. The membranes were periodically cleaned and the recovery of water permeability of the ABM membranes was found to be greater than 90%. The quality of the water permeate exceeded the World Health Organisation standard for drinking water. Compared with a commercial RO membrane, the ABM membrane only needed half of the applied pressure to achieve the same water flux, suggesting the advantage and potential of ABM membranes to reduce energy consumption in practical applications.

3 **Salt-Excluding Artificial Water Channels in Lipidic Systems.**

Istvan Kocsis[1] *and Mihai Barboiu*[2]
(1)1Adaptive Supramolecular Nanosystems, Institut Europeen des Membranes, Montpellier, France, (2)Adaptative Supramolecular Nanosystems Group, Institut Européen des Membranes, Montpellier, France

Synthesis of High-Performance Biologically-Inspired Nanofiltration Membranes for Water Treatment.

Priyesh Wagh[1], *Isabel Escobar*[2], *Xinyi Zhang*[3], *and Yinan Wei*[4]
(1)University of Kentucky, Lexington, KY, (2)Department of Chemical and Materials Engineering, University of Kentucky, Lexington, KY, (3)Chemistry Department, University of Kentucky, Lexington, KY, (4)Department of Chemistry, University of Kentucky, Lexington, KY

Aquaporins are membrane protein water channels present in cells, and they restrict the passage of contaminants including bacteria, viruses, minerals, proteins, DNA, dissolved gases, salts, detergents, and even protons without preventing the passage of water. Small molecules such as urea and boric acid, the removal of which is inefficient by conventional membranes are also rejected. Biomimetic membranes, an innovation in water filtration technology attempts to replicate a natural process occurring at the cell level; specifically, the highly-selective and efficient transport of different molecules across a cell membrane. Therefore, aquaporins have received worldwide attention because of their potential to form biomimetic membranes with high flux and selectivity for water reuse and desalination. However, challenges involved in the incorporation of aquaporin proteins in membranes limit their applicability. One of them is to attach aquaporins to the membranes without chemically altering or damaging the aquaporins during the binding to the membrane. The second challenge is to design and prepare an assembly that allows biomimetic membranes with aquaporins to sustain hydraulic water pressure gradients without losing their integrity and performance. The overarching objective of this project is to form a biomimetic membrane made of unaltered aquaporins dispersed in a polymeric membrane selective layer and capable of operation under high hydraulic pressure. To this end, polybenzimidazole (PBI) membranes were surface modified with treated aquaporins in order to achieve higher water flux and selectivity. Membranes modified with aquaporins displayed lower flux declines and higher flux recovery values after backwash as compared to unmodified PBI membranes. Also, modified membranes showed improved rejection values for both protein and salt solutions of different concentrations. However, there was leakage of ions between the channels. Therefore, in order to improve the rejection of protons, ions and other impurities, aquaporin channels were aligned with the direction of water flow. Functional groups were installed on AqpZ for covalent attachment to the polymer matrix so that the proteins could be immobilized to the membranes and aligned in the direction of the flow.

DESALINATION AND POTABLE WATER PRODUCTION I

4 **Modeling Low-Pressure Nanofiltration Membranes for Softening and Pretreatment.**

Omar Labban[1], *Chang Liu*[2], *Tzyy Haur Chong*[2], *and John Lienhard V*[1]
(1)Department of Mechanical Engineering, Massachusetts Institute of Technology, Cambridge, MA, (2)School of Civil and Environmental Engineering, Nanyang Technological University, Singapore, Singapore

SMALL Flexible Reverse Osmosis System for Off-GRID Water Purification and Desalination.

Tae Lee[1], *Anditya Rahardianto*[2], *and Yoram Cohen*[2]
(1)Chemical and Biomolecular Engineering, UCLA, Los Angeles, CA, (2)Chemical and Biomolecular Engineering, University of California, Los Angeles, Los Angeles, CA

small Flexible REVERSE OSMOSIS SYSTEM for off-grid Water Purification and Deslaination

Authors: ***Tae Lee***[#], *Anditya Rahardianto, Yoram Cohen*

Chemical and Biomolecular Engineering Department and Institute of the Environment and Sustainability, Water Technology Research Center, University of California Los Angeles,

Abstract

Various remote communities around the world are confronted with impaired local water supplies. In such communities establishing and maintaining reliable distributed water treatment systems for the production of safe drinking water is a major challenge. Water treatment systems for deployment in remote communities must possess the following attributes: a) provide a physical barrier for reliable contaminant removal, b) high operational flexibility to handle geographical and temporal variability of feed water quality and other operational requirements (e.g., water recovery), c) amenable to process automation, d) compact foot-print capable for fixed or mobile deployment, and e) standardized system configuration (for ease of maintenance and operator training). In this regard, membrane-based (nanofiltration (NF) and low-pressure reverse osmosis (RO)) water treatment is most suitable for meeting the above needs. Accordingly, a novel flexible RO/NF (FRO) system platform was developed with enhanced small-system energy efficiency and high operational flexibility (allowing a wide water recovery range of 10-95%), while minimizing plant footprint. Advanced system control was implemented that enables autonomous, self-adaptive operation of FRO over a wider range of feed water salinity (e.g., 500 mg/L – 35,000 mg/L) while maintaining energy-optimal conditions. A key feature of FRO is the ability to switch (on demand) between various operational modes (e.g. batch, continuous, and semi-batch modes) depending on feed water conditions and treatment targets using a single system platform. Furthermore, the present unique FRO system design enables continuous operation (without interruptions of permeate productivity) and without the need for a typical high pressure pump (>150 psi) used in single or multi-stage RO systems. System performance was then analyzed via process models that were validated with both laboratory and field data for brackish water and sea water desalination, as well as for nitrate removal from impaired potable groundwater sources. Both experimental and modeling results demonstrated that the innovative FRO system platform allows for flexible operation over a wide range of product water recovery, while maintaining high energy efficiency (i.e., low energy consumption suitable for use with renewable energy). Therefore, FRO is particularly attractive for distributed deployment in rural or isolated communities where high water recovery is critical for various water sources of different salinity levels.

Use of Electrodeionization in Small Scale Desalination.

Jamie Hestekin[1] *and Alexander Lopez*[2]
(1)Ralph E. Martin Department of Chemical Engineering, University of Arkansas, Fayetteville, AR, (2)Chemical and Biological Engineering, University of Colorado at Boulder, Boulder, CO

Currently reverse osmosis dominates desalination because of low cost, reliability, and choices in the market place. However, there are many new deslatination opporunities, especially in brackish water application, where the size and scale make it possible to consider other technology. One of these technologies is electrodeionization. Electrodeionization is a technique in which we use wafers to remove ions to low concentration through an electricially driven process. Because the cost is related to the amount of ions being moved this technology can be cost competitive at low to medium ion concentration and low to medium scale. The purpose of this presentation is to explore places where electrodeionization can be economically feasible and show the fundamental scientific challenges with the technology.

5 **A Module-Scale Computational Fluid Dynamics Model to Evaluate Hybrid Osmotically-Driven Desalination Systems.**

Lori Jones *and Andrea Achilli*
Environmental Resources Engineering, Humboldt State University, Arcata, CA

7 **Hybrid Pressure Retarded Osmosis-Membrane Distillation (PRO-MD) Process for Osmotic Power and Clean Water Generation.**

Gang Han*, Jian Zuo, and Tai-Shung Chung*
Chemical and Biomolecular Engineering, National University of Singapore, Singapore, Singapore

9 **Novel Thermal Efficiency-Based Model for Geothermal Membrane Distillation.**

Johan Vanneste*, John Bush, Christopher Marks, and Tzahi Y. Cath*
Civil and Environmental Engineering, Colorado School of Mines, Golden, CO

DESALINATION AND POTABLE WATER PRODUCTION II

Water Filtration Using Naturally-Occurring Membranes in Plant Xylem.

Rohit Karnik[1] *and Krithika Ramchander*[2]
(1)Massachusetts Institute of Technology, Cambridge, MA, (2)Mechanical Engineering, MIT, Cambridge, MA

Low-cost, accessible, point-of-use technologies for water filtration have the potential to greatly reduce the global burden of waterborne illnesses. Drawing inspiration from fluidic transport in plants, we explored the possibility of using plant xylem tissue for low-cost water filtration. The xylem tissue conducts sap in plants and has evolved porous membranes that provide low resistance to flow while arresting the spread of cavitation-induced bubbles. We demonstrate that plant xylem in the sapwood of coniferous trees can be used to effectively remove E. coli bacteria from water at high flow rates. However, a major challenge associated with the use of xylem for water filtration is the drop in permeability and deterioration in rejection ability due to the structural changes induced during drying, which is crucial in the context of transportation and shelf-life. We have developed methods to preserve the structural integrity of the xylem and to minimize the negative impacts on filtration characteristics due to drying. Further, we found that the permeability after drying is a strong function of filter length and have studied ways to design the filter to achieve an optimal balance between flow rate and rejection ability. We have also demonstrated gravity-driven filtration through the xylem and conducted fouling studies to gain insights into filter lifetime and variation in flow characteristics over time. Fabrication of the filter is simple, and the wood required for fabrication of a gravity-driven filter with a flow rate exceeding 1 L/h costs less than 1 cent and weighs only 8 grams. This work suggests the potential of plant xylem for realizing inexpensive, locally-manufacturable, and disposable point-of-use water purification devices.

10 **Development of Highly Permeable Thin Film Composite Polyamide Membranes for Water Desalination through Reverse Osmosis with Polydopamine/Molecular Sieve Particles Layer.**

Pinar Cay Durgun *and Mary Laura Lind*
School for Engineering of Matter, Transport, and Energy, Arizona State University, Tempe, AZ

11 **Ion and Methanol Sorption and Transport in a Cation Exchange Membrane Based on Crosslinked Sulfonated Polystyrene.**

Michele Galizia[1]*, Francesco Maria Benedetti*[2]*, Donald R. Paul*[3]*, and Benny D. Freeman*[1]
(1)McKetta Department of Chemical Engineering, The University of Texas at Austin, Austin, TX, (2)Department of Civil, Chemical, Environmental and Materials Engineering (DICAM), University of Bologna, Bologna, Italy, (3)Chemical Engineering, University of Texas at Austin, Austin, TX

12 **Controlling Water Content of Cross-Linked Polymer Membranes and Its Effect on Ion Sorption Properties.**

Eui-Soung Jang[1]*, Don R. Paul*[2]*, and Benny D. Freeman*[1]

(1)McKetta Department of Chemical Engineering, The University of Texas at Austin, Austin, TX, (2)Department of Chemical Engineering, The University of Texas at Austin, Austin, TX

Recent Achievements in the Zeta Potential Analysis of Membranes for Desalination.

Thomas Luxbacher[1], *Bryan D. Coday*[2], *Tzahi Y. Cath*[2], *and Andriy Yaroshchuk*[3]
(1)Material Characterization, Anton Paar, Graz, Austria, (2)Civil and Environmental Engineering, Colorado School of Mines, Golden, CO, (3)ICREA and Department of Chemical Engineering, Polytechnic University of Catalonia, Barcelona, Spain

Surface charge plays a major role in solute rejection and fouling of polymer membranes used for nanofiltration and reverse osmosis processes. Surface charge is accessible through the membrane zeta potential, which has gained severe interest starting in the early 1990s. Since then a variety of flat sheet membranes but also of hollow fiber membranes have been characterized in terms of the surface zeta potential.

However, what has been neglected in most of these studies was the possible influence of membrane porosity on the zeta potential. Furthermore, the majority of membrane zeta potential data was obtained by using a standard aqueous solution with an ionic strength of 0.001-0.01 mol/l. The boundary conditions for zeta potential analysis thus differ significantly from those experienced by NF and RO membranes for seawater or brackish water desalination, where the membranes are exposed to high salinity conditions.

In this contribution, we address the effects of membrane porosity and water salinity on the zeta potential analysis. Recently, evidence for the contribution of the porous membrane body to the zeta potential was confirmed for flat sheet membranes used in microfiltration. We extend the new approach, i.e., the streaming current measurement at varying distances between samples of a flat sheet membrane, to TFC NF and RO membranes. Although the nonwoven support dominates the total thickness of such membranes, we find little evidence for an influence of its porous structure.

For the second topic, we report on the impact of high ionic strengths on the membrane zeta potential. Streaming potential measurements on a TFC RO membrane were conducted in a wide range of salt concentration. Zeta potential calculated from streaming potential was then extrapolated to an environmentally relevant ionic strength (up to 1 mol/l). Results revealed that the membrane zeta potential approaches a finite equilibrium value at high salinity.

13 **Environmental and Economic Assessment of Fertilizer Drawn Forward Osmosis and Nanofiltration Hybrid System for Desalination of Mine Impaired Water.**

Jung Eun Kim[1], *Sherub Phuntsho*[1], *Ko Kyong Shon*[1], *and Laura Chekli*[2]
(1)Faculty of Engineering and Information Technology, University of Technology Sydney, Sydney, Australia, (2)University of Technology Sydney, Sydney, Australia

DESIGN AND PROCESSING OF INORGANIC MEMBRANES

Development of Zeolite Membrane Products in a Thin Sheet Form.

Wei Liu[1] *and* ***Lorna Mao***[2]
(1)Pacific Northwest National Laboratory, Richland, WA, (2)Molecule Works Inc., Richland, WA

Developing low-cost zeolite membranes is important for a range of industrial applications. In this presentation, design principles and preparation methods of zeolite membranes in a thin-sheet form will be introduced. A zeolite membrane layer is directly grown on a flexible, 50μm-thin metal support sheet with strong adhesion and without defects. The resulting zeolite membrane sheet has mechanical flexibility and strength like a thin metal sheet. The new membrane product development will be elaborated with NaA-type, MFI-type, and Faujasite-type zeolite frameworks – the three commonly-used zeolite materials. Characterization results of micro-structures and molecular sieving functions will be discussed. These thin zeolite membrane sheets show excellent gas and molecular separation performances, such as air dehumidification, alcohol dehydration, CO_2 separation, and ethanol enrichment. The thin membrane sheet makes it possible to achieve low-cost, high throughput production and to make zeolite membrane modules of membrane area packing density comparable to polymeric membrane sheets.

Highly Permeable N_2/CH_4 Separation SAPO-34 Membranes Synthesized By Diluted Gels and Increased Crystallization Temperature.

Zhaowang Zong[1], *Xuhui Feng*[1], *Yi Huang*[2], *Zhuonan Song*[2], *Rongfei Zhou*[3], *Shaojun James Zhou*[4], *Moises Carreon*[1], *Miao Yu*[5], *and Shiguang Li*[4]
(1)Chemical & Biological Engineering, Colorado School of Mines, Golden, CO, (2)Department of Chemical Engineering, University of South Carolina, Columbia, SC, (3)State Key Laboratory of Materials-Oriented Chemical Engineering, College of Chemistry and Chemical Engineering, Nanjing Tech. University, Nanjing, China, (4)Gas Technology Institute, Des Plaines, IL, (5)Chemical Engineering, University of South Carolina, Columbia, SC

Synthesis conditions were optimized for synthesizing SAPO-34 membranes towards N_2/CH_4 separation. Compared to previous SAPO-34 membrane preparations, synthesis gels were diluted 1.3–2.7 times and an increased crystallization temperature was used. As a result, a 3–4 fold increase in N_2 permeance was observed in this study. The membranes separated N_2 from CH_4 with N_2 permeance as high as 1300 GPU and selectivity of 7.4 at 23 °C for a 50/50 N_2/CH_4 mixture. Furthermore, membranes were reproducible and effectively separated gas mixtures having low N_2 content which are relevant compositions of natural gas. More specifically, membranes separated N_2 from CH_4 with N_2 permeance as high as 1160 GPU and selectivity of 7.1 at 23 °C for an 18.5/81.5 N_2/CH_4 mixture. The resultant SAPO-34 membranes have potential for economic N_2 rejection in natural gas processing.

15 **Macro-Porous Alpha-Alumina Support Tubes for Thin Defect-Free Membranes, Made By Centrifugal Colloidal Casting.**

Ralph Bauer
Department of Materials Science and Engineering, The Ohio State University, Columbus, OH

N/A **Fabrication of CNT Hollow Fiber Membranes and Its Applications.**

Heeyoung Kim
Civil and Environmental Engineering, KAIST, , Korea, Republic of (South)

16 **Large-Area Fabrication of Graphene-Based Membranes By Shear Alignment of Discotic Nematic Liquid Crystals of Graphene Oxide.**

Abozar Akbari[1], *Dibakar Bhattacharyya*[2], *and Mainak Majumder*[1]
(1)Mechanical and Aerospace Engineering, Monash University, Clayton, Australia, (2)Department of Chemical and Materials Engineering, University of Kentucky, Lexington, KY

17 **Nitrogen Selectivity of Molybdenum Metallic Membranes.**

Kyoungjin Lee[1], *Simona Liguori*[1], *Ralph Bauer*[2], *Hendrik Verweij*[2], *and Jennifer Wilcox*[1]
(1)Department of Energy Resources Engineering, Stanford University, Stanford, CA, (2)Department of Materials Science and Engineering, The Ohio State University, Columbus, OH

19 **Process Design of Natural Gas Treatment Using DDR-Type Zeolite Membrane.**

Hiroaki Hasegawa[1], ***Keiichi Nishida***[1], *Shuichi Oguro*[1], *Yasushi Fujimura*[1], *Kenji Yajima*[2], *Makiko Niino*[2], *Manabu Isomura*[2], *and Toshihiro Tomita*[2]
(1)JGC Corporation, Yokohama, Japan, (2)NGK Insulators, Ltd., Nagoya, Japan

DESIGN AND PROCESSING OF POLYMERIC AND ORGANIC MEMBRANES II

Electrospun Nanofiber Membranes for Water Treatment and Reuse.

Jeffrey R. McCutcheon[1], *Ngoc Bui*[2], *Liwei Huang*[3], *Mustafa Al-Furaiji*[4], *Basma I. Waisi*[5], *and Maqsud R. Chowdhury*[1]
(1)Department of Chemical and Biomolecular Engineering, University of Connecticut, Storrs, CT, (2)Physical and Life Sciences, Lawrence Livermore National Laboratory, Livermore, CA, (3)Department of Chemical & Biomolecular Engineering, University of Connecticut, Storrs, CT, (4)Inorganic membranes, University of Twente, Enschede, Netherlands, (5)Department of Inorganic membrane, University of Twente, Enschede, Netherlands

Water treatment and reuse is becoming an increasingly important issue for all regions of the world regardless of their level of water abundance or aridity. Membrane technology has provided a versatile platform for developing a host of water treatment technologies that enable us to tap sources for water that only decades ago would have never been considered as viable sources of water (seawater, produced water, wastewater). Continued innovation in the water treatment membrane field continues today as new materials are sought that leverage important structure-property relationships that push the barrier of membrane performance. One material in particular, nanofiber nonwovens, has recently achieved attention as a possible platform for improving membrane performance because of their intrinsically high permeability, low resistance to mass transfer, and high surface area. In our work, we explore the use of polymer nanofiber nonwovens made through electrospinning for use across a wide range of membrane applications. These include:

- *Electrospun supported thin film composite membranes for forward osmosis (FO):* Electrospun materials are demonstrated to have excellent mass transfer properties which reduce resistances to solute transfer in forward osmosis.

- *Nanofiber membranes for membrane distillation (MD):* Electrospun nonwovens have properties of high porosity and small pore size, making them suitable for MD applications. We demonstrate high flux performance of these membranes under very modest temperature gradients using high salinity feeds.
- *High surface area activated carbons:* Using electrospun precursors, we have established protocols for making high surface area carbon materials in a nonwoven configuration that are appropriate for organic removal in sorption applications.
- *Use of commercial nanofiber materials for membrane applications:* We have explored the use of commercial nanofiber nonwovens from DuPont in both FO and MD applications. This demonstrates that commercial nanofiber nonwovens, which are made roll-to-roll, are applicable to these emergent membrane disciplines. Availability of these materials commercially could aid in commercialization efforts.

In all, this work demonstrates the promise of using nanofibers across a number of key membrane disciplines in the water treatment space.

22 **A New Method of Fabricating Charge Mosaic Membranes By Inkjet Printing.**

Peng Gao[1], *Aaron Hunter*[1], *and William Phillip*[2]
(1)Department of Chemical and Biomolecular Engineering, University of Notre Dame, Notre Dame, IN, (2)Chemical and Biomolecular Engineering, University of Notre Dame, Notre Dame, IN

23 **Thin-Film Composite Membranes with Regular, Sub-Micron Surface Patterns: Fabrication and Characterization.**

Yifu Ding[1], *Sajjad Maruf*[2], *and Alan R. Greenberg*[3]
(1)Department of Mechanical Engineering, University of Colorado at Boulder, Boulder, CO, (2)Mechanical Engineering, University of Colorado at Boulder, Boulder, CO, (3)Department of Mechanical Engineering, University of Colorado Boulder, Boulder, CO

24 **Carbon Nanotube Multilayer Polyelectrolyte Charge Mosaic Membranes.**

Ghazaleh Vaseghi[1] *and Glenn Lipscomb*[2]
(1)Chemical Engineering, University Of Toledo, Toledo, OH, (2)Department of Chemical and Environmental Engineering, University of Toledo, Toledo, OH

Robust Outer-Selective Thin-Film Composite Polyethersulfone Hollow Fiber Membranes with Low Reverse Salt Flux for Renewable Salinity-Gradient Energy Generation.

Zhen Lei Cheng[1], *Xue Li*[1], *Ying Da Liu*[1], *and Tai-Shung Chung*[1,2]
(1)Department of Chemical and Biomolecular Engineering, National University of Singapore, Singapore, Singapore, (2)Water Desalination & Reuse (WDR) Center, King Abdullah University of Science and Technology, Thuwal, Saudi Arabia

This study reports outer-selective thin-film composite (TFC) hollow fiber membranes with extremely low reverse salt fluxes and robustness for harvesting salinity-gradient energy from pressure retarded osmosis (PRO) processes. Almost defect-free polyamide layers with impressive low salt permeabilities were synthesized on top of robust polyethersulfone porous supports. The newly developed TFC-II membrane shows a maximum power density of 7.81 W m^{-2} using 1 M NaCl and DI water as feeds at 20 bar. Reproducible data obtained in the 2nd and 3rd runs confirm its stability under high hydraulic pressure differences. Comparing to other PRO membranes reported in the literature, the newly developed membrane exhibits not only the smallest slope between water flux decline and ΔP increase but also the lowest ratio of reverse salt flux to water flux. Thus, the effective osmotic driving force could be well maintained even under high pressure operations. For the first time, the effect of feed pressure buildup induced by feed flowrate was evaluated towards PRO performance. A slight increment in feed pressure buildup was found to be beneficial to water flux and power density up to 10.06 W m^{-2} without comprising the reverse salt flux. We believe this study may open up new perspectives on outer-selective PRO hollow fiber membranes and provide useful insights to understand and design next-generation outer-selective TFC hollow fiber membranes for osmotic power generation.

Elucidating the Reverse Osmosis Polyamide Formation Mechanism from Pseudo-Stop-Motion Imaging.

Caleb V. Funk[1] *and Matthew Reichert*[2]
(1)R&D, Dow Water & Process Solutions, Edina, MN, (2)The Dow Chemical Company, Edina, MN

The formation mechanism of reverse osmosis polyamide membranes has long eluded the scientific community, primarily due to an inability to observe the formation, which occurs both too quickly and on too small a scale. Extensive data has been gathered over the years to understand the effects of recipe and processing conditions on the performance of these membranes, but as the technology approaches performance limits, understanding the intricacies of film synthesis may be the key to fine-tuning an established chemistry to maximize separation efficiency.

For the first time, methods of freezing in time polyamide at various stages of formation have been developed. Custom-designed casting and imaging methods provide a pseudo-stop-motion view of polyamide film growth and insight into the forces that lead to the typical polyp structures of RO membranes. The structures have been shown to be much less random than they appear, and through well-controlled pilot-scale membrane manufacturing have been manipulated to illustrate a new mechanistic theory.

Evidence of a relationship between structure and parameters such as interfacial tension, amine pK_a and temperature will be presented. Further evidence based on manipulation of downstream process conditions solidifies the argument for a "balloon inflation" growth mechanism. This evidence will be presented in the context of three common polyamide structures along with application to the future of reverse osmosis membrane development.

26 **Facilitated Transport Membranes in the Solid State Containing Surface-Activated Ag Nanoparticles for Olefin/Paraffin Separation.**

Young Rae Kim, Gi Hyeon Moon, and ***Yong Soo Kang***
Energy Engineering, Hanyang University, Seoul, Korea, Republic of (South)

DESIGN AND PROCESSING OF POLYMERIC AND ORGANIC MEMBRANES III

Plasticization of Glassy Polymeric Membranes in Gas Separations: Assessment and Prediction.

Matteo Minelli and ***Giulio C. Sarti***
Department of Civil, Chemical, Environmental and Materials Engineering (DICAM), University of Bologna, Bologna, Italy

The permeability behavior of glassy membranes as upstream pressure increases shows different trends, including the monotonous decline, a monotonous increase, as well as an initial decline followed by a subsequent increase after a minimum permeability value is reached. The minimum value, whenever present, occurs at a pressure conventionally indicated as the plasticization pressure. The plasticization phenomenon has not been inspected in detail thus far, in terms of material property variations; current physical interpretation associates to the plasticization pressure the onset of a generic softening of the polymer matrix, which is considered responsible of the observed increase in permeability and decrease in selectivity.

On the other hand, by considering several different glassy membranes we have observed that the different behaviors observed for gas permeability can be described by considering only a solution-diffusion model in which molecular mobility varies with penetrant concentration through an exponential law, with two adjustable parameters only. Diffusivity is taken as the product of molecular mobility and a thermodynamic factor, which is calculated by using the NELF model for thermodynamic properties of the glassy phase.

It is observed that by fitting the only two adjustable parameters of the model to the initial branch of the permeability-pressure curve it is possible to predict the plasticization pressure, if existing, by using only the solubility diffusivity model without introducing any additional physical phenomenon.

The agreement observed between model calculations and experimental data of CO_2 permeability in various glassy polymers is very satisfactory. That allow us to offer a deeper insight on the so-called plasticization phenomenon. By considering at the same time the permeability isotherm, the solubility isotherm and the concentration and swelling profiles in the membrane, we can show that in some cases the plasticization pressure takes place after part of the membrane has turned into rubbery phase; while in other cases, e.g. for Matrimid polyimide, plasticization pressure takes place when the entire membrane is still glassy but with a polymer swelling sufficient for a permeability increase.

Finally, it is observed that all parameters used have a defined physical meaning which lead to good general correlations with both polymer and penetrant properties, based on which permeability predictions can be obtained.

Solvent and Thermal History Effects on Gas Transport in TR Polymers and Their Precursors.

Michelle Dose[1], *Benny D. Freeman*[1], *and Donald Paul*[2]
(1)McKetta Department of Chemical Engineering, The University of Texas at Austin, Austin, TX, (2)Department of Chemical Engineering, The University of Texas at Austin, Austin, TX

HAB-6FDA polyimide was synthesized by a two-step polycondensation method with chemical imidization. Dense polymer films were solution cast from tetrahydrofuran (THF), acetone, and N,N-dimethylacedamide (DMAc). Films cast from THF and acetone were cast in a solvent saturated glove bag then dried at 200°C for 24 h. Two separate films were cast from DMAc in a vacuum oven using different heating and vacuum protocols, with one utilizing a gas purge and the other a temperature ramp under partial vacuum. Each precursor film was used to prepare thermally rearranged (TR) polymers. The rearrangement reaction was performed at 400°C and resulted in a 54% conversion to the polybenzoxazole TR product. Pure CH_4 and CO_2 gas permeability measurements were made on both the precursor and the TR polymer. While the ideal CH_4/CO_2 selectivity remained constant for all four polyimide films, the different casting procedures resulted in a 40% difference in permeability coefficients, with the film prepared from DMAc under partial vacuum having the lowest permeability coefficients for both gases and the DMAc film cast using a gas purge having the highest. The respective TR polymer films exhibited the same trend in permeability with a 37% variation with only minimal differences in selectivity. These trends have been attributed free volume variations due to the different casting techniques and the inability to thermally "reset" the history of the polyimide precursor.

Molecular Design of Nanofiltration Membranes for the Recovery of Phosphorus from Urban Wastewater.

Zhiwei Thong[1], *Yue Cui*[2], *Yee Kang Ong*[3], *and Tai-Shung Chung*[1]
(1)Chemical and Biomolecular Engineering, National University of Singapore, Singapore, Singapore, (2)Department of Chemical and Biomolecular Engineering, National University of Singapore, Singapore, Singapore, (3)Department of Chemical & Biomolecular Engineering, National University of Singapore, Singapore, Singapore

With the rapid depletion of mineral phosphorus, the recovery of phosphorus from sewage sludge will become increasingly important in the near future. However, the presence of various contaminants such as heavy metals in the sewage sludge only serves to complicate the issue, and the challenge lies in the separation of phosphorus from the heavy metals to produce fertilizers of high quality. Among various available methods, nanofiltration (NF) has been demonstrated to be a feasible and promising option when the sewage sludge undergoes acidic dissolution (pH <2). Since the performance of commercially available thin film composite (TFC) NF membranes reported thus far has great rooms for improvement, the development of highly permeable positively charged NF membranes are recommended. To this aim, a NF membrane which is desirable for phosphorus recovery was fabricated via interfacial polymerization of polyethyleneimine (PEI) and trimesoyl chloride (TMC) on a porous polyethersulfone (PES) membrane substrate. Through an optimization of the interfacial polymerization process which involves varying the molecular weight of PEI and the concentration of TMC, the resultant membrane displays a low molecular weight cut-off (MWCO) of 170 Da with a reasonably high pure water permeability (A) of 6.4 LMH/bar. The newly developed NF membrane can effectively reject a wide variety of heavy metal ions such as Cu, Zn, Pb and Ni (>93%) while demonstrating a low phosphorus rejection of 19.6% at 10 bar using a feed solution of pH 2. Thus, up to 92% of the feed phosphorus may be recovered using this newly developed NF membrane at a permeate recovery of 90%. This is a highly competitive value for the recovery of phosphorus.

Interfacial Polymerization for Sugar / Ionic Liquid Separations.

S. Ranil Wickramasinghe[1], *Xianghong Qian*[2], *Alexandru Avram*[3], *and Pejman Ahmadiannamini*[1]
(1)Department of Chemical Engineering, University of Arkansas, Fayetteville, AR, (2)Department of Biomedical Engineering, University of Arkansas, Fayetteville, AR, (3)University of Arkansas, Fayetteville, AR

Ionic Liquids (IL) are tunable green solvents for a variety of chemical and biochemical processes. However, the economic viability of utilizing these green solvents depends on developing processes for their recycle and reuse. ILs are outstanding solvents for biomass fractionation and sugar conversion processes. However, separating saccharides from ILs is generally very difficult due to the similarity of their molecular weight. Here a selective skin layer containing 3-aminophenylboronic acid has been successfully synthesized via interfacial polymerization on the surface of asymmetric polyethersulfone (PES) membranes. The fabricated composite membranes show superior ionic liquid/sugar selectivity at competitive permeability. Moreover, the selectivity can be tuned by varying the monomer composition, the polymerization time and post treatment conditions. While [EMIM]Cl and [BMIM]Cl are able to selectively permeate through the membranes, over 90% of mono- and oligosaccharides are rejected. The compositional, chemical and morphological properties of these nanofiltration membranes are characterized and correlated with their separation characteristics.

Organic Solvent Nanofiltration with Novel Polymeric Membranes.

Kamalesh K. Sirkar[1], *Xianghong Qian*[2], *Yucheng Hu*[1], *Jasmine Kaur*[1], *John Chau*[1], *and Hongbo Du*[2]
(1)Otto York Department of Chemical, Biological and Pharmaceutical Engineering, New Jersey Institute of Technology, Newark, NJ, (2)Department of Biomedical Engineering, University of Arkansas, Fayetteville, AR

A very large variety of membranes polymeric or otherwise have been investigated over the last two decades for organic solvent nanofiltration (OSN). The materials and structures used to make OSN membranes include among others the following: poly (dimethylsiloxane) (PDMS); mixed matrix membranes (MMMs) of PDMS with zeolites and other fillers; asymmetric integrally skinned polyimide (PI) membrane crosslinked with aromatic or aliphatic diamines; polyaniline; polypyrrole; interfacially polymerized polyamide w/o carbon nanotubes; carbon-based membranes including graphene. One of the weaknesses of most polymeric membranes is their varying tendencies for swelling with demanding organic solvents aprotic or otherwise; this gets reflected in the solute rejection behaviors of such polymers. To this end we have started studying membranes from a particular class of polymers for OSN. We report preliminary results of solvent flux and solute rejection for such membranes. We also report results of molecular dynamic simulations of transport in one such polymeric membrane.

27 **Comparison of Performance Between Laboratory Scale and Industrial Scale Membranes.**

Sneha Chede[1], *Peter Griffiths*[2], *Tequila Harris*[2], *Nicholas Linck*[3], *Matthew Weisenberger*[3], *and Isabel Escobar*[4]
(1)Chemical and Environmental Engineering, The University of Toledo, Toledo, OH, (2)Georgia Institute of Technology, Atlanta, GA, (3)University of Kentucky, lexington, KY, (4)Department of Chemical and Materials Engineering, University of Kentucky, Lexington, KY

DESIGN AND PROCESSING OF POLYMERIC AND ORGANIC MEMBRANES I

28 **Manipulating Nanopore Selectivity in Lyotropic Liquid Crystal Polymer Nanofiltration Membranes for the Recovery of Targeted Organic Solutes.**

Sarah M. Dischinger[1], *Richard D. Noble*[2], *and Douglas L. Gin*[2]
(1)Chemical and Biological Engineering, University of Colorado Boulder, Boulder, CO, (2)Department of Chemical and Biological Engineering, University of Colorado, Boulder, CO

29 **Polymeric Micro-Spherical Particles Grafted on Water Treatment Membranes: Mechanism of Formation and Applications.**

Roni Kasher[1], *Clil Regev*[2], *Rozalia Feinstein*[3], *Chad Small*[4], *Sophia Belfer*[5], *and Moshe Herzberg*[6]
(1)Ben Gurion University, Sde Boker, Israel, (2)Chemistry, Ben-Gurion University of the Negev, Midreshet Ben-Gurion, Israel, (3)Zuckerberg Institute for Water Research, Ben-Gurion University of the Negev, Midreshet Ben-Gurion, Israel, (4)Chemical Engineering, Yale University, New Haven, CT, United States Minor Outlying Islands, (5)Desalination & Water Treatment , Zuckerberg Institute for Water Research, Ben-Gurion University of the Negev, Sde Boker, Israel, (6)Department of Desalination and Water Treatment, Zuckerberg Institute for Water Research, Jacob Blaustein Institutes for Desert Research, Ben Gurion University, Midreshet Ben Gurion, Israel

Surface Characterization of Membrane Surfaces Nanostructured with Tethered Hydrophilic Polymers.

Soomin Kim *and Yoram Cohen*
Chemical and Biomolecular Engineering, University of California, Los Angeles, Los Angeles, CA

Membrane surface modification with end-tethered hydrophilic polymers is an effective approach for reducing fouling propensity and improving cleaning efficiency of RO and

UF membranes. It has been suggested that, in addition to chemistry, swelling of the polymer layers play a crucial role in minimizing foulant attachment, yet detailed characterization of swollen tethered polymer chains is still needed in order to elucidate the mechanism by which swelling minimizes fouling. In this study, AFM was used to characterize swelling of surface tethered polymer chains by utilizing poly(acrylic acid) (PAAc) and poly(2-hydroxyethyl methacrylate) (PHEMA) chains end-grafted onto polysulfone (PSf) membrane surrogate surfaces via atmospheric pressure plasma-induced graft polymerization (APPIGP). Impact of polymer chain swelling on surface topography of PAAc and PHEMA layers was evaluated by AFM imaging in various solvents (i.e., air, deionized water, saline water). The thickness (or extension) of the swollen polymer chains was assessed by determining contour length distributions of the tethered polymer chains via single molecule force spectroscopy performed in aqueous solutions of different salinity. The adhesion force measurements were performed with a functionalized colloidal probe in order to quantify fouling potential of the polymer layers of different swelling degrees. The results of the present study provide a direct measure of the potential effectiveness of tethered chains for membrane fouling reduction. In this regard, direct surface force measurement can guide the optimization of surface modification for synthesizing high performance low fouling membranes.

Assessing Biofouling Resistance of a Polyamide Reverse Osmosis Membrane Surface-Modified with a Zwitterionic Polymer.

Maria Bass[1], *Miguel Marre*[2], *Maria Piatkovsky*[3], *Mathias Ulbricht*[4], *Moshe Herzberg*[3], *and Viatcheslav Freger*[1]
(1)Wolfson Department of Chemical Engineering, Technion - Israel Institute of Technology, Haifa, Israel, (2)D.I.I.A.R. Environmental Engineering Department, Politecnico di Milano, Technical University of Milan, Milano, Italy, (3)Department of Desalination and Water Treatment, Ben-Gurion University of the Negev, Midreshet Ben Gurion, Israel, (4)Lehrstuhl für Technische Chemie II, Universität Duisburg-Essen, Essen, Germany

Reverse osmosis (RO) is a technology that can help increase clean water supply in today's world. However, mitigation of biofouling of RO membranes is a remaining challenge towards a wider use of this technology. It was suggested that a promising way to control membrane biofouling could be surface modification, e g., using low-fouling zwitterionic polymers. Published studies demonstrated that zwitterionic coatings can indeed improve the resistance of polyamide (PA) membranes to the initial bacteria adhesion, however, no long-term experiments with real water effluents were conducted. In this report a commercial PA RO membrane was surface-modified with a zwitterionic polymer and its propensity to biofouling was tested in both short-term bacteria adhesion experiments and longer filtration tests conducted using real treated wastewater effluents spiked with a small level of nutrients. The obtained results showed that, initially, there was a clear improvement in the biofouling resistance of the modified membranes and their permeation flux remained stable, in contrast to the non-modified counterpart. However, eventually, the permeabilities of the two membranes declined to a similar degree. The results indicate that antifouling coatings might not promise a better membrane performance in long term. The analysis of the biofilms grown on the pristine and the modified membranes suggested that adaptation capabilities of biofilms overcame the favorable changes in surface properties of the membrane achieved by the modification. The presented results emphasized the importance of long-term filtration experiments as an ultimate test for assessing biofouling resistance of the modified desalination membranes.

30 **UV and EB Cross-Linked Polysulfone Based Membranes for Versatile Applications.**

Ivo Vankelecom[1] *and* ***Veysi Altun***[2]
(1)Centre for Surface Chemistry and Catalysis, K.U. Leuven, Heverlee, Belgium, (2)Bioengineering, KU Leuven, Leuven, Belgium

Design of Hydrophilic Conductive MD Membranes for Fouling Mitigation, and Direct Surface Heating.

Alexander Dudchenko[1] *and David Jassby*[2]
(1)Chemical and Environmental Engineering, University of California, Riverside, Riverside, CA, (2)Department of Chemical and Environmental Engineering, University of California, Riverside, Riverside, CA

Membrane distillation (MD) is promising technique for treatment of waters contaminated with high salt concentration, such as ground water, produced water, and brines generated from desalination. Like all membrane processes, MD suffers from fouling that leads to performance degradation. The hydrophobic nature of the MD membrane surface allows for adsorption of NOM and other organic foulants that clog membrane pores, and can lead to membrane wetting, irreversibly degrading their performance. Here, we present a method of synthesizing composite MD membranes with hydrophilic or hydrophobic conductive carbon nanotube thin film surfaces layered on a hydrophobic MD support. We demonstrate that the hydrophilic surface reduces fouling caused by humic acids, while still maintaining the same performance as a bare MD membrane. Finally, we demonstrate a novel method of generating the thermal driving force in the MD process by heating the carbon nanotube thin film surface directly via joule heating. The direct heating of the MD surface eliminates the need to heat the bulk feed solution, allows for high recovery in single pass, and could significantly reduce the complexity and construction cost of MD plants.

DESIGNS AND PROCESSING OF COMPOSITE AND HYBRID MEMBRANES I

31 **Development of a Composite Membrane with a Liquid-Barrier Chemical-Resistant Polymer Matrix to Recover Water from Urine-Containing Wastewater.**

Afsaneh Khosravi[1]*, Andrew Dopilka*[2]*, Katelyn Kline*[3]*, and Mary Laura Lind*[1]

(1)School for Engineering of Matter, Transport, and Energy, Arizona State University, Tempe, AZ, (2)School for Engineering of Matter, Transport and Energy, Arizona State University, TEMPE, AZ, (3)School for Engineering of Matter, Transport and Energy, Arizona State University, Tempe, AZ

32 **How to Enhance ZIF-8 Growth on Polymers?.**

***Eva Barankova**, Xiaoyu Tan, Pradeep Neelakanda, Mahendra Kumar, and Klaus Viktor Peinemann Sr.*
AMPM, KAUST, Thuwal, Saudi Arabia

33 **Mixed-Matrix Membranes Containing Engineered Nanoporous Materials for Biogas Purification.**

Tae-Hyun Bae
Chemical and Biomedical Engineering, Nanyang Technological University, Singapore, Singapore; School of Chemical and Biomedical Engineering, Nanyang Technological University, Singapore, Singapore

35 **Metal-Rich Polymer Membranes By Cips: From Metal Ions to Crystals.**

***Luis Francisco Villalobos**[1], Suzana Nunes[2], and Klaus-Viktor Peinemann[3]*
(1)Advanced Membrane and Porous Materials Center, King Abdullah University of Science and Technology, Thuwal, Saudi Arabia, (2)Biological and Environmental Science and Engineering Division, King Abdullah University of Science and Technology (KAUST), Thuwal, Saudi Arabia, (3)Advanced Membranes and Porous Materials Center, King Abdullah University of Science and Technology, Thuwal, Saudi Arabia

36 **The Enhanced Performance of Iron-Oxide and Silica-Coated-Iron-Oxide Nanocomposite Membranes during the Filtration of Oily Emulsion.**

Ikechukwu A. Ike*, John D. Orbell, and Mikel Duke*
Institute for Sustainability and Innovation, Victoria University, Melbourne, Australia

Porous Solid Supports for Dual-Phase Liquid/Solid Membranes.

Patrick G. Campbell[1]*, Maira R. Ceron*[1]*, and Marcus Worsley*[2]
(1)Materials Science Division, Lawrence Livermore National Laboratory, Livermore, CA, (2)Lawrence Livermore National Laboratory, Livermore, CA

Dual phase membranes consisting of a liquid phase contained within a solid support show promise in a variety of applications including high-temperature CO2 separations and fuel cells. The solid support phase can simply host the active liquid phase or can participate in the separation by supplying electrons or ions. In many cases, the solid material must be inert to corrosive chemical environments at high temperature, while the pore morphology and material properties need to be optimized to suit the particular application. Lawrence Livermore National Laboratory has developed expertise in the synthesis of novel porous materials over the last four decades. These efforts have mainly focused on creating ultra-low-density carbon, metal-oxide, and inorganic materials, but recently we have turned our attention to creating support materials for molten-salt-based CO2 separation membranes. This contribution will highlight our recent developments in this area.

Funding was provided by Lawrence Livermore National Laboratory Directed Research and Development (LDRD) Grant 16-LW-013. This work was performed under the auspices of the U.S. Department of Energy by LLNL under Contract DE-AC52-07NA27344.

37 **Sub–10 Nm Polyamide Nanofilms with Ultrafast Solvent Transport for Molecular Separation.**

Andrew G. Livingston[1]*,* ***Santanu Karan***[2]*, and Zhiwei Jiang*[2]
(1)Chemical Engineering, Imperial College London, London, United Kingdom, (2)Department of Chemical Engineering, Imperial College London, London, United Kingdom

DESIGNS AND PROCESSING OF COMPOSITE AND HYBRID MEMBRANES II

Electrospun Composite Bipolar Membranes.

M. Cunningham, R. Wycisk, C. Shen, and ***P. Pintauro***
Vanderbilt University, Nashville, TN

Bipolar membranes, typically a laminate of an anion-exchange and cation-exchange membrane, have the unique capability to split water at a potential as low as 0.8 V while conventional electrolysis requires at least 1.2 V. Such membranes are employed in a number of important industrial processes, including the production of mineral acids, the recovery of organic acids from fermentation processes, pH control, and ion-exchange resin regeneration. The use of bipolar membranes in electrodialysis separations eliminates unwanted salt efflux and disposal problems and offers significant savings in electrical energy consumption. Recently, bipolar membranes have also been examined for use in self-humidifying hydrogen/air fuel cells, where the recombination of electrochemically generated hydroxide ions and protons at the bipolar junction produces water that improves membrane hydration, with higher power output at reduced feed gas humidity conditions.

We report here for the first time on the design, fabrication, and testing of a new bipolar membrane that was made by polymer nanofiber electrospinning, where the anion-exchange polymer is quaternized poly(phenylene oxide) and the cation-exchange polymer is sulfonated poly(etherether ketone). These films have an increased interfacial area for water splitting at the junction of anion and cation exchange polymers. This morphology is important for two reasons: (1) it improves the attachment (adhesion) of the two ion-exchange layers so that ballooning/delamination is minimized or eliminated and (2) the increased bipolar junction surface area will lead to a better distribution of the water splitting reaction and lower membrane voltage drop for the same operating current density. The procedure for membrane fabrication will be described and the performance of electrospun films will be compared to bipolar membranes in the literature.

Influence of Nanoparticle Processing on PES Casting Solution Viscosity.

Lauren F. Greenlee[1] *and Nikki S. Rentz*[2]
(1)Ralph E. Martin Dept. of Chemical Engineering, University of Arkansas, Fayetteville, AR, (2)Applied Chemicals and Materials Division, National Institute of Standards & Technology, Boulder, CO

Nanoparticle incorporation into porous polymeric membranes has received continued attention with the concurrent continued development of reactive metallic and metal oxide nanoparticles for water treatment applications. Nanoparticles comprised of metals such as iron, nickel, palladium, silver, aluminum, and copper have been shown to remove both organic and inorganic water contaminants, and novel multi-metallic nanoparticle combinations and morphologies continue to be reported. Iron nanoparticles (Fe NPs) remain at the forefront of efforts focused on water remediation due to the low cost of the metal and the versatility of Fe NPs in addressing a wide variety of water contaminants. The incorporation of these nanoparticle materials into a support material such as a polymeric membrane immobilizes the nanoparticles, preventing movement within the treatment system, and may also provide enhanced contaminant removal. In particular, there is an interest in embedding nanoparticles in phase inversion polymeric membranes, where the nanoparticles can be incorporated into the membrane casting solution and retained in the solid membrane phase when the membrane polymer precipitates.

In this work, the incorporation of iron nanoparticles in a polyethersulfone membrane was studied. Nanoparticle stabilizer type, casting solution additives, and nanoparticle processing by centrifugal or magnetic separation were varied, and specific membrane characteristics were evaluated. Specifically, membrane casting solution viscosity, cast membrane thickness, pure water flux, and internal morphology were evaluated. While the addition of the additives ethanol and polyvinylpyrrolidone to a polyethersulfone-dimethylacetamide solution causes an expected increase in viscosity, the addition of nanoparticles can cause an increase or decrease in viscosity depending on the ligand stabilizer used during nanoparticle synthesis and casting solution additive concentration. Viscosity can also be affected by nanoparticle separation method, but again, changes are also dependent on casting solution composition. Varying changes in membrane thickness are observed and can be correlated to viscosity. Pure water flux decreases for all samples when nanoparticles are added, but the extent of the change is affected by casting solution composition. Internal morphology can partially explain the decrease in flux for nanoparticle-embedded membranes.

Membrane Separation of CO2 from Syngas Using PDMS and Mixed Matrix Membranes Oral.

F. Handan Tezel[1] *and* ***Lauren Rose***[2]
(1)Department of Chemical and Biological Engineering, University of Ottawa, Ottawa, ON, Canada, (2)Chemical and Biological Engineering, University of Ottawa, Ottawa, ON, Canada

Carbon dioxide (CO_2), being one of the leading global greenhouse gas emissions, can be important to remove from other gaseous components in downstream processing. In processes such as gasification or dry CH_4 reforming, CO_2 along with syngas (CO and H_2) are the primary products. In order for syngas to be used in other processes, such conversion into methanol or ammonia, these components must be separated. Polymeric membranes have been proven to selectively separate CO_2 over syngas, due to the favoured transport mechanism. Along with these materials, mixed matrix membranes are also proving to be novelty separation methods, combining the adsorption and membrane separation mechanisms to further improve their performances. In this study, commercial pure PDMS membranes, along with PDMS mixed matrix membranes including activated carbon and zeolites were investigated for CO_2 removal from syngas

38 **Iron Nanoparticle Functionalized Membrane for PCB Degradation from Water.**

Hongyi Wan[1], *Minghui Gui*[1], *Lindell Ormsbee*[2], *and Dibakar Bhattacharyya*[1]
(1)Department of Chemical and Materials Engineering, University of Kentucky, Lexington, KY, (2)Department of Civil Engineering, University of Kentucky, Lexington, KY

39 **Novel Supra-Molecular Organic Framework (SOF) Based Mixed Matrix Membranes (MMMs) for CO2/CH4 Separation.**

Zoban V. Singh[1], *Richard Noble*[1], *Douglas L. Gin*[1], *Wei Zhang*[2], *LiLi Tan*[2], *and Matthew G. Cowan*[3]
(1)Department of Chemical and Biological Engineering, University of Colorado, Boulder, CO, (2)Department of Chemistry, University of Colorado, Boulder, CO, (3)Department of Chemistry & Biochemistry, University of Colorado, Boulder, CO

40 **Fouling Control Using Temperature Responsive Membranes Activated Using Superparamagnetic Iron Oxide Nanoparticles.**

Sneha Chede[1], *Geoffrey Bothun*[2], *and Isabel Escobar*[3]
(1)University of Kentucky, Lexington, KY, (2)Chemical Engineering, University of Rhode Island, Kingston, RI, (3)Department of Chemical and Materials Engineering, University of Kentucky, Lexington, KY

41 **Tuning the Carboxylic Groups of the Thin-Film Composite Membranes and Their Effects on the Performance.**

Jianzhong Xia[1], *Shuren Chou*[2], *Yong Lin*[3], *Qinglei Zhang*[3], *Wenjuan Peng*[3], *and Yili Chen*[3]
(1)R&D Center, Beijing OriginWater Technology, Beijing, China, (2)R&D Center, Beijing Origin Water Co., Ltd., Beijing, China, (3)BeijingOriginWater, Beijing, China

43 **Extended Field Trials of Sweep Modules for Carbon Capture with Flue Gas from a Power Plant.**

Lloyd S. White[1], *Karl D. Amo*[2], *Tony Wu*[3], *and Timothy C. Merkel*[4]
(1)39630 Eureka Drive, Membrane Technology & Research, Newark, CA, (2)MTR, Newark, CA, (3)National Carbon Capture Center, Southern Company, Wilsonville, AL, (4)Membrane Technology & Research, Newark, CA

ENERGY APPLICATIONS (FUEL CELLS, BATTERIES, BUILDINGS)

44 **Analysis of Flux Reduction Behaviors of PRO Hollow Fiber Membranes: Experiments, Mechanisms, and Implications.**

Junying Xiong
Department of Chemical and Biomolecular Engineering, National University of Singapore, Singapore, Singapore

Membranes for Air-to-Air Water Vapor Exchangers.

Ryan Huizing, *Frankie Wong, and Hao Chen*
dPoint Technologies, Vancouver, BC, Canada

Energy recovery ventilation (ERV) is an emerging application for polymeric membranes used in energy efficient building ventilation systems. In these enthalpy exchange devices, heat and moisture are transported through a membrane between fresh supply ventilation air streams and stale building exhaust air streams. Membranes for these devices must have high water vapor permeance and good selectivity of water vapour over indoor air contaminants.

This presentation will discuss ongoing membrane research and development efforts at dPoint Technologies. Experimental results on multi-layer membrane structures based on several substrate platforms are presented, demonstrating the contribution of substrate and selective layer resistances to transport in these air-to-air transport devices. Several selective layer polymers are also discussed and results are reported in terms of water vapour permeability and selectivity for water vapour over gases and volatile organic compounds. The thermal effects on permeability in these polymers and interactions of water and water vapour with these polymers are also discussed. Specifically development of the "Gryphon" series of membranes will be discussed, these blended membranes are designed to have low VOC transport under humid air conditions. Results from mixed gas studies demonstrate how high humidity impacts selectivity when using various polymers. Focus areas for future membrane, exchanger, and process research will be discussed.

46 **Highly Proton-Conductive Polyelectrolyte Membranes with Greatly Reduced Water Swelling.**

Joseph Aboki, Shuangjiang Luo, and ***Ruilan Guo***
Department of Chemical and Biomolecular Engineering, University of Notre Dame, Notre Dame, IN

47 **Polymer Blend Electrolyte Membrane Containing Hydroxyl Group for Long Term Durable Proton Exchange Membrane.**

Bo Hyun Kim[1,2]*, So Young Lee*[1]*, Jong Hyun Jang*[1]*, Dirk Henkensmeier*[1]*, Sung Jong Yoo*[1]*, Ki Wan Bong*[2]*, and Hyoung-Juhn Kim*[1]
(1)Fuel Cell Research Center, Korea Institute of Science and Technology (KIST), Seoul, Korea, Republic of (South), (2)Department of Chemical and Biological Engineering, Korea University, Seoul, Korea, Republic of (South)

Graphene Quantum Dot Sensitization and Nitrogen Doping of Ordered Mesoporous TiO2 Thin Films for Visible Light Driven Water Splitting.

Syed Islam[1]*, Namal Wanninayake*[2]*, Allen Reed*[2]*, Doo Young Kim*[2]*, and Stephen E. Rankin*[1]
(1)Department of Chemical and Materials Engineering, University of Kentucky, Lexington, KY, (2)Department of Chemistry, University of Kentucky, Lexington, KY

The optical and electronic properties of TiO_2 thin films provide tremendous opportunities in several applications including photocatalysis, photovoltaics and photoconductors for energy production. Despite many attractive features of TiO_2, critical challenges include the innate inability of TiO_2 to absorb visible light and the fast recombination of photoexcited charge carriers. In this study, we prepared ordered mesoporous TiO_2 films co-modified by graphene quantum dot sensitization and nitrogen doping (GQD-N-TiO_2) for hydrogen production from photoelectrochemical water splitting under visible light irradiation. First, cubic ordered mesoporous TiO_2 films were prepared by a surfactant templated sol-gel method. Then, TiO_2 films were treated with N_2/Ar plasma for the incorporation of substitutional N atoms into the lattice of TiO_2. GQDs were prepared by chemically oxidizing carbon nano-onions. The immobilization of GQDs was accomplished by reacting carboxyl groups of GQDs with amine groups of N-TiO_2 developed by the prior immobilization of (3-aminopropyl)triethoxysilane (APTES). Successful immobilization of GQDs onto N-TiO_2 was probed by UV-Vis, FT-IR, zeta potential and contact angle measurements. The nitrogen-doped TiO_2 (N-TiO_2) showed the significant reduction of band-gap (3.5 eV to 2.88 eV) and the unprecedented enhancement (240 times) of photocatalytic hydrogen production visible light illumination. The GQD-immobilized N-TiO_2 film showed about 2 times enhancement compared to N-TiO_2, showing the important role of GQDs. This outstanding enhancement is attributed to the high surface area of mesoporous films and synergistic effects of nitrogen doping and GQD sensitization resulting in enhanced visible light absorption, efficient charge separation and transport.

48 **Application of Trimethylamine–Carbon Dioxide in Osmotic Heat Engine for Power Generation.**

Lingling Xia, *Jason T. Arena, Jian Ren, and Jeffrey R. McCutcheon*
Department of Chemical and Biomolecular Engineering, University of Connecticut, Storrs, CT

ENVIRONMENTAL EMISSION CONTROLS

Present Situation of Air Pollution and Control Strategy in China.

Yunfa Chen
Institute of Process Engineering, Chinese Academy of Sciences, Beijing, China

For recent years, people in most northern and eastern parts of China have suffered from the heavily polluted haze. Besides natural climate variations, emissions from transportation and industrial sources have been the main contributors. In this presentation, current situation of industrial air pollution emissions in China will be summarized and applications of pollution control technologies will be critically reviewed with emphasis on the catalytic oxidation of volatile organic compounds (VOCs) and potential opportunities for porous ceramic membrane filters.

Post-Combustion Flue Gas Separation Membrane Performance Testing Skid.

Victor A. Kusuma[1], *Surendar Venna*[2], *Ganpat J. Dahe*[1], *Shan Wickramanayake*[3], *Kevin P. Resnik*[4], *David Hopkinson*[5], *and Christina Myers*[6]

(1)National Energy Technology Laboratory, Pittsburgh, PA, (2)West Virginia University, 395 Evansdale Drive, PO Box 6102, Morgantown, WV 26506, USA, Morgantown, WV, (3)US DOE /NETL/AECOM, Pittsburgh, PA, (4)URS - US DOE/NETL, Pittsburgh, PA, (5)National Energy Technology Laboratory, Morgantown, WV, (6)NETL/US DOE, Pittsburgh, PA

We have recently constructed and deployed a constant pressure/variable volume membrane gas separation performance testing system at the National Carbon Capture Center (NCCC). The skid-mounted system was intended to perform fully automated testing of bench-scale experimental membranes using real flue gas obtained from Alabama Power's Gaston Unit 5, a pulverized coal fired boiler. Currently the unit is optimized to deliver up to 50 cc/min flue gas at up to 2 bar to small membrane samples (around 2 cm^2 in area) in both hollow fiber and flat sheet form, but the unit itself is amenable to future upgrades to test larger membrane modules. In this talk, we will present the unit's capabilities and share some preliminary results from testing several mixed matrix thin film composite hollow fiber membranes on actual flue gas with humidity and minor contaminants present. These lab-scale membranes were tested over short testing periods, quantifying the carbon dioxide, oxygen, nitrogen and water vapor permeances, and we expect to continue testing more membranes over longer periods during the summer.

N/A **The Removal of Elmental Mercury from Flue Gas By Different Materials.**

Yinghong Wu
Institute of Process Engineering, Chinese Academy of Sciences, Beijing 100190, People's Republic of China, Beijing, China

Dust-Removal Technique Combined with Dust Pre-Charge and Electric Ceramic Tubes Loaded with High Voltage.

Haidi Liu
State loboraotry of multiphase complex system, The Institute of Process Engineering, CAS, Beijing, China

Abstract:A new dust elimination technique combined with dust pre-charge and electric ceramic tubes loaded with high voltage was investigated. During the operation of this technique, the dust was firstly pre-charged by pre-charge appliance with DC negative corona (-20 kV), and then the gas containing dust was forced to penetrated the wall of porous electric ceramic tubes loaded with negative high voltage (-5000 V) to undergo dust elimination. It was revealed by the results that, this new technique obviously decreased the pressure drop of the ceramic tubes during the de-dust process. With the new technique, the increasing speed of the pressure drop was only 0.18 times to that with normal technique in same conditions during the same operation time. The increasing speed of the pressure drop under the new technique was only 0.29 times to that with normal technique in same conditions under the same dust layer mass. The results were high possibly caused by the fact that the electrostatic repulsion force between the dust particles gives rise to loosely compacted dust layer, furthermore, the electrostatic repulsion between the electric porous ceramic tubes and the pre-charged dust particles bring about very loose attachment between the dust layer and tube surface, as a result, the dust layer was easy to remove and its pressure drop was greatly decreased. This new technique was expected to obviously reduce the pressure drop of ceramic tube in industrial de-dust operation, and prolong its regeneration period.

Preparation and Application of Ceramic Membranes for Hot Metallurgical Gas Cleaning.

Kaiqi Liu[1], *Yunfa Chen*[2], *and Haidi Liu*[3]
(1)State Key Laboratory of Multi-phase Complex Systems, Institute of Process Engineering, Chinese Academy of Sciences, Beijing, China, (2)Institute of Process Engineering, Chinese Academy of Sciences, Beijing, China, (3)State loboraotry of multiphase complex system, The Institute of Process Engineering, CAS, Beijing, China

Alumina-based ceramic filtration membranes were prepared on a large-pore silicon carbide supports using alumina and mullite particles as the raw materials by spray-coating, followed by dip-coating for modification. The preparation process, including the powder pre-treatment, the formation and sintering of membranes, was systematically studied. Optimization of the raw materials ratio was found to be necessary to avoid crack forming during sintering and application. The microstructure, pore properties and gas permeation flux of prepared membranes were investigated. The preferred sintering temperature for the membranes was in the range of 1190–1220 °C based on the results of SEM, pore-structure and pressure drop. Low cost alumina-based membrane tubes with average pore size 6.7μm and pressure drop of 470Pa under the filtration velocity of 1 m/s were obtained. The length, outer diameter, and wall thickness of the membrane tubes were 1600mm, 60 mm and 10 mm, respectively. A dust filter equipped with 1760 resulting membrane tubes (capacity:45000Nm^3/h) was built at Huaian Special Steels (East China) to clean metallurgical sintering flue gas. Efficiencies of filtration of sintering flue gas were found to be greater than 99.95% which gave rise to very low dust content in the outlet fume gas (<10 mg/Nm^3). These results make the ceramic membrane filter very promising for hot and corrosive industrial exhaust gas filtration applications.

Functionalized Cellulose Nanofiber Membranes for Heavy Metals Removal from Impaired Water.

Nithinart Chitpong[1] *and Scott M. Husson*[2]
(1)Department of Chemical and Biomolecular Engineering, Clemson University, Clemson, SC, (2)Chemical and Biomolecular Engineering, Clemson University, Clemson, SC

In this project, high-performance, nanofiber-based ion-exchange membranes were prepared for removal of heavy metals from impaired waters. Our hypothesis was that such membranes would have high metal ion capacities resulting from large surface area-to-volume ratios and high water permeability relative to barrier-type membranes such as reverse osmosis and nanofiltration. The nanofiber membranes were produced by electrospinning cellulose acetate nanofiber mats, which were strengthened by thermal-mechanical annealing and then converted to regenerated cellulose nanofiber mats. Subsequent surface modification transformed the membranes into ion-exchange membranes for removal of trace metal ions in water.

Cellulose acetate was selected as the starting material because it has been used widely to electrospin fibers due to simple processing. The electrospinning formulation and conditions yielded fiber mats having a fiber diameter in submicron range and a thickness range of 160 to 330 μm. Following annealing, the nanofiber mats were converted to regenerated cellulose, which is a low-fouling material that can be modified using convenient surface chemistry to introduce the ion-exchange ligands.

Ion exchange using carboxylate anion ligands is a potential strategy to capture cadmium and nickel ions found in industrial wastewaters. Thus, the regenerated cellulose nanofiber mats were modified by grafting poly(itaconic acid) (dicarboxylic acid side groups) and poly(acrylic acid) (monocarboxylic acid side groups) to the surfaces of individual nanofibers that had been modified with poly(glycidyl methacrylate).The success of poly(itaconic acid) and poly(acrylic acid) grafting was confirmed by attenuated total reflectance Fourier-transform infrared spectroscopy.

Direct-flow filtration measurements showed that the nanofiber membranes have water permeabilities of 400-1600 $L/m^2/bar/h$, much higher than reverse osmosis or nanofiltration membranes. The single-component ion-exchange isotherms were measured at constant pH for cadmium, nickel, and calcium ions. Competitive sorption measurements were made using environmentally relevant concentrations of these ions to determine the selectivity of the membranes for cadmium ions. Experimental isotherms for Cd-Ca and Cd-Ni were compared to model predictions. The multi-component isotherms for Cd-Ca and Cd-Ni systems were modeled using the competitive Langmuir model with Langmuir parameters obtained by fitting single-component isotherms.

Results show that the polymer-modified nanofiber membranes have outstanding permeability and high selectivity for heavy metals over a common competing anion present at much higher concentration. The nanofiber membrane adsorbers bind cadmium ions with static capacities that are similar to and, in some cases, superior to commercial ion-exchange resins and activated carbon.

50 **Photocatalytic Membrane Reactors for High Throughput Virus Removal and Inactivation.**

Bin Guo[1], *Brian Starr*[1], *Irene Xagoraraki*[1], *and* ***Volodymyr V. Tarabara***[2]
(1)Civil and Environmental Engineering, Michigan State University, East Lansing, MI,

(2)Department of Civil and Environmental Engineering, Michigan State University, East Lansing, MI

GAS SEPARATION I - NOVEL MEMBRANE MATERIALS

Preparation and Characterization of a Hydrophobic Metal-Organic Framework Membrane.

***Jian Liu**, Nathan Canfield, and Wei Liu*
Pacific Northwest National Laboratory, Richland, WA

A hydrophobic metal-organic framework (MOF) UiO-66-CH_3 is prepared and its solvothermal stability is investigated in comparison to UiO-66. It is confirmed that the MOF stability is enhanced by introduction of the two methyl groups, while the water adsorption is reduced. Given its hydrophobicity and stability, UiO-66-CH_3 is proposed as an attractive membrane material for gas separation under moisture conditions. The UiO-66-CH_3 membrane is prepared on a 50μm-thick porous Ni support sheet for the first time by use of a secondary growth method. It is found that uniform seed coating on the support is necessary to form a continuous membrane. In addition to growth time and temperature, presence of a modulator in the growth solution is found to be useful controlling parameters for hydrothermal membrane growth on the seeded support. A dense, inter-grown membrane layer is formed under 140 °C and 160 °C for 24 h with modulator addition. The membrane surface comprises 500 nm octahedral crystals, which are supposed to grow out of the original 100 nm spherical seeding crystals. The separation characteristics of resulting membranes are characterized with pure CO_2, air, CO_2/air mixture, and humid CO_2/air mixture. CO_2 permeance as high as 1.9E-06 mol/m^2/s/Pa at 31°C is obtained. Unlike the hydrophilic zeolite membranes, CO_2 permeation through this membrane is not blocked by the presence of water vapor in the feed gas. The results suggest that this MOF framework is a promising membrane material worth to be further investigated for separation of CO_2 and other small molecules from humid gas mixtures.

55 The Latest Development of Thin-Sheed Zeolite Membranes for Building Applications

Wei Liu
Pacific Northwest National Laboratory, Richland, WA

Effects of Pyrolysis Conditions on Gas Separation Properties of 6FDA/Detda:Daba(3:2) Derived Carbon Molecular Sieve Membranes.

Shilu Fu[1], *Graham Wenz*[2], *Edgar S. Sanders*[3], *Sudhir Kulkarni*[4], *and William Koros*[5]
(1)process engineering group, American Air Liquide DRTC, Newark, DE, (2)School of Chemical and Biolmolecular Engineering, Georgia Institute of Technology, Atlanta, GA, (3)Air Liquide Advanced Separation, Newport, DE, (4)Air Liquide, Newark, DE, (5)Chemical & Biomolecular Engineering, Georgia Institute of Technology, Atlanta, GA

***Corresponding author. Tel.: +1 4043852845; fax: +1 4043852683. E-mail address: wjk@chbe.gatech.edu (W.J.Koros).**

A previously reported polyimide, 6FDA/DETDA:DABA(3:2), was pyrolyzed under different protocols to produce carbon molecular sieve (CMS) dense film membranes for separation of important gas pairs, including pure gases CO_2/CH_4, O_2/N_2 and mixture gases 50% CO_2/50% CH_4 and 50% C_3H_6/50% C_3H_8. This study investigated the effects of pyrolysis temperature, O_2 doping, and precrosslinking on the separation performance of 6FDA/DETDA:DABA(3:2) CMS membranes. Comparing to the precursor membranes, separation performance of all 6FDA/DETDA:DABA(3:2) CMS membranes improves significantly with both higher permeability and selectivity. The CMS film pyrolyzed with a novel method, i.e., 800 ^{o}C with precrosslinking, shows very attractive separation performance with CO_2 and O_2 permeability of 4678 Barrer and 683 Barrer, CO_2/CH_4, O_2/N_2 selectivity of 71.5 and 8.0. Sorption measurements provide insight into the pore size distributions among these CMS membranes. The effects of these three parameters during pyrolysis on the CMS membranes gas separation performance are summarized and compared, hypotheses regarding the structure changes brought about by changing these parameters are also provided.

High-Performance Zeolitic Imidazolate Framework ZIF-8 Membranes Fabricated By Interfacial Microfluidic Processing in Polymeric Hollow Fibers.

Kiwon Eum[1], *Chen Ma*[1], *Ali A. Rownaghi*[2], *Christopher W. Jones*[1], *and Sankar Nair*[1]
(1)School of Chemical & Biomolecular Engineering, Georgia Institute of Technology, Atlanta, GA, (2)School of Chemical and Biomolecular Engineering, Georgia Institute of Technology, Atlanta, GA

High-performance ZIF-8 Membranes for Propylene Separation Fabricated by Interfacial Microfluidic Processing in Polymeric Hollow Fibers

Kiwon Eum[1], *Chen Ma*[1], *Ali Rownaghi*[3], *Christopher W. Jones*[1,2], *and Sankar Nair*[1]*

[1]School of Chemical & Biomolecular Engineering, Georgia Institute of Technology, Atlanta, GA

[2]School of Chemistry and Biochemistry, Georgia Institute of Technology, Atlanta, GA

[3]Department of Chemical & Biochemical Engineering, Missouri University of Science and Technology, Rolla, MO

Metal Organic Frameworks (MOFs) are nanoporous materials consisting of organic linkers coordinated to metal ions in crystalline structure. They are potentially attractive as energy-efficient gas separation materials and membranes.[1] The MOF material ZIF-8 has great potential for use in C_3H_6/C_3H_8 separation membranes. However, industrial propylene separation relies on energy-intensive high-pressure distillation due to the lack of a high-performance membrane that can meet operating requirements.[2] Most of the reported ZIF membranes are fabricated by non-scalable techniques or develop defects when challenged at higher pressures (5-10 atm). Previously, we have reported a novel membrane fabrication technique called Interfacial Microfluidic Membrane Processing (IMMP) that has potential to overcome scalability and economic feasibility issues.[3]Here, we will discuss in detail the unprecedented high performance of ZIF-8 hollow fiber membranes achieved by rational engineering of the IMMP technique, to yield membranes with excellent selectivities and permeances at ambient conditions as well as elevated pressures and temperatures of industrial interest.

*Corresponding author: sankar.nair@chbe.gatech.edu

[1] H.-C. Zhou, J. R. Long, O. M. Yaghi, *Chemical Reviews*2012, 112, 673.

[2] B. R. Pimentel, A. Parulkar, E.-k. Zhou, N. A. Brunelli, R. P. Lively, *ChemSusChem*2014, 7, 3202.

[3] A. J. Brown, N. A. Brunelli, K. Eum, F. Rashidi, J. R. Johnson, W. J. Koros, C. W. Jones, S. Nair, *Science* 2014, 345, 72.

Separation of Olefins and Paraffins Using Stable Customized Amorphous Fluoropolymer (CAF) Membranes.

Sudip Majumdar, *Yosuke Koizumi, Ken Loprete, Kenneth J. Pennisi, Stuart Nemser, Andrew Feiring, and Ning Shangguan*
Compact Membrane Systems, Wilmington, DE

The oil and gas industry expends 250 trillion BTU/year separating olefins and paraffins (e.g., propylene from propane) through the use of costly and energy intensive distillation columns, a technology that is little changed in the last 40 years. The separations of these materials are some of the costliest, most energy intensive, and most technically difficult separations in the industry due to the very similar size and nature of the molecules being separated. The ability to separate these materials using a stable membrane would provide substantial economic benefit to petrochemical processes and dramatically decrease the energy required.

Using customized amorphous fluoropolymer (CAF), Compact Membrane Systems has developed membranes which have shown high propylene flux and propylene/propane selectivity over a period of greater than 180 days. The permeance and selectivity combination of the membranes are above the industry Robeson curve. While we have previously shown data for propylene/propane separation, we have recently demonstrated the ability to separate C2s (as well as C3s) obtaining very high olefin permeance (~270 GPU) and selectivity (>70) for ethylene/ethane separation. In addition, membrane stability was evaluated in presence of poisons such as hydrogen sulfide, acetylene and hydrogen. We will discuss these results as well as some of our economic modeling showing very low payback times and positive IRRs for a variety of scenarios governing petrochemical processes.

56 **Organo-Soluble Polybenzoxazoles (PBOs) As Gas Separation Membranes: New Synthesis and Gas Transport Properties.**

Ashish Kushwaha, Shuangjiang Luo, Jennifer Wiegand, and ***Ruilan Guo***
Department of Chemical and Biomolecular Engineering, University of Notre Dame, Notre Dame, IN

N/A **Membrane.**

Vicki Chen
School of Chemical Engineering, The University of New South Wales, Sydney, Australia

Influence of Humidity on Sulfonyl-Containing Polybenzimidazoles for Gas Separations.

Joshua D. Moon[1]*, Michele Galizia*[1]*, Hailun Borjigin*[2]*, Ran Liu*[2]*, Pellegrino Musto*[3]*, Pietro La Manna*[3]*, Judy S. Riffle*[4]*, Benny D. Freeman*[1]*, and Donald R. Paul*[1]
(1)McKetta Department of Chemical Engineering, The University of Texas at Austin, Austin, TX, (2)Macromolecules and Interfaces Institute, Virginia Tech, Blacksburg, VA, (3)Institute of Polymers, Composites and Biomaterials, National Research Council of Italy, Pozzuoli, Italy, (4)Chemistry, Virginia Tech, Blacksburg, VA

Due to their high chemical and thermal stability, polybenzimidazoles (PBIs) are promising candidates for use as gas separation membranes for high temperature applications. A series of sulfonyl-containing PBIs based on the 3,3',4,4'-tetraaminodiphenylsulfone (TADPS) monomer were recently synthesized and shown to have enhanced solubilities in common organic solvents relative to commercial PBI (Celazole™), thus improving their processability. Celazole™ and TADPS-based PBIs show a strongly hydrophilic behavior, and can uptake up to 25% water by mass at room temperature. Due to the potential effects of water uptake on both mechanical and gas transport properties, this study seeks to understand the fundamentals of the interactions between water and PBIs. Water vapor gravimetric sorption and dilation measurements have been performed at 35 °C, and a significant hysteresis was observed during desorption experiments, due to swelling and plasticization. The specific volume of the water-polymer mixture as well as the partial molar volume of water in the glassy polymer have also been calculated. Furthermore, a separate FTIR analysis showed that strong hydrogen-bonding interactions occur between water molecules and the PBI backbone.

GAS SEPARATION II - CO2 SEPARATION

Crosslinked Polyimide Hollow Fiber Membranes for CO2 Removal from Natural Gas: From Laboratory to Field (Featured talk).

Daniel Chinn[1], *Nitesh Bhuwania*[1], *Shabbir Husain*[2], *Stephen Miller*[1], *Claire Daniel*[2], *and Katherine Bell*[2]
(1)Chevron, Richmond, CA, (2)Chevron, Houston, TX

Polymeric hollow fiber membranes can remove CO_2 from natural gas with compact modular systems that have a small footprint and low operating cost, making them especially attractive for offshore applications. While these membranes have demonstrated attractive performance in selectivity and productivity, the sensitivity of the polymer membrane to impurities and harsh conditions continues to be a drawback. These impurities, such as C_3+ hydrocarbons, as well as a high partial pressure of CO_2, due to its enhanced solubility in glassy polymers, can lead to swelling (plasticization) of the polymer, resulting in a loss of selectivity. This has led to considerable research in polyimides containing crosslinkable groups within the chains which, upon crosslinking, have shown significantly more resilience to degradation by impurities or high CO_2 pressure. Such crosslinkable groups include carboxylic acids which can then form chemically crosslinked bonds with multifunctional alcohols. Chevron's proprietary XLP (Cross-Linked Polyimide) hollow fiber membrane technology, developed in a joint program with Georgia Tech, offers both high flux and high selectivity, and attractive robustness for CO_2/CH_4 separation. This presentation will cover these aspects of XLP membranes:

1. The evolution of XLP membranes from university R&D to industrial bench-scale
2. How advanced gas-separation membranes, such as XLP create value from an end-user perspective
3. Performance testing of XLP membranes, from clean laboratory environments to harsh field conditions
4. Commercialization aspects of XLP and areas for further performance optimization

57 Pilot-Scale Development of a PEEK Hollow Fiber Membrane Contactor Process for Post-Combustion CO2 Capture.

Shiguang Li[1], *Travis Pyrzynski*[1], *Naomi Klinghoffer*[1], *Timothy Tamale*[1], *James Aderhold*[1], *S. James Zhou*[1], *Howard S. Meyer*[1], *Yong Ding*[2], *and Benjamin Bikson*[2]

(1)Gas Technology Institute, Des Plaines, IL, (2)Air Liquide Advanced Separations, Woburn, MA

58 **High Ionic Liquid Content Membranes for Effective Biogas (CO2/CH4) Separation.**

Karel Friess
Institute of Chemical Technology, Prague 6, Czech Republic

High Performance UV-Irradiated PIM-1-Based Mixed-Matrix Membranes for Carbon Capture.

Lin Hao[1], *Kuo-Sung Liao*[1], *and Tai-Shung Chung*[2]
(1)Chemical & Biomolecular Engineering, National University of Singapore, Singapore, Singapore, (2)Chemical and Biomolecular Engineering, National University of Singapore, Singapore, Singapore

A series of mixed-matrix membranes (MMMs) consisting of zeolitic imidazolate framework-71 (ZIF-71, Zn(dcIm)$_2$) and polymer of intrinsic microporosity (PIM-1) were investigated before and after ultraviolet (UV) irradiation for various gas separations. Well compatible MMMs are formed between ZIF-71 nanoparticles and PIM-1. The incorporation of ZIF-71 considerably enhances the gas permeability without compensating the gas pair selectivities of O_2/N_2, CO_2/N_2 and CO_2/CH_4. The PIM-1 comprising 30 wt% ZIF-71 exhibits a 154% increase in CO_2 permeability. FESEM and PAS confirm the formation of an ultrathin dense layer at the membrane surface after the UV treatment. The UV-irradiated PIM-1 has CO_2/CH_4, CO_2/N_2 and O_2/N_2 selectivities of 34.1, 29.8 and 6.7, respectively. The UV-treated MMMs possess much higher permeability compared to that of pristine UV-irradiated PIM-1 with similar gas pair selectivities. Under mixed gas tests, the UV treated PIM-1 containing 20 wt% ZIF-71 has a CO_2 permeability of around 2000 barrers and a CO_2/CH_4 selectivity of 32. This performance outperforms PIM-1 and UV treated PIM membranes. The UV-PIM-1/ZIF-71 membranes have superior separation performance with both high permeability and selectivity for CO_2/CH_4 or other gas pair separation.

CO2 Capture from Power Plant Flue Gas By Polaris Membranes: Update on Field Demonstration Tests.

Jay Kniep[1], *Carlos Casillas*[1], *Ken Chan*[2], *Don Fulton*[1], *Jurgen Kaschemekat*[2], *Jennifer Ly*[2], *Tim Merkel*[3], *Vincent Nguyen*[3], *Zhen Sun*[1], *Xuezhen Wang*[1], *Xiaotong Wei*[3], *and L. S.*

White[4]
(1)MTR, Newark, CA, (2)Membrane Technology and Research, Inc., Menlo Park, CA, (3)Membrane Technology and Research, Inc., Newark, CA, (4)Membrane Technology and Research, Inc (MTR), Menlo Park, CA

Working with the US Department of Energy, MTR has developed new membranes and process designs to recover CO_2 from power plant flue gas. MTR Polaris membranes have CO_2 permeances ten times higher than standard commercial membranes, which greatly reduces the cost of a membrane capture system. These membranes are combined with a novel process design that uses incoming combustion air to sweep membranes and recycle CO_2 to the boiler. Design calculations estimate that this membrane process can capture 90% of the CO_2 in flue gas as a supercritical fluid using approximately 25% of the plant power, at a cost of $40-$50/ton of CO_2 captured. This translates to an increase in the levelized cost of electricity of about 50%.

In 2010, MTR completed a three-month field demonstration of the membrane CO_2 capture process with Arizona Public Service (APS) at their Cholla coal-fired power plant. The pilot system used commercial-scale membrane modules, and removed 90% CO_2 from a post-FGD flue gas slipstream containing 1 ton CO_2/day. In late 2011, the system was moved to National Carbon Capture Center (NCCC) operated by Southern Company in Birmingham, AL, for further demonstration. Overall, the system has an accumulated run time of over 10,000 hours, and the membrane modules have exhibited stable performance. With experience learned from these tests, MTR has designed a scale-up small pilot unit to capture 20 ton CO_2/day from a power plant slipstream (equivalent to 1 MW_e-scale power generation). The system was initially commissioned in August 2014 and testing concluded in June 2015. During testing, advanced modules demonstrating higher packing density, lower pressure drop, and cost savings were validated. Current work involves fully integrating the small pilot unit with Babcock & Wilcox's (B&W) 1.8 MW_{th} coal fired research boiler in Barberton, OH. The field demonstration to be conducted at B&W will be the first time membrane operation with real flue gas and boiler operation with CO_2-laden air have been conducted in a single test. Technical results to date for the field tests, and future plans on technical development will be discussed in this presentation.

Bench Scale Testing of Next Generation Hollow Fiber Membrane Modules.

David Hasse[1], *Alex Augustine*[2], *Sudhir Kulkarni*[3], *Trapti Chaubey*[2], *Dean Kratzer*[1], *Shilu Fu*[4], *and Edgar S. Sanders*[5]
(1)Research & Development, Air Liquide, Newark, DE, (2)Research and Development, Air Liquide, Newark, DE, (3)Air Liquide, Newark, DE, (4)process engineering group, American Air Liquide DRTC, Newark, DE, (5)Air Liquide Advanced Separation, Newport, DE

Air Liquide (AL) is developing a next generation polyimide membrane material (PI-2) for application in our cold membrane hybrid process for CO_2 capture. Combining the cold membrane operation with an integrated CO_2 compression and purification unit significantly reduces the overall cost of CO_2 capture. The current work is aimed at developing the PI-2 material to a commercial scale, then validating its performance under real flue gas conditions.

Previous work (including grants DE-FE0004278 and DE-FE0013163) showed promising results for the cold membrane hybrid process utilizing the commercial available AL membrane bundles (PI-1 material). The Levelized Cost of Electricity (LCOE) for 90% CO_2 capture from a 550 MWe net coal power plant was projected to increase by 48 – 53%. The cold membrane technology was demonstrated at the National Carbon Capture Center with a 0.3 MWe field-test unit incorporating two 12" OD PI-1 commercial membrane bundles. That system exceeded its performance targets with stable membrane operation for over 500 hours. Finally, the new PI-2 membrane fiber was spun at lab scale and showed significant performance improvements: up to 10 times the CO_2 permeance and similar CO_2/N_2 selectivity as the existing commercial fiber (3 – 5 times the potential productivity per bundle).

The current project seeks to develop this next generation PI-2 material to a technology readiness level of 5 via manufacturing scale-up and testing with 300 - 1,000 Nm^3/h of real flue gas, 0.1 - 0.3 MWe equivalents. Air Liquide manufacturing methodology will be applied to fabricate prototype membrane modules. Once the manufacturing methods prove reliable and performance is demonstrated in line with expectation, commercial scale bundles will be fabricated. Those commercial scale bundles will be performance tested with real flue gas at the National Carbon Capture Center. Finally, simulation studies of novel hybrid processes have begun that take advantage of the next generation fiber performance. The advantages and drawbacks of several arrangements will be discussed.

The current development work is supported by the Department of Energy National Energy Technology Laboratory under Award Number DE-FE0026422 (J. Figueroa, Program Manager).

60 **Dual CO2- and N2- Selective Intrinsically Microporous Hydroxyl-Functionalized Triptycene-Based Polyimide for Natural Gas Upgrading.**

Fahd I Alghunaimi[1], *Bader Ghanem*[1], *Nasser Alaslai*[1], *Mohsin Mukaddam*[1], *and Ingo Pinnau*[2]
(1)Advanced Membranes and Porous Materials Center, KAUST, Thuwal, Saudi Arabia, (2)Advanced Membranes and Porous Materials Center, King Abdullah University of Science and Technology, Thuwal, Saudi Arabia

GAS SEPARATION III - NEW CONCEPTS

Physical Aging of Perfluoropolymers in Ultrathin Film Composite Membranes and Its Interpretation Using the Free Volume Model.

Milad Yavari[1], *Sajjad Maruf*[2], *Yifu Ding*[3], *and Haiqing Lin*[4]
(1)Chemical and Biological Engineering, University at Buffalo, Buffalo, NY, (2)Mechanical Engineering, University of Colorado at Boulder, Boulder, CO, (3)Department of Mechanical Engineering, University of Colorado at Boulder, Boulder, CO, (4)Department of Chemical and Biological Engineering, University at Buffalo, The State University of New York, Buffalo, NY

Thin film composite (TFC) membranes for gas separation are comprised of a selective layer of glass polymers, which are usually subjected to physical aging, i.e. permeability decreases with time. The aging behavior has been ascribed to the rearrangement of polymer chains in the thin film, resulting in changes in free volume (f) and glass transition temperature (T_g). Our objective is to investigate the effect of physical aging on gas permeance, selectivity and T_g of the selective layer in the TFC membranes.
Two-layer TFC membranes consisting of perfluoropolymers of Teflon® AF or Hyflon® AD at various film thicknesses (50 – 400 nm) on polyethersulfone ultrafiltration support were prepared. Pure-gas permeances of CH_4, N_2, H_2 and CO_2 at 35 °C were measured and monitored for aging over 1000 hr. Apparent T_g of the thin selective layer was measured over time by a new thermo-mechanical method, Nano Thermal Analysis. In general, aging decreases gas permeances and increases T_g, and the rate of changes is more significant for thinner films. For example, after aging for 2000 hr, N_2 permeance decreases from 1000 gpu to 550 gpu by 45% and T_g increases from 160 °C to 172 °C in a TFC membrane with 100-nm thick Teflon® AF1600. In this talk, we will compare the aging behavior and gas permeance reduction of the TFC membranes with those of freestanding thin films. More importantly, for the first time we have developed a correlation between the gas permeance reduction and T_g increase using the free volume model to explain the physical aging behavior.

61 **New Membrane Based Hybrid Unit Operation for Gas Separation.**

Ferruccio Doghieri[1], *Matteo Minelli*[1], *Giulio Sarti*[2], *Francesco Miccio*[3], *Valentina Medri*[3], *and Elena Landi*[3]
(1)Department of Civil, Chemical, Environmental and Materials Engineering (DICAM), University of Bologna, Bologna, Italy, (2)DICAM, University of Bologna, Bologna, Italy, (3)Institute of Science and Technology for Ceramics, National Research Council of Italy, Faenza, Italy

62 **Steric Hindrance Effect Demonstrated in Polyvinylamine Facilitated Transport Membranes for CO_2 Separation.**

Zi Tong[1] *and W.S. Winston Ho*[1,2]
(1)William G. Lowrie Department of Chemical and Biomolecular Engineering, The Ohio State University, Columbus, OH, (2)Department of Materials Science and Engineering, The Ohio State University, Columbus, OH

Reverse Selective Glassy Polymers for Higher Hydrocarbon Removal from Natural Gas.

Justin T. Vaughn, *Daniel J. Harrigan, Benjamin J. Sundell, John A. Lawrence III, and Tatiana Pilyugina*
Boston Downstream Research Center, Aramco, Cambridge, MA

Saudi Arabia owns one of the largest proven reserves of natural gas in the world. Much of this resource is utilized for domestic energy production; however, other opportunities exist for natural gas monetization, such as fuel gas conditioning and recovery of C_3^+ hydrocarbons for feedstock and petrochemical applications. As such, Saudi Aramco has strong interest in developing gas separation membranes for improving the recovery of C_3^+ hydrocarbons from methane in existing liquid recovery units (LRU). In this case, reverse selective membranes are needed in order to retain methane on the high pressure so that recompression into transmission pipelines can be avoided. Recently, a series of novel glassy polynorbornenes was developed that demonstrates improvement in mixed gas C_3^+/C_1 selectivity relative to commercial PDMS under field relevant conditions. Through a systematic study in the transport properties of these materials under industrially-relevant feed streams, the effect of polymer structure as well as crosslinking on membrane performance has been elucidated. This talk will discuss mixed gas transport results of these polynorbornenes compared to commercial PDMS, demonstrating the promise of these glassy membranes for reverse selective membrane applications.

64 **Transferring Membrane Polymer Properties into Gas Separation Process Performance.**

Markus Priske[1], *Tymen Visser*[2], *Kah Peng Lee*[3], *Joerg Balster*[3], *and Goetz Baumgarten*[2]
(1)Evonik Corporation, Theodore, AL, (2)Evonik Resource Efficiency GmbH, Marl, Germany, (3)Evonik Fibres GmbH, Schoerfling, Austria

Pressure-Vacuum Swing Permeation: A Novel Process for Gas Separations.

Xianshe Feng[1], *Darren Lawless*[2], *and Glenn Towe*[3]

(1)Chemical Engineering, University of Waterloo, Waterloo, ON, Canada, (2)Applied Research & Innovation, Humber Institute of Technology, Toronto, ON, Canada, (3)Imtex Membranes Corp., Mississauga, ON, Canada

A novel process mode for gas separation by means of membranes was presented. It involved cyclic pressure-vacuum swings for feed pressurization and permeate evacuation, thereby increasing the transmembrane pressure difference and the feed to permeate pressure ratio. The feed pressurization and permeate evacuation were preferably accomplished using a single pump capable of pressurization and suction. Unlike transient permeation processes reported in the literature which were based on the differences in sorption uptake rates or desorption falloff rates, this process was based on the selective permeability of the membrane for separations. The dynamic gas permeation process associated with the pressure-vacuum swing permeation was analyzed to elucidate the working principle of the process, and a parametric study was carried out to evaluate the effects of design and operating parameters on the separation performance. To exemplify the feasibility and effectiveness of the process, air separation for oxygen enrichment and CO2 separation from lanfill gas were investigated as representative applications. It was shown that the separation efficiency was enhanced by means of the pressure-vacuum swing, and both product purity and product throughput higher than those from conventional steady-state permeation could be obtained.

MEMBRANE REACTORS AND CONTACTORS

Polymeric, Catalytic, Conductive, Porous Hollow Fiber Membrane Cathode for Water and Energy Recovery from Wastewater Treatment.

NM Srivatsa Bettahalli[1]*, Krishna P Katuri*[2]*, Xianbin Wang*[3]*, Stefan Chisca*[1]*, Suzana Pereira Nunes*[1]*, and Pascal Saikaly*[2]
(1)Biological and Environmental Sciences & Engineering Division (BESE), King Abdullah University of Science & Technology (KAUST), Thuwal, Saudi Arabia, (2)Water Desalination and Reuse Center (WDRC), King Abdullah University of Science & Technology (KAUST), Thuwal, Saudi Arabia, (3)Nanofabrication Core Lab, King Abdullah University of Science & Technology (KAUST), Thuwal, Saudi Arabia

Polymeric porous hollow fiber membrane (HFM) is regarded as revolutionary solution in the wastewater industry for recovery of clean water at low cost. Further advancing research towards the energy recovery using microbial electrochemical principle has been conceptualized by precisely modifying polymeric porous HFMs to become catalytic and electrically conductive. The polymeric catalytic conductive HFMs were used as a cathode in anaerobic electrochemical membrane bioreactor (AnEMBR) for recovery of both water/energy from low organic strength wastewaters (such as sewage). In turn this mechanism also reduces biofouling by membrane scouring through the generation of H_2 bubbles on the surface of the membrane and/or generation of electrical field. The advantage of such membranes in biological waste water treatment has been reported for polymeric flat sheet [1] and inorganic HFM [2].

In this study, we have investigated the development of multilayer polymeric catalytic and conductive porous HFM cathodes using atomic layer deposition (ALD) technique. These fibers were compared with the benchmark Pt-coated (0.5 mg Pt/cm^2) carbon cloth cathodes (control reactor) generally used in microbial electrochemical systems research. They were used as cathode in AnEMBR and operated in electrolysis mode under applied potential (-0.1V vs Ag/AgCl). The e^- and H^+ generated at the anode ultimately get reduced to H_2 on the surface of cell cathode. The performance of the AnEMBR in terms of fouling propensity was compared with a control reactor, which was operated as an anaerobic MBR (AnMBR). The AnEMBr fouled significantly less than that of AnMBR due to the indigenous H_2 evolution on the conductive fibers surface. Their dual functionality and architecture open the window for sustainable harnessing of resources from wastewaters (i.e. pollution to purification) and mitigating the unavoidable phenomenon of biofouling in membrane processes.

References

[1] L. Malaeb, K.P. Katuri, B.E. Logan, H. Maab, S.P. Nunes, P.E. Saikaly, *A Hybrid Microbial Fuel Cell Membrane Bioreactor with a Conductive Ultrafiltration Membrane Biocathode for Wastewater Treatment*, Environmental Science & Technology, 47 (2013) 11821-11828.

[2] K.P. Katuri, C.M. Werner, R.J. Jimenez-Sandoval, W. Chen, S. Jeon, B.E. Logan, Z. Lai, G.L. Amy, P.E. Saikaly, *A novel anaerobic electrochemical membrane bioreactor (AnEMBR) with conductive hollow-fiber membrane for treatment of low-organic strength solutions*, Environmental Science and Technology, 48 (2014) 12833-12841.

High-Temperature Membrane Reactors: An Effective Approach for Equilibrium-Limited Reactions.

Nazanin Entesari, *Zhongtang Li, and Theodore Tsotsis*
University of Southern California, Los Angeles, CA

Reaction and separation are considered as two of the most crucial steps in chemical plants. Performance of each of these two steps can strongly affect the profitability of a given process by affecting its raw materials and energy demand and consumption. In an effort to improve the overall performance of the process, reaction and separation can be combined into a multifunctional Membrane Reactor (MR) that offers a platform for simultaneously carrying out the reaction and separation steps in one single stage. Such reactors typically consist of a catalyst bed in contact with a high-temperature membrane that either enables the removal in situ of one or more products (extractive MR) from the system or the controlled dosing of one or more of the reactants (contactor MR). Replacing the conventional two-stage reaction and separation systems with such a multifunctional reactor increases the process synergy, reduces the by-product formation as well as amount of reactants consumed. Additionally, a considerable reduction in the overall energy costs is to be expected [Sanchez and Tsotsis, 2002; Soltani et al., 2013]. While there are several reports on the applications of MR for both gas and liquid phase reactions, further studies are still required to characterize and optimize their performance, and to move towards commercialization of these multifunctional reactors [Sanchez and Tsotsis, 2002; Soltani et al., 2013].

In this talk, we will present two different examples of high-temperature MR currently under investigation by our Group. The first is a extractive MR that is used for methanol (MeOH) synthesis, and the other a contactor MR that is used for carbon (CO_2) capture and utilization (CCU). Synthesis of MeOH from syngas is a severely equilibrium-limited reaction, which means that a large fraction of the unreacted syngas feed, must be recycled to make this equilibrium-limited process economically feasible. The ability of MR to increase the per-pass yield makes them a optimal choice for process intensification [Soltani et al., 2013], particulalry for coal-deruved and biomass-derived syngas from air-blown gasifiers that contains considerable amounts of inert N_2.

Power plants are responsible for a large fraction of the man-made carbon-dioxide emissions, and are major contributors to the green house effect. Thus, reducing the amount of CO_2 emissions from such plants is of the utmost importance. The key challenge here is the dilute CO_2 concentration in the flue-gas, which make the conventional approaches such as direct sequentration technically as well as economically inefficient. Coupling of reaction and separation in a contactor MR, provides the driving force for the separation which moderates the challenge of separating dilute CO_2 from flue-gas, and can also produce valuable products (e.g., synthetic natural gas, alcohols, etc.) which can be used as fuels to reduce the overall energy demand of the plant, and to improve its efficiency.

During our talk, we will present the two multifunctional reactor configurations that are being presently experimentally investigated, and will discuss the influence of reaction conditions and separation efficiency on the final yield of the reaction. The different types of high-temperature membranes that have been investigated will be described, and the influence of membrane characteristcs on overall system performance will be described. Efforts to optimize MR performance both experimentally and numerically will be furher outlined.

References:

1. Sanchez Marcano, J.G., and Tsotsis, T. T., "Catalytic Membranes and Membrane Reactors," Wiley VCH, 2002.
2. Soltani, S., Sahimi, M., and Tsotsis T.T., "Catalytic Membrane Reactors: A Brief Overview", *Encyclopedia of Membrane Science and Technology,* Eric. M.V. Hoek and Volodymyr V. Talabar Editors, John Wiley & Sons, 2013.

67 **Bio-Mimetic Hollow Fiber Membrane Contactor Process for Low-Cost High-Purity Oxygen Production.**

Travis Pyrzynski[1], *Naomi Klinghoffer*[1], *Howard Meyer*[1], *Shiguang Li*[1], *Mahdi Fathizadeh*[2], *and Miao Yu*[3]
(1)Gas Technology Institute, Des Plaines, IL, (2)Department of Chemical Engineering, University of South Carolina, Columbia, SC, (3)Chemical Engineering, University of South Carolina, Columbia, SC

Influence of Microporous Membrane Properties on VMD and DCMD.

Lin Li *and Kamalesh K. Sirkar*
Otto York Department of Chemical, Biological and Pharmaceutical Engineering, New Jersey Institute of Technology, Newark, NJ

A variety of microporous hydrophobic flat sheet membranes of polyvinylidene fluoride (PVDF) and expanded-polytetrafluoroethylene (e-PTFE) were studied to evaluate the influence of membrane properties on their performance in desalination by vacuum membrane distillation (VMD) and direct contact membrane distillation (DCMD) processes. The membrane thickness was varied between 23 μm to 125 μm; the pore size was varied from 0.05 μm to 0.45 μm. The porosity was generally high in the range of 0.7-0.8. VMD experiments were performed over a hot brine temperature range of 65 °C to 85 °C for various feed flow rates and various vacuum levels in a rectangular chlorinated polyvinyl chloride (CPVC) cell. Membrane properties, such as the maximum pore size, tortuosity as well as the liquid entry pressure (LEP) were experimentally determined. Water vapor fluxes were predicted and compared using two models: the Knudsen diffusion and the dusty-gas model (DGM). The deviation between two models is within 1.3 %. Knudsen diffusion is the dominant regime in VMD transport since Knudsen number, K_n, for all membrane were larger than 1 at all temperatures. The boundary layer heat transfer resistance in the membrane cell and the membrane surface temperature were determined from experimental data via Wilson plot. Good agreements of membrane mass transfer coefficients and water vapor fluxes were found between the DGM simulations and the experimental results (deviation within 5%). Same membranes were studied in a circular stainless steel cell and the CPVC cell for DCMD behavior using the Wilson plot method. Experimental and predicted values of the membrane mass transfer coefficients and water vapor fluxes have been compared for brine temperature range of 65°C - 85°C and distillate temperature at 25°C. Pore size distribution (PSD) did not matter if the membrane pore size variation falls entirely in the Knudsen diffusion or the transition regime. However, for membranes having nominal pore size 0.1 μm, K_n 1, PSD played an important role in DCMD. The effect of membrane thickness on water vapor flux and thermal efficiency were also simulated and compared with the experimental results.

68 **Development of Multibore Hollow Fiber Membranes for Water Deoxygenation.**

Jincai Su
Life Sciences & Chemical Technology, Ngee Ann Polytechnic, Singapore, Singapore

69 **Fabrication & Performance Study of Pd-Au PSS Supported Membrane Reactor Via Natural Gas Steam Reforming.**

Bryce Anzelmo[1], *Jennifer Wilcox*[2], *Simona Liguori*[2], *Ivan Mardilovich*[3], *and Ed Ma*[4]
(1)Energy Resources Engineering, Stanford University, Stanford, CA, (2)Department of Energy Resources Engineering, Stanford University, Stanford, CA, (3)Department of Chemical Engineering, Worcester Polytechnic Institute, Worcester, MA, (4)Chemical Engineering, Worcester Polytechnic Institute, Worcester, MA

MEMBRANE TRANSPORT: THEORY AND CHARACTERIZATION TOOLS

N/A **Transport of Organic Compounds in DCMD.**

Judith Winglee
Duke University, Durham, NC

N/A **New Insights into Salt Transport Mechanism in Membrane Desalination.**

Viatcheslav Freger[1], *Noga Fridman-Bishop*[2], *and Vesselin Kolev*[2]
(1)Wolfson Department of Chemical Engineering, Technion - Israel Institute of Technology, Haifa, Israel, (2)Technion - Israel Institute of Technology, Haifa, Israel

New Insights into Salt Transport Mechanism in Membrane Desalination

Viatcheslav Freger*, Noga Fridman-Bishop, Vesselin Kolev,

[1]*Wolfson Department of Chemical Engineering, Technion - Israel Institute of Technology, Haifa 32000, Israel*

Despite >50 years of industrial membrane desalination, salt transport in membranes (NF and RO) and membrane structure are still insufficiently understood. Our group has studied the mechanism of membrane transport using two basic methods: theoretical analysis using analytical models and molecular dynamics and experimental measurements using filtration data and impedances spectroscopy of genuine and model membrane films. Some open fundamental questions we tried to address were:

1) Why polyamide has been the best material for RO/NF membranes

2) What is the actual bottleneck of salt transport in RO/NF: a nanopore or a dense film?

3) What is the charge of the membrane and how does it uptake salt?

4) How much salt uptake and salt transport are related;

5) How much role ion-specific effects (including H^+ and OH^-) play in salt sorption and transport;

6) How separation of complex ion mixtures (including pH changes) should be modelled etc.

Even though some question still remain open, we seem to have obtained (partly surprising) answers to some of the above questions, which will be presented in the talk.

Numerical Simulations of Concentration Polarization and Mineral Scaling in Reverse Osmosis Systems.

Nils Tilton *and Steven DeCaluwe*
Mechanical Engineering, Colorado School of Mines, Golden, CO

Fluid flow and mass transport in reverse osmosis (RO) systems are difficult to simulate numerically due to the coupling between the transmembrane filtrate flow, pressure, and osmotic pressure. This coupling is particularly challenging for real-world systems in which the local filtrate flow varies both spatially and temporally, and is further complicated for systems in which filtrate flow is also affected by mineral scaling and membrane fouling. Consequently, previous simulations of RO tend to focus on two dimensional steady cases. We present a numerical method specifically tailored to simulating RO. Using the method, we simulate the transient formation of concentration polarization layers as a function of feed Reynolds number, solute concentration, and transmembrane pressure. From the computed flow fields, we investigate the importance of advection and diffusion of solutes both normal and tangential to the membrane surface. Contrary to the thin-film approximation, we find that normal diffusion can be balanced by both normal and tangential advection within the concentration polarization layer. Finally, we use our method to simulate the transient decline in filtrate flux due to the precipitation of sparingly soluble salts on the membrane surface.

Membrane Porous Structure and Its Impact on Permeability and Selectivity.

Sourav Mondal[1], Maria Bruna[1], Ian Griffiths[1], and ***Guy Z. Ramon***[2]
(1)Mathematical Institute, University of Oxford, Oxford, United Kingdom, (2)Civil & Environmental Engineering, Technion - Israel Institute of Technology, Haifa, Israel

Membranes comprise complex porous structures, either as the actual separation medium, e.g. in the case of Dialysis, Micro- and Ultrafiltration or, in the case of composites used for gas separation, Nanofiltration and Reverse Osmosis, where an underlying porous support can impact the overall performance of the membrane. Such structural features have a wide impact on membrane performance, ultimately dictating the water permeability and solute selectivity as well as concentration polarization, propensity for and severity of fouling.

The porous structures can be characterized by their pore-size distribution as well as their spatial distribution, which can further be divided between the membrane surface distribution vs the depth distribution. Here, we theoretically probe the relative importance of these distributions for several illustrative cases: pressure-driven flow as impacted by membrane surface pore locations and depth morphologies, the obstruction presented by a support layer to diffusion through the overlying thin-film, and the selectivity of a porous membrane as affected by a porosity gradient. In particular, we examine the difference between periodic and random spatial distributions of pores.

Separation membranes come in many morphological forms, mostly dependent on the materials used as well as the fabrication method and conditions. Identifying advantageous porous structures would benefit design and motivate the development of better control over the formation of a given morphology.

71 Twisted Hollow Fiber Membranes for Enhanced Mass Transfer.

Seyed Pouria Motevalian[1]*, Ali Borhan[1], Hongyi Zhou[2], and Andrew L. Zydney[1]*
(1)Chemical Engineering, The Pennsylvania State University, University Park, PA, (2)Membrane and Separations Technology, GE Global Research, Niskayuna, NY

72 How Do Hydrodynamic Conditions in Spiral Wound Membrane Elements Translate into Shear Forces Effective for Fouling Control at Membrane Surface?.

Sepideh Jankhah[1] *and Pierre P. Bérubé[2]*
(1)Sterlitech Corporation, Kent, WA, (2)Civil Engineering, University of British Columbia, Vancouver, BC, Canada

Prediction of Vapor Recovery from Light Gas with Mesoporous Membranes Using Dynamic Mean Field Theory.

***Ashutosh Rathi**, David Ford, and Peter A. Monson*
Chemical Engineering, University of Massachusetts Amherst, Amherst, MA

Mesoporous membranes have great potential for the separation of condensable species such as in the recovery of volatile organic compounds from air or the removal of carbon dioxide from flue gas. Selectivity and permeance of membranes in these processes are influenced by mechanisms like preferential adsorption, surface flow, and capillary condensation, but predictive correlations relating these mechanisms to mesopore properties such as geometry and surface chemistry are absent. Atomistic simulations provide a pathway to such correlations but are computationally intensive. Dynamic mean field theory (DMFT), a lattice-based density functional theory that successfully describes adsorption equilibrium and uptake dynamics in mesoporous materials, is a less intensive approach that provides qualitative and semi-quantitative insight. We have recently adapted DMFT to study non-equilibrium steady-state transport in mesopores and published a computational analysis of a permporometry experiment, in which a light gas permeates through a membrane in the presence of an essentially static condensable vapor. In this work, we use DMFT to study the recovery of a condensable vapor from a light gas, a situation in which the vapor is under a substantial partial pressure gradient. Model parameters are chosen to be relevant to experimental work in the literature on recovery of volatile organic compounds (VOCs) from air. Even in a simple slit pore, the results show interesting new phenomena like capillary condensation that is confined to the part of the pore closest to high pressure (feed side) of the membrane. We explore how modifying the geometry of the pore influences the selectivity and permeance, which will be useful to researchers designing mesoporous membrane materials to perform such separations.

MEMBRANES FOR BIO-SEPARATIONS AND DEVICES

Purification of Oncolytic Viruses - a New Challenge for Membrane Processes.

*Shabnam Shoaebargh[1] and **David R. Latulippe**[2]*
(1)Department of Chemical Engineering, McMaster University, Hamilton, ON, Canada, (2)Chemical Engineering, McMaster University, Hamilton, ON, Canada

Oncolytic viruses (OVs) are a class of cancer therapies that are currently going through clinical trials on their way to full approval. They are attractive given their inherent specificity for disease targets. The upstream production procedure of these bio-therapeutic molecules vectors results in a high amount of cell debris, host cell proteins, and nucleic acids that must be removed to meet the guidelines of regulatory authorities. Thus, there is a need for robust, high-yielding manufacturing downstream processes (DSP) to improve the speed of production, along with the yields and purity of OV products. One of the primary challenges during production is the final sterile filtration step – some recent studies have reported an over 80% loss in the vector titer. The objective of this study is to develop new filtration processes that can achieve the required purity requirements but at the higher throughputs and lower losses that are needed for full-scale production of OVs.

For this work, a purified lot of the Maraba virus (MG1) was provided by the Mulcahy Family Manufacturing Facility (at the Ottawa Hospital Research Institute) to study the filtration performance in a dead-end (i.e. normal flow) configuration. The Maraba virus is from the rhabdovirus family and features a unique bullet-shaped, enveloped structure. The effects of formulation buffer conditions (salt concentrations, buffer type, buffer concentrations, and pH values), membrane type, and the length of the holding time before the filtration step were studied. Our hypothesis is that virus aggregation is the dominant reason for the high losses reported in recent studies. The transmembrane pressure (TMP) profile was used to monitor for membrane fouling and the number of plaque forming units (PFUs) was used to assess the virus concentration. Future work will use unpurified lots of the same Maraba virus solution to assess the performance of different membrane filters under actual DSP conditions.

Voltage Activated Membrane Platforms for Drug Delivery and Bioseparation.

Bruce J. Hinds
Chemical and Materials Engineering, University of Kentucky, Lexington, KY

Voltage Activated Membrane Platforms for Drug Delivery and Bioseparation

B.J. Hinds

Department of Materials Engineering, University of Washington, Seattle WA

bjhinds@engr.uky.edu

An important challenge for the membrane community is to mimic the dynamic activity of natural protein channels that outperform by orders of magnitude man-made systems based on pore size and coarse chemical selectivity. To mimic protein channel pumping on a robust engineering membrane platforms applied bias can be used to actuate charged gatekeepers and induce ionic pumping. Described here are two platforms of Carbon nanotube membranes and Anodized Aluminum Oxide (AAO) with nm-thick electrodes at pore entrances/exits. Carbon nanotubes have three key attributes that make them of great interest for novel membrane applications 1) atomically flat graphite surface allows for ideal fluid slip boundary conditions and extremely fast flow rates [*1,2*] 2) the cutting process to open CNTs inherently places functional chemistry at CNT core entrance for chemical selectivity and 3) CNT are electrically conductive allowing for electrochemical reactions and application of electric fields gradients at CNT tips. The CNT membrane, with tips functionalized with charged molecules, is a nearly ideal platform to induce electro-osmotic flow with high charge density at pore entrance and a nearly frictionless surface for the propagation of plug flow. Through diazonium electrochemical modification we have successfully bound anionic surface charge to CNT tips and along CNT cores. High electro-osmotic flows of 3 cm/s-V at are seen that are 10,000 fold faster than in conventional nanoporous materials[*3,4*] and are consistent with pressure driven flow enhancements. Use of the electro-osmotic phenomenon for responsive/programmed transdermal drug delivery devices for nicotine addiction [*5*]. Another approach is to mimic natural protein channel transport cycles with binding/transport/release/reset events. Porous alumina (AAO) membranes have top and bottom electrodes coated with thin Au layers with pore dimension tuned to match protein dimensions. At this thin layer at pore entrances, Ni-ETA is able to bind to hys-tag residue on target protein, as is commonly employed in chromatography. A binding voltage pulse attracts anionic target protein to top electrode and blocking the pore, while repelling the cationic imidazole release agent. The second voltage cycle attracts cationic release agent to top of membrane while pumping anionic target protein to bottom permeate and resetting the pumping cycle. This system was able to successfully mimic natural membrane transporter cycles and the separation efficiency of 1cm2 of membrane was comparable to convention 1cm3 volumes used in chromatography [*6*]

1 "Nanoscale hydrodynamics: Enhanced flow in carbon nanotubes" Majumder, M.; Chopra, N.; Andrews, R; Hinds, B.J *Nature* **2005**, 438, 44.

2 'Mass Transport through Carbon Nanotube Membranes in three different regimes: ionic diffusion, gas, and liquid flow' Mainak Majumder, Nitin Chopra, B.J. Hinds *ACS Nano* **2011** 5(5) 3867-3877

3 'Highly Efficient Electro-osmotic Flow through Functionalized Carbon Nanotubes Membrane' Ji Wu, Karen Gerstandt, Mainak Majunder, B.J. Hinds, *RCS Nanoscale* **2011** 3(8) 3321-28

4 " Programmable transdermal drug delivery of nicotine using carbon nanotube membranes" J. Wu, K.S. Paudel, C.L. Strasinger, D. Hamell, Audra L. Stinchcomb, B. J. Hinds *Proc. Nat. Acad. Sci.* **2010** 107(26) 11698-11702.

5 "Electrophoretically Induced Aqueous Flow through sub-Nanometer Single Walled Carbon Nanotube Membranes" Ji Wu, Karen Gerstandt, Hongbo Zhang, Jie Liu, and B. J. Hinds *Nature Nano* **2012** 7(2) 133-39

6 "Dynamic Electrochemical Membranes for Continuous affinity protein separation", Z. Chen, T. Chen, X. Sun and B.J. Hinds. Advanced Functional Materials **2014** 24(27) 4317-23

Developing High Performance Responsive HIC Membrane Adsorbers.

***Xianghong Qian**[1], Zizhao Liu[2], and S. Ranil Wickramasinghe[2]*
(1)Department of Biomedical Engineering, University of Arkansas, Fayetteville, AR,
(2)Department of Chemical Engineering, University of Arkansas, Fayetteville, AR

Fast convective mass transport, low pressure drop and better accessibility to large biomolecules make membrane adsorbers an attractive alternative to packed-bed chromatography used in the downstream process. Responsive hydrophobic interaction chromatography (HIC) membranes are investigated for protein purification in a bind/elute mode. The surface of these responsive HIC is grafted with polymeric ligands with tunable hydrophobicity. Unlike traditional HIC ligands, responsive HIC ligands can switch between hydrophobic state (under high salt) and hydrophilic state (under low salt) for protein binding and elution respectively reducing the possibility of irreversible binding.

Thermo-responsive poly (N-vinyl caprolactam) (PVCL) ligands have been grafted on the surface of regenerated cellulose membrane (0.45 μm pore size) via atom transfer radical polymerization (ATRP). The conformations and degrees of hydration/dehydration of these responsive ligands are concentration and salt specific. Systematic studies have been performed to understand and optimize the chromatographic behavior of these responsive HIC membranes using BSA and IgG as model proteins. In addition, the effects of copolymerization with other monomers such as hydroxyethyl methacrylate, N-(2-hydroxypropyl) methacrylamide and hydroxyethyl acrylamide on binding capacity and recovery were also investigated. Our results suggest that the 3-D conformation, monomer property and composition all strongly affect the binding capacity and recovery.

73 **Rapid Uranium Isotopic Analysis Using Ultrafiltration and Alpha Spectroscopy.**

Christine Duval[1], *Timothy DeVol*[2], *and Scott M. Husson*[3]
(1)Department of Chemical and Biomolecular Engineering, Clemson University, Clemson, SC, (2)Department of Environmental Engineering and Earth Science, Clemson University, Anderson, SC, (3)Chemical and Biomolecular Engineering, Clemson University, Clemson, SC

74 **A Method to Evaluate Membrane Protein Compatibility with Biomimetic Membranes.**

Tingwei Ren[1], *Mustafa Erbakan*[2], *Mariusz Grzelakowski*[3], *Peter J. Butler*[4], *and Manish Kumar*[5]
(1)Chemical Engineering, the Pennsylvania State University, State College, PA, (2)Department of Biosystem Engineering, Bozok University, Yozgat, Turkey, (3)Applied Biomimetic Inc., Cincinnati, OH, (4)Bioengineering, Penn State University, University Park, PA, (5)Chemical Engineering, The Pennsylvania State University, University Park, PA

75 **Fabrication of Ultrathin ZnO Nanocrystal-Nanocellulose Composite Paper.**

Jingda Wu *and Lih Lin*
Electrical Engineering, University of Washington, Seattle, WA

N/A **Concentrating Membrane Proteins without Concentrating Detergent.**

Hasin Feroz
Department of Chemical Engineering, Pennsylvania State University, University Park, PA

MICRO, ULTRA, AND NANO-FILTRATION APPLICATIONS II

77 **Comparison of Polymeric and Ceramic MF/UF Membranes for the Treatment of Produced Water.**

Hooman Vatankhah
CEE, Colorado School of Mines, Golden, CO

79 **Desalination of Brackish Inland Surface Water Using Nanofiltration: Fouling Control and Water Quality Improvements.**

Shankar Chellam[1] *and Mutiara Ayu Sari*[2]
(1)Department of Civil Engineering, Department of Chemical Engineering, Texas A&M University, College Station, TX, (2)Civil Engineering, Texas A&M University, College Station, TX

80 **Influence of Chemical Coating Combined with Nanopatterning on Alginate Fouling during Nanofiltration.**

Steven Weinman *and Scott M. Husson*
Chemical and Biomolecular Engineering, Clemson University, Clemson, SC

81 **The Impact of Slickwater Fracturing Fluid Composition / Fracturing Interactions on Membrane Fouling of Flowback Water.**

Boya Xiong[1], *Manish Kumar*[2], *Andrew L. Zydney*[3], *Travis Tasker*[1], *and William Burgos*[1]
(1)Department of Civil and Environmental Engineering, Pennsylvania State University, University Park, PA, (2)Department of Chemical Engineering, Pennsylvania State University, University Park, PA, (3)Chemical Engineering, The Pennsylvania State University, University Park, PA

82 **Removal of Human Adenovirus 40 in Membrane Bioreactors: Roles of Membrane Pore Size, Fouling, Pressure Relaxation and Membrane Cleaning.**

Volodymyr V. Tarabara[1], *Irene Xagoraraki*[2], *and Ziqiang Yin*[2]
(1)Department of Civil and Environmental Engineering, Michigan State University, East Lansing, MI, (2)Civil and Environmental Engineering, Michigan State University, East Lansing, MI

Filtration of Fine Particles in Atmospheric Aerosol with Electrospinning Nanofibers.

Zhong Zhang
National Center for Nanoscience and Technology, Beijing, China

In this study, a needleless electrospinning was studied, and the equipment was set up through theoretical simulation and a large number of experiments. The polylactic acid (PLA) nanofibers were prepared for filtering the aerosols by electrospinning. The PLA nanofibers mats were used as the middle layer to design the composite filter membranes. The atmospheric aerosols were used as dust sources in filtration test. The results showed that the filtration efficiency of the composite filter media was increased along with the thickness of nanofibers mats which was controlled by the collection time during the electrospinning. The filtration efficiency can be improved obviously by compositing with a thin layer of nanofibers. Moreover, the size distributions of atmospheric aerosols were monitored by the scanning mobility particle sizer spectrometer and the optical particle sizer. The size of particles in atmospheric aerosols was primarily distributed in range of less than 1 μm. It showed different changes with the mass concentrations of PM 2.5 (particulates with an aerodynamic diameter of ≤ 2.5 μm) for different sizes of fine particles. The amount of fine particles (less than about 60 nm) was decreased while the larger ones (> 60 nm) increasing along with the mass concentrations of PM 2.5 in atmospheric aerosols. That was due to the formation of the secondary atmospheric aerosols.

Removing Mercury (II) from Wastewater By Polymer-Enhanced Ultrafiltration.

Yifeng Huang *and Xianshe Feng*
Chemical Engineering, University of Waterloo, Waterloo, ON, Canada

Mercury (II) is one of the most strictly regulated contaminants released from various anthropogenic emission sources to the aquatic environment. When transformed to an organic form of methylmercury, it will be the most toxic, persistent and bioaccumulative form of mercury in nature. This study deals with removal of mercury (II) from wastewater by polymer-enhanced ultrafiltration (PEUF) using polyvinylamine as the mercury-binding polymer. A mercury removal rate as high as 99% was obtained, which was otherwise impossible to achieve with conventional ultrafiltration. Over the feed mercury concentration range tested (0 - 50 ppm), the PVAm dosage used did not affect the mercury rejection significantly, while water flux was reduced at a higher dosage of PVAm. A flux vs pressure relationship typical of ultrafiltration of macromolecular solutes was observed because of the presence of the amine polymer in the system, and the limiting flux appeared to follow the gel layer formation model. The fouled membrane surface was cleaned periodically with dilute chloric acid to recover the membrane permeability. Mercury removal with the PEUF was also tested with a simulated chlor-alkali wastewater that contained mercury and other chemicals (i.e., sodium chloride and sulphate), and the accompanying compounds in the feed solution was shown to influence the performance of PEUF for mercury removal.

Comparative Study of NF-RO Membranes for Sustainable Desalination of Produced Water and Frac Flowback Wastewater Following Biological Pretreatment.

Stephanie M. Riley[1], *Victoria B. . Frank*[2], *Kathryn Newhart*[3], *and Tzahi Cath*[1]
(1)Civil and Environmental Engineering, Colorado School of Mines, Golden, CO, (2)Chemical Engineering, Colorado School of Mines, denver, CO, (3)Colorado School of Mines, denver, CT

Development of unconventional oil and gas (O&G) resources is reshaping the energy landscape of the United States. Technological advances in hydraulic fracturing and horizontal drilling have made previously unrecoverable resources economically viable. However, extraction from these formations requires hundreds of thousands of gallons of clean water per well and generates substantial (and unavoidable) volumes of wastewater. These wastewaters (e.g., hydraulic fracturing flowback and produced water (PW)) contain varying concentrations of total dissolved solids (TDS), ranging from 10,000 mg/L to more than 200,000 mg/L, oil, grease, hydrocarbons, and various organic and inorganic matter. Consequently, a multi-barrier approach is required to effectively treat O&G waste streams for beneficial reuse.

Biological and membrane processes are promising methods to treat O&G wastewaters. Biological technologies have proven effective at removing organic material from a variety of waste streams including domestic waste streams, landfill leachate, and oily wastewaters; tight/dense membranes are most established in desalination of seawater and brackish waters. For sustainable membrane treatment, optimizing water recovery, and reducing fouling propensity, pretreatment targeting removal of organic constituents and suspended solids is critical. Integration of biologically-active filtration (BAF), ultrafiltration (UF), and nanofiltration/reverse osmosis (NF/RO) results in comprehensive removal and degradation of contaminants, reducing membrane fouling propensity due to BAF pretreatment, and producing a high quality permeate suitable for reuse applications such as hydraulic fracturing fluid and irrigation.

The current study evaluates the performance of NF/RO membranes following BAF and UF pretreatment of Denver-Julesburg (DJ) basin PW and frac flowback wastewater. The main objectives include (a) optimizing BAF and UF pretreatment for maximum removal of organic compounds, (b) evaluating NF/RO fouling propensity and membrane performance based on solids, inorganic, and organic rejection, and water flux, and (c) identifying favorable NF/RO membranes for treatment of PW and flowback, based on membrane characterization and performance. Bench-scale experiments with BAF have been conducted to acclimate a biofilm and optimize degradation of organic matter. In addition, experiments with UF were conducted to further reduce the turbidity of the BAF effluent and prepare it for NF/RO desalination. Ongoing NF/RO experiments compare the performance of various membranes (e.g., NF90, NF270, BW30 (DOW Chemical Company, Midland, MI), and NFX (Synder Filtration, Vacaville, CA)). Following rejection tests, membrane properties for each NF/RO membrane are being investigated using scanning electron microscopy (SEM) and atomic force microscopy (AFM) (morphology), contact angle (hydrophobicity), zeta potential (surface charge), and Attenuated Total Reflectance Fourier Transform Infrared Spectroscopy (ATR-FTIR) (surface chemistry) to evaluate variations in fouling.

MICRO, ULTRA, AND NANO-FILTRATION APPLICATIONS I

85 Fouling and Scaling Behavior of Charged Nanofiltration Membranes and the Influence of Graphene Quantum Dot Surface Modification.

Andrew Colburn[1]*, Namal Wanninayake*[2]*, Doo Young Kim*[2]*, and Dibakar Bhattacharyya*[1]
(1)Department of Chemical and Materials Engineering, University of Kentucky, Lexington, KY, (2)Department of Chemistry, University of Kentucky, Lexington, KY

Kam Agarwal
Patheon Inc, Mississauga, ON, Canada

86 **Real-Time Fouling Monitoring and Characterization for Effective Operational Control of Ultrafiltration As Pretreatment for Sewater Desalination.**

Han Gu[1], *Larry Gao*[2], *Anditya Rahardianto*[3], *Panagiotis D. Christofides*[4], *and Yoram Cohen*[3]
(1)Chemical & Biomolecular Engineering, University of California, Los Angeles, Los Angeles, CA, (2)Chemical & Biomolecular Engineering, UCLA, Los Angeles, CA, (3)Chemical and Biomolecular Engineering, University of California, Los Angeles, Los Angeles, CA, (4)Department of Chemical and Biomolecular Engineering and Department of Electrical Engineering, University of California, Los Angeles, Los Angeles, CA

88 **Application of Ceramic Membranes in Treatment of Oil Sands Produced Water.**

Saviz Mortazavi[1], *André Yves Tremblay*[2], *Yves Thibault*[1], *Charbel Atallah*[2], *Matthew Hudder*[1], *Alex Doiron*[1], *Joanne Gamage McEvoy*[1], *and Derek Smith*[1]
(1)CanmetMINING, Natural Resources Canada, Ottawa, ON, Canada, (2)Department of Chemical and Biological Engineering, University of Ottawa, Ottawa, ON, Canada

Treatment of Surface Water By a Hybrid Coagulation and Immersed Ultrafiltration Membrane for Potable Water Production.

Esper Jacob Ncube
Process Technology, Rand Water, Johannesburg, South Africa

In recent years, surface water purification with ultrafiltration membrane has become an attractive alternative to conventional treatment. Some of the advantages of ultrafiltration membrane compared to conventional water treatment process include none or less use of chemical agents, good quality of produced water independent of feed water quality, good removal efficiency towards microorganisms, less production of sludge, compact process, and easy automation. Originally, ultrafiltration was used to remove turbidity and microorganism from good-quality surface water. Presently the main goal of this technology is for the removal of microorganisms, natural organic matter (NOM), dissolved organic matter (DOC) and other components. This trend can be attributed to the improved cost effectiveness of membrane systems as compared to conventional treatment technologies such as filtration and to progressively more stringent water quality regulations.

This study examined the effect of hybrid process coagulation and ultrafiltration membrane performance for potable water production. An ultrafiltration system with a capacity of $24m^3/d$ was constructed to treat raw water taken from the impounded source water supplying the largest water utility in South Africa. A polyvinylidene (PVDF) hollow fibre ultrafiltration membrane was immersed in the process tank as the main component in the filtration system. The membrane area was $37.6\ m^2$ with a design membrane flux of $25\ L/\ m^2.h$ and a maximum membrane flux of $45\ L/\ m^2.h$. It was operated on the outside – In mode with a pore size of 0.04 μm nominal. Surface water contained turbidity of approximately 100NTU. Ferric chloride was used as the primary coagulant. The system performance was evaluated by measuring dissolved organic carbon (DOC), ultraviolet absorbance at 254 nm (UV_{254}) and membrane permeate flux. Also residual concentration of metal ion (Fe) in permeate was determined.

The results showed that regardless of the raw water turbidity, the membrane permeate water turbidity was always below 0.1 NTU. The hybrid system removed DOC more efficiently than the conventional water treatment process, and the removal percentage by the hybrid system was in the range of 34–54%. The membrane permeability improved greatly with the inclusion of citric acid as part treatment chemical for chemical cleaning in place (CIP). Pre-treatment with coagulation not only improved the treated water quality, but also enhanced membrane performance, that is, the permeate flux increased and flux decline was reduced.

Ultrafiltration of surface water with PVDF membranes was proven very useful for drinking water production. The results of this study have shown that ultrafiltration membrane can be applied to impounded water treatment for public supply from the viewpoint of treatment efficiency and effectiveness of contaminant rejection.

89 **Role of Permeate Flux and Specific Membrane-Foulant-Solute Affinity Interactions (°¼ΔG°½_slm) in Transport of Trace Organic Solutes through Fouled Nanofiltration (NF) Membranes.**

Oranso T. Mahlangu[1], *Klaas V. K. M. Schoutteten*[2], *Arnout D'Haese*[2], *Bhekie B. Mamba*[3], *and Arne R. Verliefde*[4]
(1)Nanoscience and Water Sustainability (NanoWS), University of South Africa, Johannesburg, South Africa, (2)Ghent University, Ghent, Belgium, (3)School of Science, Engineering and Technology, University of South Africa, Johannesburg, South Africa, (4)Applied Analytical and Physical Chemistry, University of Ghent, Ghent, Belgium

90 **Behavior of Oil Droplets at the Membrane Surface during Crossflow Microfiltration of Oil-Water Emulsions.**

Emily Tummons[1], *Volodymyr Tarabara*[1], *Anthony G. Fane*[2], *and Jia Wei Chew*[3]
(1)Civil and Environmental Engineering, Michigan State University, East Lansing, MI, (2)Singapore Membrane Technology Centre, Nanyang Environment and Water Research Institute, Nanyang Technological University, Singapore, Singapore, (3)School of Chemical and Biomedical Engineering, Nanyang Technological University, Singapore, Singapore

Mixed Matrix Membrane for Protein Separation (oral).

Junfen Sun
Donghua University, Shanghai, China

In this study, strontium-substituted nano hydroxyapatite (Sr-HAP) was synthesized by ultrasonic method. The surfaces of Sr-HAP particles were grafted with acrylic acid (AAc) by low temperature plasma technology and Sr-HAP grafted with AAc (Sr-HAP-AAc) was obtained. The dual layer polyether sulfone (PES)/HAP mixed matrix membranes (MMMs) were prepared by an immersion phase separation process. A particle free PES membrane layer was co-cast with the active support layer, which led to formation of dual layer MMMs. The active support layer consisted of Sr-HAP-AAc particles embedded in macroporous PES. The dual layer MMMs showed separation of protein via diffusion as well as adsorption. The bovine serum albumin (BSA) was used as a model protein. The BSA adsorption capacity of MMMs increased with extending equilibrium time and almost reached stable at 20 h. The BSA adsorption capacity of dual layer MMMs was 48.6 mg/(g MMMs). The dual layer MMMs with top PES layer kept good filtration and adsorption properties.

MOLECULAR SEPARATION AND PERVAPORATION

92 **Organic Solvent Reverse Osmosis: Formation of Solvent-Resistant Asymmetric Carbon Molecular Sieve Hollow Fiber Membranes.**

Dong-Yeun Koh[1], *Melinda L. Jue*[1], *Benjamin A. McCool*[2], *Harry W. Deckman*[2], *and Ryan P. Lively*[1]
(1)School of Chemical & Biomolecular Engineering, Georgia Institute of Technology, Atlanta, GA, (2)ExxonMobil Research and Engineering, Annandale, NJ

PEEK Hollow Fiber Composite Gas Separation Membranes Applied in Minimal Pretreatment and Condensing Environments.

Yong Ding, *Benjamin Bikson, James Macheras, and Joyce Nelson*
Air Liquide Advanced Separations, Woburn, MA

Poly(ether ether ketone), PEEK, is a semi-crystalline, high performance engineering polymer, well known for its solvent/chemical resistance and a high temperature operating capability. The excellent chemical and physical characteristics of PEEK make it an ideal candidate for the preparation of the next generation of polymeric membranes that combine inorganic membrane robustness with polymeric membrane performance and price.

PEEK hollow fiber membranes were successfully developed and commercialized by melt extrusion of compatible PEEK/polyimide blend, . The polyimide in the blend serves as a pore former and is quantitatively removed to produce the target porous PEEK membrane. Membrane morphology and pore size is controlled by the blend composition and processing conditions. The methodology is highly flexible and provides for preparation of both flat sheet and hollow fiber membranes. PEEK exhibits thermo-mechanical properties superior to almost all engineering polymers currently in use in membrane preparation.

A family of composite hollow fiber membranes based on PEEK have been developed and commercialized for variety of natural gas purification applications, such as BTU reduction, H2S and CO2 removal. PEEK based hollow fiber membranes are extremely robust. The membrane separation performance is not affected by the presence of organic liquids, such as NGLs. Only coalescing filter pretreatment is required for all commercial PEEK membrane systems. The elimination of the extensive pretreatment significantly reduces the capital cost and maintenance cost. One example of the use of the PEEK hollow fiber membrane is for H2S reduction. Application of the PoroGen™ membrane portfolio to multiple industrial application will be presented.

94 **Spent IPA Dehydration for in-House Reclamation.**

Yu (Ivy) Huang[1], *Richard Baker*[2], *Yan Gao*[3], *Ramin Daniels*[4], *and Xuezhen Wang*[2]
(1)Membrane Technology & Research, Inc., Menlo Park, CA, (2)MTR, Newark, CA, (3)Membrane Technology & Research, Newark, CA, (4)Membrane Technology and Research, Inc., Newark, CA

95 **Aromatic Polyimide and Crosslinked Thermally Rearranged Poly(benzoxazole-co-imide) Membranes for Isopropanol Dehydration Via Pervaporation.**

Yi Ming Xu[1], *Ngoc Lieu Le*[2], *Jian Zuo*[1], *and Tai-Shung Chung*[1]
(1)Chemical and Biomolecular Engineering, National University of Singapore, Singapore, Singapore, (2)Biological and Environmental Sciences and Engineering, King Abdullah University of Science and Technology, Thuwal, Saudi Arabia

97 **A New Molecular Approach to Tuning Separation Properties of Carbon Molecular Sieves.**

Graham Wenz[1] *and William J. Koros*[2]
(1)School of Chemical and Biolmolecular Engineering, Georgia Institute of Technology, Atlanta, GA, (2)School of Chemical & Biomolecular Engineering, Georgia Institute of Technology, Atlanta, GA

98 **Electric Field-Assisted Membranes for Molecular Separation.**

Oleg Polyakov
Thermosolv LLC, Laramie, WY

Highly-Efficient Immersion Engineering of PEO Membranes Toward Superior CO2 Separation Performance.

Lu Shao *and Shuai Quan*
Harbin Inst Technol, Harbin, China

Herein, we designed an extremely versatile strategy to significantly improve gas transport properties of PEO-containing cross-linked membranes based on their specific swelling properties. Low-molecular-weight poly (ethylene glycol) (PEG) was deliberately incorporated into synthesized swellable pure PEO cross-linked membrane via immersing PEO membrane into PEG-water solution and membranes with various PEG loadings are obtained. The imbedded PEG molecules can be located between polymer chains to improve chain mobility and increase chain distance, resulting in the increase of fractional free volume (FFV), which is in favor of gas diffusion. For example, the flexible incorporation of low-molecular-weight poly(ethylene glycol) dimethyl ether (PEGDME) into the swollen network membrane greatly improved the gas transport performance and the CO_2 permeability was increased by **550%**. The separation performance was correlated nicely with the microstructure of the membranes. This study may provide useful insights about the formation and mass transport behaviour of highly-efficient polymeric membrane applicable to clean energy purification and CO_2 capture, and possibly bridge the material-induced technology gap between academia and industry. Most importantly, our discovered strategy has great potential for treating composite membranes and regenerating "aged" membranes to realize clean energy purification and CO_2 capture by membrane technology.

NAMS AWARD SESSION

99 **Optimizing Hollow Fiber Membranes and Modules for Osmotic Processes: From Models to Membranes to Modules.**

Jian Ren *and Jeffrey McCutcheon*
Department of Chemical and Biomolecular Engineering, University of Connecticut, Storrs, CT

101 **Fundamental Studies of Ion Sorption and Transport in Ion-Exchange Membranes.**

Jovan Kamcev*, Benny D. Freeman, and Donald R. Paul*
McKetta Department of Chemical Engineering, The University of Texas at Austin, Austin, TX

102 **Antibody Purification from Cell Culture Supernatant Using New Multimodal Membranes.**

Juan Wang[1] *and Scott M. Husson*[2]
(1)Chemical and Biomolecular Engineering Department, Clemson University, Clemson, SC, (2)Chemical and Biomolecular Engineering, Clemson University, Clemson, SC

103 **Poly(ionic liquid)/Ionic Liquid Ion-Gel Membranes for CO_2/Light-Gas Separations.**

Matthew G. Cowan[1]*, William M. McDanel*[2]*, Trevor K. Carlisle*[3]*, Zoban V. Singh*[4]*, Douglas L. Gin*[4,5]*, and Richard D. Noble*[2]

(1)Department of Chemistry & Biochemistry, University of Colorado, Boulder, CO, (2)Dept. of Chemical and Biological Engineering, University of Colorado at Boulder, Boulder, CO, (3)Dept. Chemical and Biological Engineering, University of Colorado at Boulder, Boulder, CO, (4)Department of Chemical and Biological Engineering, University of Colorado, Boulder, CO, (5)Dept. of Chemistry & Biochemistry, University of Colorado at Boulder, Boulder, CO

Engineering Nanostructured Ion-Containing Block Polymers for Membrane Separation Applications.

Matthew D. Green
Chemical Engineering, Arizona State University, Tempe, AZ

Membrane-mediated transport has industrial and commercial applications in food packaging, water and gas separation, energy storage and harvesting, performance clothing, sensors, and many others. However, decoupling conflicting constraints is commonly necessary; for example, water purification and gas separation membranes frequently suffer from the permeability-selectivity trade off. Ion-containing block polymers offer a design strategy to decouple transport from thermomechanical performance: membranes that form transmembrane or bicontinuous microphase separated morphologies enable transport through the ionomer phase while the surrounding microphase(s) provide robust mechanical properties. This presentation will focus on strategies for preparing and the performance of ion-containing block polymer membranes with bicontinuous morphologies and percolated transport pathways.

104 **Membrane Performance Characterization and Process Prediction in Osn: Challenges, Achievements and Perspective.**

Patrizia Marchetti[1], *Binchu Shi*[2], *Dimitar Peshev*[3], *and Andrew Livingston*[1]
(1)Chemical Engineering, Imperial College London, London, United Kingdom, (2)Evonik MET Ltd., London, United Kingdom, (3)University of Chemical Technology and Metallurgy, Sofia, Bulgaria

NOVEL MEMBRANE MATERIALS I

106 **Polymer Nanofilms with Enhanced Microporosity By Interfacial Polymerisation for Molecular Separations.**

Maria F. Jimenez Solomon[1], Qilei Song[1], Kim E. Jelfs[2], Marta Munoz-Ibanez[1], and ***Andrew G. Livingston[3]***
(1)Department of Chemical Engineering, Imperial College London, London, United Kingdom, (2)Department of Chemistry, Imperial College London, London, United Kingdom, (3)Chemical Engineering, Imperial College London, London, United Kingdom

Novel Polynorbornenes for Gas Separation Membranes.

Benjamin J. Sundell*, John A. Lawrence III, Daniel J. Harrigan, Justin T. Vaughn, and Tatiana Pilyugina*
Boston Downstream Research Center, Aramco, Cambridge, MA

Saudi Arabia produces approximately half of its domestic electricity from natural gas. With nearly 300 trillion standard cubic feet in proven natural gas reserves, a small fraction of those reserves has low BTU content. Gas separation membranes have the potential to help monetize these low BTU content reserves, but there are still challenges with membrane technologies for these applications. To address these challenges, a systematic series of novel polynorbornenes with varying substituents were synthesized. Structural evidence via NMR and FTIR demonstrate the role that substituent *endo-exo* isomerism has on resulting polymeric microstructure, yield, and permeation properties. Substitution also enables covalent crosslinking via tunable sol-gel routes that helps to stabilize the membranes. This talk will describe how X-ray diffraction has been used to provide a unique perspective into the porosity and crosslinking of these materials and to demonstrate that pendant substitution can be a successful technique to increase intermolecular spacing, a paradigm for further improving selectivities. Finally, both pure and mixed gas permeation studies have revealed that substitution can double gas separation selectivities compared to unmodified polynorbornene while retaining high gas permeabilities, further demonstrating the promise of these novel membranes for gas separation applications.

107 **Highly Permeable Peptide-Appended Pillar[5]Arene Artificial Water Channel in Block Copolymer Membranes.**

Yuexiao Shen[1]*, Tingwei Ren[2], Daniel Tsai[2], Ian Sines[2], and Manish Kumar[3]*
(1)Department of Chemical Engineering, The Pennsylvania State University, University Park, PA, (2)Pennsylvania State University, (3)Department of Chemical Engineering, Pennsylvania State University, University Park, PA

Zwitterion-Containing Polymer Additives for Fouling Resistant Ultrafiltration Membranes: Choosing the Right Chemistry and Architecture.

Ayse Asatekin
Department of Chemical Engineering, Massachusetts Institute of Technology, Cambridge, MA

Fouling is likely the most significant obstacle to the wider use of membranes in many processes, especially when the feed contains large quantities of organic contaminants such as proteins, natural organic matter, and oil. Zwitterions, which have equal numbers of positively and negatively charged groups, are very hydrophilic and strongly resist fouling. Blending a zwitterion-containing polymer with a commodity polymer such as polyvinylidene fluoride (PVDF) during membrane manufacture by non-solvent induced phase inversion (NIPS) is a promising approach to making fouling resistant ultrafiltration (UF) membranes. It does not require any post-processing steps and can be plugged directly into existing membrane manufacturing facilities. A handful of studies demonstrate varying degrees of fouling resistance when zwitterion-containing amphiphilic copolymers are blended with PVDF to prepare membranes. However, we do not know how copolymer chemistry, architecture, and blending ratio with the base polymer affect final membrane properties. There are no guidelines on how to design and select a zwitterion-containing copolymer for optimal performance. In this study, we seek to identify the effect of several copolymer features on the performance of membranes prepared from its blends with PVDF: Zwitterion chemistry, hydrophobic monomer chemistry, copolymer composition (i.e. zwitterionic/hydrophobic monomer ratio), copolymer architecture, and blend composition. We synthesized several copolymers varying these parameters, and evaluated the morphology, permeance, selectivity and fouling resistance of resultant membranes. We have found design rules to direct the selection of best polymer structures and architectures for this use. First, we found that unlike past reports, a higher hydrophilic/zwitterionic monomer content in the copolymer does not necessarily lead to better performance. Copolymers with high zwitterion contents macrophase separate from PVDF during the NIPS process. The resultant membrane has poor performance. Second, we can identify the copolymer compositions where this problem will be avoided using thermal analysis of copolymer/PVDF blends. If the copolymer leads to a shift in the glass transition temperature of PVDF, this predicts good anchoring of the copolymer and improved fouling resistance. Third, even very small quantities of zwitterion-containing copolymers can lead to significant increases in permeance and exceptional fouling resistance. As little as 5 wt% zwitterionic copolymer can lead to 6x higher flux with little change in rejection, and show essentially complete resistance to irreversible fouling. Finally, this effect is more prominent with comb-shaped copolymers with the zwitterions are arranged as short side-chains attached to the hydrophobic backbone, in comparison with random copolymers of the same repeat units. This information is crucial for membrane companies considering such copolymer additives to improve the performance of PVDF membranes, enabling rational design of zwitterion-containing polymers for membrane applications. We are currently further exploring the responsive and self-cleaning properties of these membranes, and their performance in realistic applications such as wastewater treatment.

Crosslinked Polytriazole Membranes for Organophilic Filtration.

Stefan Chisca[1], *Gheorghe Falca*[2], *Valentina Musteata*[1], *Cristiana Boi*[2], *and Suzana Pereira Nunes*[1]
(1)Biological and Environmental Sciences & Engineering Division (BESE), King Abdullah University of Science & Technology (KAUST), Thuwal, Saudi Arabia, (2)University of Bologna, Bologna, Italy

Pharmaceutical and chemical industries generate a large amount of organic solvents which need to be recovered for further use, otherwise will become waste solvent. Membrane technology can be used as an alternative process to more conventional separation methods with advantage of low energy consumption.

In this study we report the facile synthesis of new crosslinked polytriazole membranes for organophilic filtration by using poly(ethylene glycol) diglycidyl ether (PEGDE), which is not toxic and cheap crosslinker. The polytriazole is highly soluble in common organic solvents due to the presence of 98% of free OH groups, which can be easily processed by a phase inversion method. The resulting membranes are stable in wide range of organic solvents and in all range of pH. The THF flux is 49 L m^{-2} h^{-1} bar^{-1} for the membrane with MWCO of 6.9 kDa and 17.5 L m^{-2} h^{-1} bar^{-1} for the membrane with a MWCO of 2.8 kDa. By using different casting solution and different crosslinking time it is possible to tailor the MWCO of crosslinked polytriazole membranes to cover the range from the lower end of ultrafiltration to upper end of nanofiltration, indicating that these membranes can be used in wide range of applications.

Acknowledgement

This research was supported by King Abdullah University of Science and Technology (KAUST)

Adsorption of Polyelectrolyte Multilayers Imparts High Monovalent/Divalent Cation Selectivity to Nafion and Fujifilm Cation-Exchange Membranes.

Yan Zhu[1] *and Merlin Bruening*[2]
(1)Chemistry, Michigan State University, East Lansing, MI, (2)Chemistry Department, Michigan State University, East Lansing, MI

Although commercial ion-exchange membranes show high selectivities between cations and anions, they have modest selectivities among monovalent and divalent cations or monovalent and divalent anions. Such selectivities among cations or among anions are important, however, in applications such as purifying salts. This presentation will describe layer-by-layer adsorption of polyelectrolytes on cation-exchange membranes to impart extreme selectivities among monovalent and divalent cations. Deposition of protonated poly(allylamine) (PAH)/poly(4-styrenesulfonate)(PSS) films on Nafion membranes leads to K^+/Mg^{2+} electrodialysis selectivities >1000. Unfortunately, the current efficiency is only 0.5, so unwanted ions (such as protons) carry 50% of the current. Moreover, the high price of Nafion membranes ($800 for 0.5 m^2) limits their practical application. We recently found that $(PAH/PSS)_5PAH$-modified Fujifilm cation-exchange membranes show a K^+/Mg^{2+} electrodialysis selectivity of ~800 with a current efficiency of 0.5. The corresponding unmodified membranes show selectivities <2. The Fujifilm substrates do not contain expensive fluorinated polymers and are thus much less expensive than Nafion. To increase current efficiency, we coated Nafion membranes with polydiallyldimethylammonium (PDADMAC)/PSS films. $(PDADMAC/PSS)_5PDADMAC$-modified Nafion membranes exhibit K^+/Mg^{2+} and Li^+/Co^{2+} selectivities > 1000, but in Li^+/Co^{2+} separations, the current efficiency is 0.6 for $(PDADMAC/PSS)_5PDADMAC$-modified Nafion and only 0.45 for $(PAH/PSS)_5PAH$-modified Nafion. The higher current efficiency may stem from swelling of PDADMAC/PSS films. However, the PDADMAC/PSS films likely delaminate at high ionic strength, so we are working on cross-linking chemistries to stabilize these membranes.

NOVEL MEMBRANE MATERIALS II

109 Sorption Enhanced Mixed Matrix Membranes for High Temperature H2/CO2 Separation.

Lingxiang Zhu, *Deqiang Yin, Shailesh Konda, Mark Swihart, and Haiqing Lin*
Department of Chemical and Biological Engineering, University at Buffalo, The State University of New York, Buffalo, NY

Abstracts:

- Abstract for NAMS 2016.docx (20.8KB) - Abstract Text

110 Fabricating Nanoporous Metallic Films As Components of Composite Membranes.

Michael Detisch[1], *Dibakar Bhattacharyya*[2], *and John Balk*[1]
(1)Dept. of Chemical and Materials Engineering, University of Kentucky, Lexington, KY, (2)Department of Chemical and Materials Engineering, University of Kentucky, Lexington, KY

Abstracts:

- Abstract Mike np_PdFe_NAMS 2016.docx (14.0KB) - Abstract Text

111 Tuning PIM-PI-OH/Z-MOF-Based Mixed-Matrix Membranes for Highly Efficient C3H6/C3H8 Separation.

Ramy Swaidan[1], *Xiaohua Ma*[2], *and Ingo Pinnau*[1]
(1)Advanced Membranes and Porous Materials Center, King Abdullah University of Science and Technology, Thuwal, Saudi Arabia, (2)KAUST

Exploring the Potential of Nanoporous Single-Layer Graphene As a Next-Generation Membrane Material.

Rohit Karnik
Massachusetts Institute of Technology, Cambridge, MA

Nanoporous graphene has potential as an ideal ultrathin membrane for high-flux, high-selectivity, chemically-resistant membrane separations. I will discuss our work on understanding mass transport and developing membranes that employ nanoporous single-layer graphene as the selective layer. Through controlled nucleation of defects *via* ion bombardment and oxidative etching, sub-nanometer pores are created in the otherwise impermeable graphene placed on a porous support. The resulting membranes exhibit selectivity between ions and between salt and small molecules. We also present strategies for the design of defect-tolerant membranes using centimeter-scale areas of single-layer graphene, which allows for measurement of water flux and solute rejection during filtration by forward osmosis. The membrane can reject multivalent ions and small molecules, and exhibits a permeance that is consistent with theoretical predictions. We also demonstrate that these membranes exhibit high mechanical strength and are able to withstand applied pressures up to 100 bar. These studies illustrate the feasibility and potential of single-layer nanoporous graphene membranes for nanofiltration, water purification, and other applications.

Separation of Chiral Molecules Using Mixed Matrix Membranes Incorporating Chiral Metal Organic Frameworks.

Bradley P. Ladewig, *Benjamin Slater, and Marine Michel*
Department of Chemical Engineering, Imperial College London, London, United Kingdom

Exciting new porous coordinated materials have recently been synthesised that incorporate chiral pore structures, which opens up the opportunity to use these materials as sorbents for the selective adsorption of enatiomerically distinct molecules from racemic mixtures. The significance of this development to the pharmaceutical and fine chemical industries is enormous, since a significant proportion of their products are chiral molecules which are usually separated using low-throughput, highly expensive elution columns (or must be synthesised using challenging chemistry which is directed towards producing enantiomerically pure products).

In this presentation, we will outline our recent results synthesising chiral metal organic framework materials using two distinct approaches; the use of chiral ligands to impart chirality to the resulting metal organic framework, and an emerging and lower-cost approach with is to post-synthetically modify non-chiral (but usually easily synthesised) subtrate materials, to impart a degree of chirality to them. These materials are tested as batch liquid-phase sorbents to screen their potential for selective sorption of a library of low molecular weight chiral molecules (primarily sugars, alcohols and amino acids). The most promising candidates, are incorporated into mixed matrix membranes and tested as membranes, with a specially- configured gas chomatograph with chiral columns used to resolve the components in the feed and permeate, to quantify the selectivity.

CNT Membranes with Sub-Nm Pores Showing Salt Rejection and Sharp Size Selectivity at High Ionic Strenghts.

Rob McGinnis[1]*, Kevin Reimund*[2]*, Kevin A. Stevens*[3]*, Joshua Moon*[4]*, Benny D. Freeman*[3]*, Jian Ren*[5]*, Lingling Xia*[5]*, and Jeffrey McCutcheon*[5]
(1)Nagare Membranes, New York, NY, (2)Nagare, New York, NY, (3)McKetta Department of Chemical Engineering, The University of Texas at Austin, Austin, TX, (4)Univ of Texas, Austin, TX, (5)Department of Chemical and Biomolecular Engineering, University of Connecticut, Storrs, CT

Abstract: We report water and gas separation performance for novel carbon nanotube (CNT) membranes consisting of the smallest (0.67 - 1.27 nm inner diameter) single wall nanotubes yet reported for CNT membranes. The CNTs traverse a non-porous, non-water-permeable, flexible polymer film, and show rejection of small salts at high ionic strength, and highly specific size selectivity for sub-1.5 nm solutes. Water flow through the membranes showed significant enhancement over that predicted by Hagen-Poiseuille flow, in agreement with previous CNT membrane studies. Gas selectivity was found to be similar to that predicted by Knudsen flow for non-condensable gases, but the permeabilities were found to be significantly higher than those predicted by Knudsen, also in agreement with previous studies of CNT membranes. CO2 permeation was found to be substantially higher than predicted. The membranes used in this study were fabricated using non-aligned bulk CNTs, orientated during membrane formation to give a membrane with tubes traversing a non-porous polymer film. Membranes of this type will allow for the use of very small diameter nanotubes previously not suitable for use in CNT membranes, offering the potential to perform separations of sub-nm solutes with angstrom-scale selectivities.

112 **Carbon Molecular Sieve Membranes for Mixed Olefins/Paraffins Separations.**

Yu-Han Chu *and William J. Koros*
School of Chemical and Biomolecular Engineering, Georgia Institute of Technology, Atlanta, GA

PLENARY LECTURE: DR. MICHAEL YANDRASITS

N/A **Fuel Cell Membranes based on Perfluorsulfonic Acids and.**

Michael Yandrasits
AL

PLENARY LECTURE: DR. WINSTON HO

N/A **Recent Developments on Membranes for CO2 Separation and Water Purification.**

Winston Ho
Chemical and Biomolecular Engineering, The Ohio State University, Columbus, OH

POSTER PRESENTATION I - ENERGY AND ENVIRONMENT

113 **High-Throughput Hollow-Fiber (HT-HF) Membrane Module – a Miniaturized, Parallel Format Approach Ideal for the Accelerated Design of Environmental Separation Processes.**

Amir S. Kazemi, Seung Mi Yoo, Raja Ghosh, and ***David R. Latulippe***
Chemical Engineering, McMaster University, Hamilton, ON, Canada

Iron Nanoparticle Functionalized Membrane for PCB Degradation from Water - Poster.

Hongyi Wan[1], *Minghui Gui*[2], *Lindell Ormsbee*[3], *and D.B. Bhattacharyya*[2]
(1)University of Kentucky, Lexington, KY, (2)Chemical and Materials Engineering, University of Kentucky, Lexington, KY, (3)Department of Civil Engineering, University of Kentucky, Lexington, KY

The development of functionalized membranes with catalytic metal nanoparticles provides an effective method of environmental remediation and wastewater treatment. Polyvinylidene fluoride (PVDF) is an excellent polymer for both chemical and mechanic resistance as well as thermal stability. By immobilizing zero-valent iron (ZVI) or bimetallic nanoparticles (NPs), functionalized PVDF membranes can be used in chlorinated organic treatment. Polyacrylic acid (PAA) was polymerized inside PVDF membrane pores to maximize iron adsorption. Adsorbed iron was reduced to form NPs in-situ in order to prevent metal ion loss, NPs aggregation and iron precipitation. Furthermore, tunable membrane pore size could be achieved by changing the environmental pH because of ionization of PAA.

This study includes three major aspects: 1. Reactivity toward model toxic chlorinated organic compounds--polychlorinated biphenyl (PCBs) in both batch and convective flow study; 2. NPs characterization of its average sizes, composition, extent of oxidation and distribution inside membrane pores; 3. Simulating pH responsive behavior and PCB reaction within the membrane.

PCBs are not biodegradable and tend to cycle between air, water and soil, which might accumulation in living organisms. PCBs dechlorination can be achieved by Fe/Pd reductive pathway. The hydrogen, generated from reaction between ZVI and water, gets activated by Pd for hydrodechlorination. In batch study, 95% of PCB 126 ([Co]=15 μM) was consumed and 67% was converted to biphenyl in 5 hours. In convective flow study, 96% of PCB 126 was consumed at 26s residence time. However, the final product biphenyl still showed toxicity. In order to degrade biphenyl, the further oxidative pathway was combined to break down the aromatic ring and eventually form organic acid. Both iron oxide with H_2O_2 method (Fenton Reaction) and persulfate method significantly reduced toxicity of the products of PCB degradation.

Fe/Pd NPs size distribution was analyzed by SEM and TEM (TEM samples were prepared by Focus Ion Beam) and the average size of NPs was 19.4±3.2nm. Iron core and discontinuous palladium shell was tested by STEM-EELS line scan and elemental mapping and extent of oxidation was studied by XRD. Furthermore, membrane permeability change from 32.2LMH/bar to 1.1LMH/bar when environmental pH change from 2 to 6.5. The pH-responsive behavior was simulated and fitted with experimental data, which is a part of PCBs reactivity model.

This research is supported by the NIEHS-SRP grant P42ES007380, and by the NSF KY EPSCOR program. Full scale PVDF membranes were provided by Nanostone/Sepro (USA).

114 **Physicochemical and Photocatalytic Properties of Carbonaceous Char and Titania Composite Hollow Fibers for Wastewater Treatment.**

David Wang
the University of Queensland, St Lucia, Australia

115 **Evaluation of Different Membrane Separation Processes for Treatment of Acid Mine Drainage and Mine Effluents.**

Saviz Mortazavi[1], *André Yves Tremblay*[2], *Georges Houlachi*[3], *Alex Doiron*[1], *and Matthew Hudder*[1]
(1)CanmetMINING, Natural Resources Canada, Ottawa, ON, Canada, (2)Chemical and Biologtical Engineering, University of Ottawa, Ottawa, ON, Canada, (3)Laboratoire des technologies de l'énergie (LTE), Institut de recherche, Hydro-Québec, Shawinigan, QC, Canada

Organic Acids and Water Membrane Separations from Bio-Oil.

George Marshall[1]*, Jamie A. Hestekin*[2]*, Yupo J. Lin*[3]*, and Michael Ackerson*[4]
(1)University of Arkansas, Fayetteville, AR, (2)Chemical Engineering, University of Arkansas, Fayettville, AR, (3)Energy Systems Division, Argonne National Laboratory, Argonne, IL, (4)Ralph E. Martin Department of Chemical Engineering, University of Arkansas, Fayettville, AR

Over 70% of the world's energy consumption is provided from fossil fuels and with their reserves depleting at a fast rate, alternative energy sources or methods are needed to provide the world's energy needs. This research hopes to aid with the alternative energy needs by making it economically feasible to produce fuel products, such as bio-diesel, from the upgrading of bio-oil obtained from the pyrolysis of biomass waste material such as sawdust. The high water and oxygenated compound content of the bio-oil make it undesirable for fuel use, but by means of a surface modified commercial membrane we hope to facilitate the separation of desired compounds from others which may be used synthesize other commercially useful products. In this research two methods were used for this separation: electrodialysis and a pressure driven system. Physically modified Nafion 117 nanofiltration pore size membranes were used for the pressure driven system in hopes of separating water and low molecular weight compounds from the bio-oil. And anion exchange membranes with added polyethylenimine groups to the membrane surface were used to remove organic acids from the bio-oil. The electrodialysis experiments results showed a decrease in conductivity giving supporting belief that some if not all of the organic acids are were removed. And the procedure to widen the Nafion 117 membrane pores also proved to be successful. With Nafion's proven chemical resistance to the bio-oil, it is expected to achieve the desired separation upon obtaining the appropriate pore diameter of the membrane.

Design Supported Ruthenium Catalysts for the Catalytic Oxidation of (chloro)-Benzene.

Jian Wang*, Xiaolong Liu, and Tingyu Zhu*
Institute of Process Engineering, Chinese Academy of Sciences, Beijing, China

Ruthenium-based catalysts are commonly known for their roles in the Fischer-Tropsch synthesis as well as the synthesis of ammonia, and these catalysts have also been applied industrially for chlorine gas production via the Deacon Process. Recently, the catalytic properties of ruthenium-based catalysts in the catalytic oxidation of light alkanes and (chloro)-aromatics have attracted significant interest due to their good activities. In this article, we present our work related to design supported ruthenium catalysts for the catalytic oxidation of (chloro)-benzene. In particular, P25-TiO_2, ZrO_2, γ-Al_2O_3 and SiO_2

supported ruthenium catalysts were synthesized and well characterized by XRD, BET, ICP, XPS and TEM. These catalysts were used for benzene oxidation at first. It was found that the Ru/P25-TiO_2 catalyst with inhomogeneous distributed RuO_2 (mainly distributed in the rutile phase of P25-TiO_2) had the best activity among these catalysts. And the total conversion temperature (T_{100}) of benzene over the Ru/P25-TiO_2 catalyst was approximately at 230 °C. Accordingly, the Ru/P25-TiO_2 catalyst was tested for chlorobenzene oxidation. Noteworthy, T_{100} of chlorobenzene over the Ru/P25-TiO_2 catalyst was below 280 °C, and the organic by-products formed during the catalytic reaction completely disappeared when T_{100} of the reactant was achieved. Hence, the Ru/P25-TiO_2 catalyst showed substantial advantages than the other reported catalysts in the catalytic oxidation of chlorobenzene. Based on the results of XPS, the active species of Ru/P25-TiO_2 was described as chlorinated RuO_2. The excellent performance of the Ru/P25-TiO_2 catalyst for chlorobenzene oxidation could be attributed to the good thermal stability of RuO_2 in the rutile phase of P25-TiO_2 and chlorine removal capacity of the active species.

116 **Non-Carbon Adsorbent and Oxidation Catalyst for Mercury Removal from Coal-Fired Flue Gas.**

Wenqing Xu[1] *and* ***Tingyu Zhu***[2]
(1)Institute of Process Engineering, Chinese Academy of Sciences, Beijing, China., Beijing, China, (2)Institute of Process Engineering, Chinese Academy of Sciences, Beijing, China

118 **Towards Conductive Chemically-Active Membranes: Flavin-Functionalized Fibers and Electrodes Active in Photo-Enhanced Oxygen Evolution.**

Madison Sloan[1], *Rupam Sarma*[2], *Shankar Miller-Murthy*[3], *and Anne-Frances Miller*[3]
(1)Chemistry, University of Kentucky, Lexington, KY, (2)Department of Chemical and Materials Engineering, University of Kentucky, Lexington, KY, (3)Chemistry, University of Kentucky, Lexington

Synthesis of Ru-Based Catalysts for Complete Oxidation of Volatile Organic Compounds.

Xiaolong Liu, *Junlin Zeng, Jian Wang, and Tingyu Zhu*
Institute of Process Engineering, Chinese Academy of Sciences, Beijing, China

In this work, low-temperature catalytic oxidation of volatile organic compounds (VOCs) over Ru-based catalysts was presented. The catalysts were prepared using the wet impregnation method, and characterized by ICP, BET, XRD, TEM, and XPS. The catalytic studies demonstrated that P25 supported catalyst contributed the highest activity. Accordingly, Ru/TiO_2(P25) was particularly studied.

Based on the characterizations, we found that ruthenium species exists as ruthenium oxides, which are often hard to observe in TEM micrographs due to their low contrast. Hence, TEM, HR-TEM, and HAADF-STEM charcterizations were conducted. It can be concluded that ruthenium species tend to be distributed on the rutile-TiO2 of P25 due to their similar crystal lattice properties, leading to an inhomogeneous distribution. The catalyst was employed in complete oxidation of different VOCs, including benzene, chlorobenzene, trichloroethene, and methyl bromide. Noteworthy, for chlorinated VOCs, Ru-based catalyst showed apparently better catalytic performance than catalysts, such as Pd or Pt. Additionally, the thermal stability test, anti-moisture test, and byproducts analysis were also conducted, revealing that Ru/TiO_2 is promising for VOCs purifications in industrial plants.

119 **Characterization of Biofouling on Forward Osmosis Membranes Used for Long-Term Produced Water Treatment.**

Kevin Chan
Civil and Environmental Engineering, Colorado School of Mines, Golden, CO

120 **Green Modification of Outer Selective P84 Nanofiltration (NF) Hollow Fiber Membranes for Cadmium Removal.**

***Jie Gao**[1], Shi Peng Sun[2], Wen-Ping Zhu[1], and Tai-Shung Chung[1]*
(1)Chemical and Biomolecular Engineering, National University of Singapore, Singapore, Singapore, (2)Singapore-MIT Alliance, National University of Singapore, Singapore, Singapore

123 **High Performances of Poly(arylene ether sulfone) with Tetra(quaternary ammonium) Group for Solid Alkaline Exchange Membrane Fuel Cell Applications.**

***Jieun Choi**[1,2], So Young Lee[3], Jong Hyun Jang[3], Dirk Henkensmeier[3], Sung Jong Yoo[3], Yung-Eun Sung[2], and Hyoung-Juhn Kim[3]*
(1)Fuel Cell Research Center, Korea Institute of Science and Technology, Seoul, Korea,

Republic of (South), (2)Department of Chemical and Biological Engineering, Seoul National University, Seoul, Korea, Republic of (South), (3)Fuel Cell Research Center, Korea Institute of Science and Technology (KIST), Seoul, Korea, Republic of (South)

Metal Oxides Film Catalyst for VOCs Degradation Derived from Hydrotalcite.

Shuangde Li, *Renliang Yue, and Yunfa Chen*
Institute of Process Engineering, Chinese Academy of Sciences, Beijing, China

Emission of volatile organic compounds (VOCs) from industrial gases is major contributors to air pollution which may pollute the atmosphere directly or indirectly as secondary pollutants, such asozone generation and photochemical smog. VOCs may be also found in non-industrial indoor air environments, which can act as irritants to the human organism and have negative health effects, and cause bad indoor air quality. Catalytic oxidation with high efficiency and less or no byproducts is recognized as an effective method for VOCs degradation. The monolithic film catalyst recently attracted much attention for their advantages of sufficient exposure of active sites with superb heat conduction and recycling. Transition metal oxides (Fe_2O_3, Cr_2O_3,CuO, Co_3O_4, NiO and MnO_2) as a potential alternative of precious metals, have received much attention for their superior properties such as low cost, high resistance to poisoning, good reducibility. Studies reveal that mixed metal oxides (MMO) own higher catalytic activities than single metal oxide, even equivalent or better than precious metal because of the synergistic effect. To design smartly catalysts with large surface area ratio, rich oxygen species and low temperature reducibility will lead outstanding catalytic activity.

Layered double hydroxides (LDH) are a class of naturally occur-ring and synthetic materials generally expressed by the formula$[M^{2+}{}_{1-x}M^{3+}{}_x(OH)_2](A^-)_{x/n}\cdot$ mH2O, in which M^{II} and M^{III} cations disperse in an ordered and uniform manner in brucite-like layers, and A^{n-} is a charge compensating anion such as NO_3^-,·CO_3^{2-}. The cation composition of LDH-related mixed oxide catalysts can be simply adjusted during precursor synthesis. After heating at moderate temperatures, the resulting mixed oxides catalysts from a topotactic transformation of LDH give finely dispersed mixed oxides of M^{II} and M^{III}metals with a sufficiently large surface area and good thermal stability. LDHs can in situ grow into thin film over varied substrate with hexagonal laminar structure stick to the substrate under proper synthetic condition. Furthermore, the synergistic effect due to the good interspersion of the oxide phases would possess some desirable features, like low temperature reducibility and more lattice oxygen for the improvement of catalytic activity. Besides, the presence of LDH matrix inhibits the agglomeration of active components and thus enhances the long-term stability.

The monolithic transitional metal oxides film catalysts were obtained from the calcination of the in situ growth of the corresponding LDHs precursor on aluminum substrates with nearly 1.32 mg/cm^2 by weight. CoMnAl-MMO sample displays the temperature of 90% benzene conversion (T_{90}) 240 °C with reaction rate 1.19 mmol g_{cat}^{-1} h^{-1} and space velocity 300,000 mL g_{cat}^{-1} h^{-1}, which is higher than CoNiAl-MMO at 275 °C and CoCuAl-MMO at 315 °C.

124 **Investigation of Ultrafiltration,Nanofiltration,Biofiltration and Ozonation for the Removal of Microcystin from Water.**

Joyner Eke, *Priyesh Wagh, and Isabel Escobar*
University of Kentucky, Lexington, KY

125 **Fundamental Studies of Ion Transport in Ion-Exchange Membranes.**

Jovan Kamcev, *Donald R. Paul, and Benny D. Freeman*
McKetta Department of Chemical Engineering, The University of Texas at Austin, Austin, TX

Characterization of Poly(ether ether ketone) Anion Exchange Membrane for Fuel Cell Application.

DeukJu Kim[1], *ChiHoon Park*[2], *and SangYong Nam*[1]
(1)Department of Materials Engineering and Convergence Technology, Gyeongsang National University, JinJu-si, Korea, Republic of (South), (2)Department of Energy Engineering, Gyeongnam National University of Science and Technology, Jinju-si, Korea, Republic of (South)

In this study, we synthesized a novel poly(ether-ether-ketone), containing di(quaternary ammonium) hydroxide groups in the polymer chain. Then we tested the alkaline-exchange properties of a membrane made from the novel polymer, for fuel-cell applications. First, we synthesized a di-hydroxide monomer with amine groups, which were then converted to ammonium functional groups by immersion in hydroxide solution. Then, we synthesized poly(ether-ether-ketone) using a synthesized monomer and bisphenol A. We controlled the mole ratio of the quaternary ammonium hydroxide groups in the polymer structure to evaluate the performance of the membrane. The performances of polymer-electrolyte membranes with differing degrees of amination were compared. The optimum composition of the membrane was determined based on the chemical and thermogravimetric analyses, mechanical properties, the behavior of water in the membrane, and the level of ion conductivity.

Techno-Economic Assessment of the Closed Loop Osmotic Heat Engine Utilized As an Energy Storage System.

Kerri Hickenbottom[1], *Johan Vanneste*[2], *and Tzahi Cath*[2]
(1)Environmental Resources Engineering, Humboldt State University, Arcata, CA,
(2)Civil and Environmental Engineering, Colorado School of Mines, Golden, CO

Osmotic power harnesses the energy of mixing between high salinity and low salinity streams to generate useful energy. The closed loop osmotic heat engine (OHE) is a membrane-based energy cycle that couples pressure retarded osmosis (PRO), an osmotically driven membrane process, with membrane distillation (MD), a thermally driven membrane process. The OHE is unique in that it can be designed to store energy that can be released as needed, for example during periods of peak electricity demand.

High power density in PRO is essential to minimize equipment costs and parasitic pumping losses. Likewise, high water flux is needed in MD to efficiently reconcentrate the diluted draw solution from the PRO process and minimize equipment costs. The objective of this study was to assess the technical and economical potential of the OHE as an energy storage system, and to evaluate several ionic organic and inorganic draw solutions as working fluids in the OHE. The performance of select working fluids was assessed in terms of PRO power density and reverse solute diffusion, and MD water flux and thermal efficiency. Potential pore wetting of the MD membrane was also evaluated. The working fluids were also assessed in terms of their potential for equipment corrosion.

Results indicate that sodium formate and calcium chloride draw solutions outperform sodium chloride draw solution (commonly used PRO draw solution) in terms of PRO power density and reverse solute diffusion, and that lithium chloride and calcium chloride outperform sodium chloride in terms of MD water flux. Results were used to perform a techno-economic analysis and make future recommendations on the most suitable working fluid for the OHE as a potential energy storage device. Experimental data and previously established MD and PRO models were used to develop an OHE system model that calculates system efficiency, net power output, and electricity generation costs. Modelling results indicate that with future system improvements, specifically implementation of next generation PRO membranes, the system could be competitive with alternative energy storage technologies.

Nanoporous Silica Membranes for Flow Battery Applications.

Gregory Newbloom *and Lilo Pozzo*
Chemical Engineering, University of Washington, Seattle, WA

Grid-scale power storage remains one of the largest challenges to wide-spread adoption of clean energy technologies with intermittent energy sources (*e.g.,* solar, wind and tidal). Flow batteries are a low cost technology with the potential to meet this need. However, ion exchange membranes that are crucial to battery performance are expensive, lack chemical stability in a strongly oxidizing environment, and have poor ion selectivity which leads to electrolyte diffusion across the membrane and performance instability. In this talk, we discuss the use of a primarily inorganic membrane structure created using liquid glass (*i.e.,* nanoporous silica). Silica is an advantageous material for membranes because of its excellent chemical stability and extremely low cost. Furthermore, the liquid glass curing process produces an average pore radius of ~ 3 nm which allows for effective size exclusion of electrolytes (*i.e.,* low permeability). In this study, small angle neutron scattering (SANS) was utilized to characterize the membrane pore structure. Proton conductivity and vanadium permeability were also determined and compared to Nafion 117. Silica membranes showed similar performance attributes to Nafion, including ion selectivity. These promising results, and routes towards performance enhancements, are discussed.

126 **Correlating Nafion Membrane Microstructure with Vanadium Redox Flow Battery Performance.**

Wei Wang
Pacific Northwest National Laboratory, Richland, WA

127 **Maximizing Yield, Cost Savings and Environmental Benefits Using Osn Membrane Technology for Efficient Purifications.**

Kumar Abhinava[1], *Markus Priske*[2], *Christian Schnitzer*[3], *Goetz Baumgarten*[3], *and Axel Kobus*[3]
(1)Evonik Corporation, Lafayette, IN, (2)Evonik Corporation, Theodore, AL, (3)Evonik Resource Efficiency GmbH, Marl, Germany

Degradation of Chlorinated Organics By Temperature Responsive Pnipam-Co-PAA Functionalized MF Membranes with Reactive Nanoparticles.

Anthony Saad[1], *Dibakar Bhattacharyya*[2], *and Minghui Gui*[1]
(1)Chemical and Materials Engineering, University of Kentucky, Lexington, KY, (2)Department of Chemical and Materials Engineering, University of Kentucky, Lexington, KY

Degradation of Chlorinated Organics by Temperature Responsive PNIPAm-co-PAA Functionalized MF Membranes with Reactive Nanoparticles

Anthony Saad (speaker), Li Xiao, Minghui Gui, Dibakar Bhattacharyya

Department of Chemical and Materials Engineering, University of Kentucky, Lexington, KY 40506

Abstract

This study is aimed at evaluating the effect of temperature on the degradation of chlorinated organics, specifically polychlorinated biphenyl. PVDF microfiltration membranes are functionalized with poly-N-isopropylacrylamide (PNIPAm), and its temperature responsive behavior is studied as it relates to water flux and partitioning of toxic pollutants. Effective pore size can be predicted based on the swelling extent of the temperature-responsive PNIPAm polymer around its LCST. PNIPAm is known to show a sharp transition to a hydrophobic state at its lower critical solution temperature (LCST) at around 32°C, making PNIPAM attractive for industrial applications. Swelling ratios of PNIPAm polymeric networks have been reported in water, and are utilized to estimate changes in both flux and pore diameter.

The entrapment of reactive Fe/Pd nanoparticles in a poly-acrylic acid (PAA) polymer domain has been reported. Here, the NPs are entrapped in the temperature responsive PNIPAm-co-PAA polymer network for dechlorination and contaminant degradation of PCBs. This study aims to evaluate PCB degradation by reactive immobilized nanoparticles in a PNIPAm functionalized PVDF MF membrane. A model is formulated to account for diffusion through the membrane, as well as reactivity of the nanoparticles. Solute concentration is predicted as a function of length through the membrane, and also as a function of time. Increasing the surrounding temperature affects inter-particle spacing and solute adsorption because of the changing PNIPAm conformation, as well as the intrinsic rate constant for the reaction. These effects are predicted, and verified by experimental degradation results. This research is supported by the NIEHS-SRP grant P42ES007380, and by the NSF KY EPSCOR program. Full-scale PVDF membranes were developed through collaborative work with Nanostone-Sepro (Oceanside, CA, USA).

Preparation of a Soluble Poly(ether ether ketone) Anion Exchange Membrane for Fuel Cell Application.

DeukJu Kim[1], *ChiHoon Park*[2], *and SangYong Nam*[1]
(1)Department of Materials Engineering and Convergence Technology, Gyeongsang National University, JinJu-si, Korea, Republic of (South), (2)Department of Energy Engineering, Gyeongnam National University of Science and Technology, Jinju-si, Korea, Republic of (South)

We synthesized a novel poly(ether-ether-ketone) containing di(quaternary ammonium) hydroxide groups. Then we tested the alkaline-exchange properties of a prepared membrane for fuel-cell applications. In other to synthesis di-hydroxide monomer with amine groups, we used formaldehyde solution and triethylamine sulotion. And then we synthesized poly(ether-ether-ketone) using a synthesized amine monomer, bisphenol A and difluorobenzophenone. We controlled the mole ratio of the quaternary ammonium hydroxide groups in the polymer. The performances of membrane with different degrees of amination were compared for the application of the anion exchange membrane.

Nanoporous Silica Membranes for Redox Flow Batteries.

Gregory Newbloom *and Lilo Pozzo*
Chemical Engineering, University of Washington, Seattle, WA

Grid-scale power storage remains one of the largest challenges to wide-spread adoption of clean energy technologies with intermittent energy sources (*e.g.,* solar, wind and tidal). Flow batteries are a low cost technology with the potential to meet this need. However, ion exchange membranes that are crucial to battery performance are expensive, lack chemical stability in a strongly oxidizing environment, and have poor ion selectivity which leads to electrolyte diffusion across the membrane and performance instability. In this poster, we discuss the use of a primarily inorganic membrane structure created using liquid glass (*i.e.,* nanoporous silica). Silica is an advantageous material for membranes because of its excellent chemical stability and extremely low cost. Furthermore, the liquid glass curing process produces an average pore radius of ~ 3 nm which allows for effective size exclusion of electrolytes (*i.e.,* low permeability). In this study, small angle neutron scattering (SANS) was utilized to characterize the membrane pore structure. Proton conductivity and vanadium permeability were also determined and compared to Nafion 117. Silica membranes showed similar performance attributes to Nafion, including ion selectivity. These promising results, and routes towards performance enhancements, are presented.

POSTER PRESENTATION II - WATER TREATMENT AND DESALINATION

129 **Comparison of Fouling Mechanisms of Four Nanofiltration Membranes: Optimal Membrane Selection for Stephenfield Regional WTP, MB, Canada.**

Ian J. Moran *and Beata H. Gorczyca*
Environmental Engineering, University of Manitoba, Winnipeg, MB, Canada

An Abrasion-Resistivity Evaluation Protocol for MBR Membranes.

Shigeyuki Mori
Kubota Corporation, Amagasaki, Japan

An Abrasion-Resistivity Evaluation Protocol for MBR Membranes

Polymeric membranes used in MBR operations have deferent chemicals, i.e., CPVC, PVDF, PES, PTFE, etc., on backers usually PET or PP. Since the membrane surface during operation is susceptible to the abrasion by the suspended solids, it's important to know the resistivity to abrasion when we design the polymeric membranes; otherwise we end up in getting the turbid filtrate. In this study we developed a new protocol for evaluating the abrasion resistivity and compared several membranes in commercial use. Since the protocol is of an accelerating nature, we can even predict the lifespan of each membrane.

Development of Reverse Osmosis Mixed Matrix Membrane Using Water Impermeable Matrix.

Afsaneh Khosravi[1], *Andrew Dopilka*[2], *Katelyn Kline*[1], *and Mary Laura Lind*[1]
(1)School for Engineering of Matter, Transport, and Energy, Arizona State University, Tempe, AZ, (2)Arizona State University, Tempe

Current osmotic processes offer the potential to increase water recovery from wastewater; however there are two problems with currently available reverse osmosis (RO) membranes. First, the polyamide-based RO membranes rapidly degrade when exposed to acid pretreated urine and urine brines. Additionally, present RO membranes have relatively low rejections (60%) of neutral solutes such as urea[1] while they have very high rejections (>99%) of ionic solutes.

Our research focuses on the development of mixed matrix membranes that solve both problems. We developed an innovative mixed matrix membrane design that integrates water-selective zeolite nanoparticles into a liquid-barrier chemically resistant polymer film.[2] The main distinguishing characteristics of our new membrane design compared with current mixed matrix membranes include: (1) the use of an impermeable polymer matrix enabling us to use a broader range of chemical resistant polymers; (2) the use of zeolite particles with specific pore size ensureing us the high rejection of the neutral molecules since water is transported through the zeolite rather than the polymer; (3) the use of latex dispersions as the polymer matrix source. Latexes are environmentally friendly water based-solutions, sharing the qualities of low volatile organic compound (VOC), low cost, and non- toxicity. In our design, the zeolite particles are anchored to a porous substrate. Then, a liquid-barrier chemically resistant polymer matrix fills the space between the zeolite nanoparticles and binds the zeolite into a versatile nanocomposite thin film.

In this talk, we will discuss our progress on development and synthesis of this novel membrane. We have successfully anchored zeolite nanoparticles to porous substrates. We have studied the effect of the polarity of solvents, leaving group of silane agents, the temperature and time of the silanization reaction on anchoring zeolite nanoparticles to porous substrates. Furthermore, we have developed a novel fundamental understanding of the latex film formation on porous substrates. We investigated the effect that the properties of the latex solution and the properties of the porous substrate have on the formation of defect free films of latexes on porous substrates.[3] Additionally, we will report on the flux and sodium chloride rejection of brackish water, as well as the material characterization, of our new mixed matrix membrane.

References

1 Yoon, Y. & Lueptow, R. M. Reverse osmosis membrane rejection for ersatz space mission wastewaters. *Water Res* **39**, 3298-3308, doi:10.1016/j.watres.2005.05.016 (2005).

2 Lind, M. L., Fink, S. & Sasaki, K., Lin, J. Nanocomposite membranes, US Patent Pending. US20160030893 A1 (2014).

3 Khosravi, A., King, J. A., Jamieson, H. L. & Lind, M. L. Latex Barrier Thin Film Formation on Porous Substrates. *Langmuir* **30**, 13994-14003, doi:10.1021/la502812d (2014).

149 **Facile One-Step Modification of Membrane Surface Using Hydrophilic Polymers for Wastewater Treatment.**

Nima Shahkaramipour, *Sankara Narayanan Ramanan, Cheng Kee Lai, Chong Cheng, and Haiqing Lin*
Department of Chemical and Biological Engineering, University at Buffalo, The State University of New York, Buffalo, NY

The Void Structure of Fully-Aromatic Polyamide and Its Influence on the Effective Thickness of the Barrier Layer and Permeability of TFC Membranes.

Yiqun Liu *and Jian Xu*
SINOPEC Beijing Research Institute of Chemical Industry, Beijing, China

The void structure of fully-aromatic polyamide and its influence on the effective thickness of the barrier layer and permeability of TFC membranes

Yiqun Liu*, Jian Xu

SINOPEC Beijing Research Institute of Chemical Industry, Beijing, China

Abstract

The thickness of the fully-aromatic polyamide (PA) barrier layer of the TFC membrane plays an important role in controlling the water permeability of the membrane. Typically this PA layer is not a homogenous dense layer, and thus the thickness of this separation layer measured with TEM or other characterization techniques is actually an apparent thickness. Our previous research [1] has proved that the PA layer has complex porous structure. Other researchers [2] also confirmed the existence of voids in the PA active layer and measured the volume fraction of these voids recently. Considering the void structure in the PA layer, the effective thickness of the PA layer through which water really penetrates during the filtration process is much thinner than the apparent thickness. In our opinion, the dense layer near the PA top surface with thickness of ca. 20 nm is the key barrier layer for PA layer possessing pores on its backside due to the connection between the voids with the backside of the PA layer.[1] In this case, the effective thickness of the PA layer with apparent thickness of several hundred nm should equal the thickness of the key barrier layer, ie. ca 20 nm.

In this study, a series of TFC membranes with different void structures in the PA layer were synthesized by only changing the concentration of m-phenylene diamine (MPD) and trimesyol chloride (TMC) respectively. The apparent thickness of the PA layer of these TFC membranes was measured by using SEM, TEM and Film Thickness Probe. The void structure of the PA layer was characterized by observing the morphology of the top surface, cross-section and backside of the PA layer. The effective thickness of the PA layer was analyzed by combining the water permeability results and the membrane void structure. Based on the membrane structure characterization and the membrane permeability performance, the mechanism of the formation of PA layer with different void structures was proposed and discussed.

References

[1] Hao Yan, Xiaopei Miao, Jian Xu, Guoyuan Pan, Yang Zhang, Yuanteng Shi, Min Guo, Yiqun Liu, The porous structure of the fully-aromatic polyamide film in reverse osmosis membranes, Journal of Membrane Science, 475 (2015) 504–510

[2] Lin Lin, ReneLopez, GuyZ.Ramon, OrlandoCoronell, Investigating the void structure of the polyamide active layers of thin-film composite membranes, Journal of Membrane Science, 497 (2016) 365–376

150 The Relation Between Structure-Performance of Thin Film Composite Membranes and the Tools Used for Their Fabrication Method.

Kelly Briceño[1], *Irakli Javakhishvili*[2], *Haofei Guo*[3], *Knud Villy Christensen*[1], *Birgir Norddahl*[1], *Søren Hvilsed*[4], *and Frank Lipnizki*[3]
(1)Department of Chemical Engineering, Biotechnology and Environmental Technology, University of Southern Denmark, Odense, Denmark, (2)Danish Polymer Centre, Department of Chemical and Biochemical Engineering, University of Southern Denmark, Kongens Lyngby, Denmark, (3)R&D department, Business Center Membrane, Alfa Laval, Nakskov, Denmark, (4)Danish Polymer Centre, Technical University of Denmark, Kongens Lyngby, Denmark

Gypsum Scaling and Membrane Integrity of Osmotically Driven Membranes: The Effect of Membrane Materials and Operating Conditions.

Yi-Ning Wang[1], *Eliisa Järvelä*[2], *Jing Wei*[3], *Hanna Kyllönen*[2], *Rong Wang*[4], *and Chuyang Y. Tang*[5]
(1)Singapore Membrane Technology Centre and School of Civil and Environmental Engineering, Nanyang Technological University, Singapore, Singapore, (2)VTT, Jyväskylä, Finland, (3)Nanyang Technological University, Singapore Membrane Technology Centre, Singapore, Singapore, (4)Singapore Membrane Technology Centre, Nanyang Environment and Water Research Institute, Nanyang Technological University, Singapore, Singapore, (5)Department of Civil Engineering, University of Hong Kong, Hong Kong, Hong Kong

The emerging thin film composite (TFC) forward osmosis (FO) and pressure retarded osmosis (PRO) membranes generally have better separation properties compared with their cellulose triacetate (CTA) counterparts. Nevertheless, their scaling performance has been rarely reported. In the current study, the phenomenon of membrane integrity loss as a result of scaling is reported for the first time for osmotically driven membrane processes (ODMPs). The results show that the TFC membrane suffered marked flux reduction during the gypsum scaling in the active-layer-facing-feed-solution (AL-FS) orientation, accompanied with the severe damage of the membrane active layer. The membrane integrity loss is attributed to the scale formation and growth in the confined space between the membrane and the feed spacer. Compared with the CTA membrane, the TFC membrane was more prone to scaling and membrane damage due to its unfavorable physiochemical properties (thin active layer with Ca^{2+} binding sites and ridge-and-valley roughness). Although antiscalant addition was shown to be effective for scaling control in AL-FS, it was ineffective in the active-layer-facing-draw-solution orientation. On the other hand, the important role of spacer in scale formation and membrane surface damage was also observed – membrane damage did not occur in the absence of feed spacer. The current study reveals the critical need for scaling control in ODMP processes with respect to the membrane integrity and flux stability. The results also have far-reaching implications for membrane and spacer design and process operation.

Dual-Functional Coating of Forward Osmosis Membranes for Hydrophilization and Anti-Microbial Resistance.

Qing Liu[1] *and Jim Yang Lee*[2]
(1)Department of Chemical and Biomolecular engineering, National University of Singapore, Singapore, Singapore, (2)Department of Chemical and Biomolecular Engineering, National University of Singapore, Singapore, Singapore

Microbial fouling and internal concentration polarization (ICP) are factors limiting the performance of forward osmosis (FO) membranes. These limitations in flat sheet thin film composite (TFC) FO membranes were mitigated by a single Ag/AgCl nanocomposite coating in this study. The coating was prepared by first adsorbing Ag^+ on the negatively charged carboxylate groups of polyacrylic acid (PAA)-polyethersulfone (PES) blend membrane substrate (support layer). Cl^- ions were introduced next to form a conformal AgCl coating in-situ. Irradiation of the coating with visible light then converted it into a composite coating with Ag on the membrane substrate exterior and AgCl in the membrane substrate interior. The Ag/AgCl-coated membranes prepared as such delivered good antimicrobial resistance concurrently with a high water flux. The antimicrobial performance could be attributed to a high loading of nano-Ag on the membrane exterior while the high water flux to the AgCl hydrophilization of the membrane interior which reduces the ICP. To the best of our knowledge, this is probably the first demonstration of a dual-functional coating for FO TFC membranes.

Fouling Resistant and Directly Heatable MD Membranes with Hydrophilic and Conductive Surface.

Alexander Dudchenko[1] *and David Jassby*[2]
(1)Chemical and Environmental Engineering, University of California, Riverside, Riverside, CA, (2)Department of Chemical and Environmental Engineering, University of California, Riverside, Riverside, CA

Membrane distillation (MD) is promising technique for treatment of waters contaminated with high salt concentration, such as ground water, produced water, and brines generated from desalination. Like all membrane processes, MD suffers from fouling that leads to performance degradation. The hydrophobic nature of the MD membrane surface allows for adsorption of NOM and other organic foulants that clog membrane pores, and can lead to membrane wetting, irreversibly degrading their performance. Here, we present a method of synthesizing composite MD membranes with hydrophilic or hydrophobic conductive carbon nanotube thin film surfaces layered on a hydrophobic MD support. We demonstrate that the hydrophilic surface reduces fouling caused by humic acids, while still maintaining the same performance as a bare MD membrane. Finally, we demonstrate a novel method of generating the thermal driving force in the MD process by heating the carbon nanotube thin film surface directly via joule heating. The direct heating of the MD surface eliminates the need to heat the bulk feed solution, allows for high recovery in single pass, and could significantly reduce the complexity and construction cost of MD plants.

N/A **An Polymeric Draw Solute for Energy Efficient Forward Osmosis Based Water Purification.**

Mi-Young Lee
Korea Advanced Institute of Science and Technology, Daejeon, Korea, Republic of (South)

152 **Facile One-Step Modification of Membrane Surface Using Zwitterionic Polymers for Wastewater Treatment.**

***Nima Shahkaramipour**, Sankara Narayanan Ramanan, Cheng Kee Lai, Chong Cheng, and Haiqing Lin*
Department of Chemical and Biological Engineering, University at Buffalo, The State University of New York, Buffalo, NY

153 **Vibrational and X-Ray Spectroscopy and Microscopy Reveal Irreversible Fouling Mechanisms of Reverse Osmosis Membranes during Water Reuse: Direct Evidence of Aluminum Coagulant Carryover.**

***Shankar Chellam** and Mutiara Ayu Sari*
Civil Engineering, Texas A&M University, College Station, TX

The AgCl-Based Mineralized Thin-Film Composites (TFC) Membrane for Forward Osmosis.

Haiyang Jin
Arizona State University, TEMPE, AZ, Armenia

The AgCl-based mineralized thin-film composites (TFC) membrane for forward osmosis

Haiyang Jin[1, 2], Tianmiao Lai[1], Huidan Yin[1], Pina Cay-Durgun[1], Afsaneh Khosravi[1], Mary Laura Lind[1] and Ping Yu[2]

[1]School for Engineering of Matter, Transport and Energy, Arizona State University, Tempe, AZ, USA

[2]College of Chemistry and Molecular Sciences, Wuhan University, Wuhan, Hubei, People's Republic of China

Forward osmosis (FO) membrane processes are emerging technologies that have demonstrated great potential in various water treatment applications. FO has many advantages including lower propensity towards fouling, lower energy, high water flux, high recovery and high quality product. However, FO still faces some critical challenges such as concentration polarization and chemical tolerance requiring further membrane development. Increasing researches have focused on preparing new FO membranes.

We synthesized novel AgCl-based mineralized thin-film composites (TFC) as forward osmosis (FO) membranes. We deposited silver chloride on the surface of the hand-cast polysulfone/polyamide (PSf/PA) membranes through surface mineralization by an alternate soaking process (ASP) in which we repeatedly dipped the membrane into aqueous solutions of silver nitrate ($AgNO_3$) and sodium chloride (NaCl). We synthesized membranes with different mineralization degrees (MD) by changing the number of ASP cycles. We use SEM, EDS, AFM, Contact angle, etc. to characterize the FO membranes. The results of MD, SEM and EDS confirmed the existence of the AgCl coating layer on membrane surface. The results of AFM and contact angle showed the mineralized membrane surfaces were smoother and more hydrophilic. We tested the FO performances of the membranes at 25 with pure water as the feed solution and 1 M NaCl solution as the draw solution. We found the mineralized membranes exhibited better water flux and salt rejection than original TFC membranes and the commercial CTA-ES FO membrane. We will choose Bovine Serum Albumin (BSA) as foulant to investigate the antifouling properties of mineralized membranes.

154 **The Influence of Surface Morphology on the Performance of Reverse Osmosis Membrane from the Perspective of Water Reclamation.**

Jianzhong Xia
R&D Center, Beijing OriginWater Technology, Beijing, China

In Situ Modification of Desalination Membranes for Improved Performance.

Katie Baransi Karkaby, *Maria Bass, and Viatcheslav Freger*
Wolfson Department of Chemical Engineering, Technion - Israel Institute of Technology, Haifa, Israel

Nanofiltration (NF) and reverse osmosis (RO) are widely used today for desalination and water treatment. In these membranes nanoscale defects and non-uniformities of the top polyamide layer may reduce overall selectivity. This study examines “caulking” such defects via concentration polarization-enhanced graft-polymerization and tuning selectivity - by filtering the solution of appropriate monomers and initiators through the membrane.[1,2] A novel improvement of this technique reported here is based on addition of a surfactant to the reactive solution, which appears to enhance polarization (via solubilisation of monomer in surfactant micelles) and adherence of graft-polymer to the membrane surface, which facilitates grafting and reduces monomer consumption.

The effect was particularly notable for a hydrophobic monomer glycidyl methacrylate (GMA) combined with a non-ionic surfactant. Surface characterization by ATR-FTIR, HR-SEM and contact angle indicated a significantly increased degree of grafting and uniformity of the coating, compared to a procedure with no surfactant added. Modified ESPA1-2521 elements had permeability comparable to commercial BWRO elements with enhanced rejection of boric acid and representative endocrine-disrupting contaminants (EDCs) compared to their rejection before modification.

The procedure was however less suitable for grafting hydrophilic zwitter-ionic polymers, which has the potential to improve resistance to fouling. For grafting of a zwitter-ionic polymer along with GMA for imparting membrane with both selectivity and antifouling properties, we are currently exploring a two-step procedure whereby GMA with added surfactant is grafted first and zwitter-ionic polymer is added on top of the poly-GMA. Tests of performance, fouling with humic acid, and surface characterization of modified membranes are underway and will be reported.

[1] Bernstein, R., Belfer, S., & Freger, V. (2012). Journal of Membrane Science, 405, 79-84.

[2] Ben-David, A., Bernstein, R., Oren, Y., Belfer, S., Dosoretz, C., & Freger, V. (2010). Journal of Membrane Science, 357(1), 152-159.

155 **Comparative Study of NF/RO Membranes for Sustainable Desalination of Produced Water and Frac Flowback Wastewater Following Biological Pretreatment.**

Stephanie M. Riley *and Tzahi Y. Cath*
Civil and Environmental Engineering, Colorado School of Mines, Golden, CO

Graphene Oxide Nanocomposite Membranes for High Flux Water Desalination.

Mahdi Fathizadeh[1], *Miao Yu*[2], *and Ngoc Tien Huynh*[1]
(1)Department of Chemical Engineering, University of South Carolina, Columbia, SC,
(2)Chemical Engineering, University of South Carolina, Columbia, SC

Among various desalination methods, reverse osmosis (RO) technology attracts great interest and has been widely applied due to its lower energy consumption. Further increasing water flux while maintaining high salt rejection would further reduce RO energy consumption. Therefore, developing novel methods to increase water flux through desalination membranes is one of the main research focuses. In this study, graphene oxide (GO) and modified GO were used as a novel additive to the thin film composite (TFC) membranes to significantly increase water flux. Permeability of TFC membrane was greatly enhanced by introducing modified-GO in interfacial polymerization reaction, resulting from the improved hydrophilicity and smoother surface. The achieved water flux of the modified TFC membranes in the best condition was around three times higher than that of the pristine TFC membrane with negligible influence on salt rejection. GO modified TFC membranes are expected to have great potential for high flux water desalination.

The Impact of Slickwater Fracturing Fluid Composition / Chemical Interactions on Membrane Fouling of Flowback Water.

Boya Xiong[1], *Selina Roman-White*[2], *Travis Tasker*[1], *Benjamin Farina*[2], *William Burgos*[1], *Manish Kumar*[2], *and Andrew L. Zydney*[2]
(1)Department of Civil and Environmental Engineering, Pennsylvania State University, University Park, PA, (2)Department of Chemical Engineering, Pennsylvania State University, University Park, PA

Although there is broad interest in the development of membrane systems for treatment of flowback water from natural gas production, there is currently little information on the critical components governing the fouling behavior of these complex fluids. Synthetic flowback water was generated by subjecting mixtures containing friction reducer, breaker, crosslinker and surfactants to the high pressures and temperatures associated with fracking operations in a 1 L Parr reactor that contained pulverized shale outcrops. Dead-end filtration experiments were performed to evaluate the impact of fluid composition and fracturing conditions on the fouling characteristics of flowback water during microfiltration (MF). Polyacrylamide (PAM) based friction reducer (FR) was found to be the most significant membrane foulant. Dynamic light scattering results suggested that the presence of breaker (ammonium persulfate) reduced the mean hydrodynamic diameter of PAM from more than 500 to 80 nm after activation with sodium tetraborate pentahydrate at high temperatures (80). This decrease in size led to a significant reduction in fouling during MF. The adsorption of organic additives within the shale formation during treatment in the Parr reactor was enhanced at high pressures (1200 psi), further reducing fouling by the recovered water. The addition of surfactants to the fracturing fluid led to the formation of a stable emulsion, which eliminated almost all of the fouling during subsequent MF treatment. These results demonstrate the significant interactions between various components used in hydraulic fracturing operations and its impact on membrane fouling, providing valuable suggestions on the development of efficient treatment strategies for flowback waters.

156 **A Fouling Comparison Study of Algal, Bacterial, and Humic Organic Matters in Seawater Desalination Pretreatment Using UF Ceramic Membranes.**

Mohammed Namazi
WDRC, KAUST, JEDDAH, Saudi Arabia

157 **Fouling Behaviour of Forward Osmosis TFC Membranes during Wastewater Reclamation Using Feed Streams Mimicking Treated Effluent.**

Machawe M. Motsa[1], *Bhekie B. Mamba*[2], *and Arne R. Verliefde*[3]
(1)Applied Analytical and Physical Chemistry, University of Gent, Gent, Belgium, (2)School of Science, Engineering and Technology, University of South Africa, Johannesburg, South Africa, (3)Applied Analytical and Physical Chemistry, University of Ghent, Ghent, Belgium

158 **Process Optimization on Clean Water Recovery from Real Reactive Dyeing Effluents.**

Burcu Kocer[1], *Mert Erkanli*[1], *Ruken Dilara Zaf*[1], *Levent Yilmaz*[2], *Ulku Yetis*[1], *and Zeynep Culfaz-Emecen*[3]
(1)Environmental Engineering, Middle East Technical University, Ankara, Turkey, (2)Chemical Engineering, Middle East Technical University, Ankara, Turkey, (3)Middle East Technical University, Ankara, Turkey

160 **Recovery and Reuse of Reactive Dyeing Wastewater By Ultrafiltration.**

Mert Erkanli[1], *Burcu Kocer*[1], *Ruken Dilara Zaf*[1], *Levent Yilmaz*[2], *Zeynep Culfaz-Emecen*[2], *and Ulku Yetis*[1]
(1)Environmental Engineering, Middle East Technical University, Ankara, Turkey, (2)Chemical Engineering, Middle East Technical University, Ankara, Turkey

162 **Sustainable Antibiofouling Properties of Thin Film Composite Forward Osmosis Membrane with Rechargeable Silver Nanoparticles Loading.**

Zhongyun Liu[1] *and Yunxia Hu*[2]
(1)Yantai Institute of Coastal Zone Research, Chinese Academy of Sciences, Yantai, China, (2)Yantai Institute of Coastal Zone Research,Chinese Academy of Sciences, Yantai, China

Silver Modified PI Membrane with Enhanced Antifouling Properties.

Dong Yan Zhang[1] *and Yan Wang*[2]
(1)Huazhong University of Science and Technology, State Key Laboratory of Material Processing and Die & Mould Technology, Wuhan, China, (2)School of Chemistry and Chemical Engineering, Huazhong University of Science & Technology, Wuhan, China

To improve antifouling and antibacterial properties of the polyimide (PI) membrane, silver particles are grafted on the membrane surface via a three-step modification route. The reaction mechanism is confirmed by Fourier Transform infra-red analysis (FTIR) and physical characterizations of the modified PI membranes. The anti-biofouling performance is evaluated by disk diffusion test and bacterial suspension immersion experiment. The antifouling effect of the silver-modified membranes is also assessed in the bovine serum albumin (BSA) aqueous solution. Both the anti-biofouling and antifouling properties of the membrane are greatly improved by the silver modification. Effects of the thermal treatment temperature on the property and silver stability of modified PI membrane are also investigated systematically.

163 **Second Interfacial Polymerization of Thin-Film Composite Polyamide Membrane with Polyethyleneimine for Forward Osmosis Applications.**

Liang Shen *and Yan Wang*
School of Chemistry and Chemical Engineering, Huazhong University of Science & Technology, Wuhan, China

164 **Mass Balance Simulation of a Full-Scale Forward Osmosis-Nanofiltration Hybrid System.**

Sherub Phuntsho*, Jung Eun Kim, and Ko Kyong Shon*
Faculty of Engineering and Information Technology, University of Technology Sydney, Sydney, Australia

Novel Thin Film Composite Forward Osmosis Membrane with Improved Water Flux and Anti-Fouling Performance with Incorporation of N-[3-(trimethoxysilyl) Propyl] Ethylenediamine (NPED) Amine Monomer.

Shu Xiong[1] *and Yan Wang*[2]

(1)School of Chemistry and Chemical Engineering, Huazhong University of Science and Technology, Wuhan, China, (2)School of Chemistry and Chemical Engineering, Huazhong University of Science & Technology, Wuhan, China

In this work, novel thin film composite (TFC) forward osmosis (FO) membranes are developed via interfacial polymerization on the polyacrylonitrile (PAN) substrate, using an organic-inorganic hybrid compound — N-[3-(trimethoxysilyl) propyl] ethylenediamine (NPED) as the amine monomer to mix with m-phenylenediamine (MPD). The PAN substrate was modified with sodium hydroxide and optimized by varying the hydrolyzation time. With the incorporation of NPED in the selective layer, the formed TFC membrane exhibits better hydrophilicity, improved water flux and fouling resistance, and therefore superior membrane performance. By varying the NPED content in the aqueous phase, the surface properties and performance of TFC membranes are investigated with various characterizations. In addition, the anti-fouling properties of the TFC membranes with different NPED content in the aqueous monomer is also studied.

Designing Anti-Oil-Fouling Membrane for Membrane Distillation: Learning from Natural Oleophobic Surfaces and Beyond.

Shihong Lin[1,2] *and Zhangxin Wang*[1]
(1)Civil and Environmental Engineering, Vanderbilt University, Nashville, TN, (2)Chemical and Bimolecular Engineering, Vanderbilt University, Nashville, TN

Membrane distillation (MD) is an emerging, thermally driven, desalination technology using micro-porous hydrophobic membranes. MD has a tremendous potential to become an economically competitive process for desalinating highly saline brine solution (such as shale gas wastewater, which reverse osmosis cannot treat), especially when low-grade thermal energy such as waste heat and geothermal energy is readily available. However, the reliance of MD on the usage of hydrophobic membrane limits its application to treating relatively “clean” saline water that is free of hydrophobic contaminants, as the strong attractive interaction between the hydrophobic contaminants and the hydrophobic membrane would lead to rapid membrane fouling and flux decline. To address the fouling problem in MD, we develop composite MD membranes with oleophobic surface coating inspired by natural surface with anti-fouling properties. Compared to a pristine hydrophobic membrane, the composite membrane showed a significantly more robust MD performance when challenged with a feed solution of oil-water emulsion. Beyond mimicking natural oleophobic surfaces in engineering surface wetting property, we further investigated the impact of other surface properties, such as surface charges, on the anti-oil-fouling performance of the MD membranes. We also conducted time-dependent contact angle measurements and oil-droplet-probe force spectroscopy to unravel the mechanism of oil-adhesion for corroborating the observations regarding membrane fouling in the MD experiments.

166 **Modified Microporous Polypropylene Flat Sheet Membrane Via Thermally Induced Phase Separation for Membrane Distillation.**

Yujie WANG[1], *Zhiping Zhao*[2], *Xi Zhenyu*[3], *Jinyi Luan*[4], *and Yang Yongqiang*[5]
(1)Beijing Research Institute of Chemical Industry, CHINA PETROLEUM & CHEMICAL CORPORATION, Beijing, China, (2)School of Chemical Engineering and the Environment, Beijing Institute of Technology, Beijing, China, (3)Environmental Protection Research Institute, Beijing Research Institute Of Chemical Industry, Beijing, China, (4)Environmental Protection Research Institute, BRICI, CHINA PETROLEUM & CHEMICAL CORPORATION, Beijing, China, (5)Environmental Protection Research Institute, SINOPEC, beijing, China

168 **Development of Anti-Fouling, Anti-Microbial Membranes for Wastewater Treatment.**

Steven Weinman[1], *Maria Bass*[2], *Viatcheslav Freger*[2], *Moshe Herzberg*[3], *and Scott M. Husson*[1]
(1)Chemical and Biomolecular Engineering, Clemson University, Clemson, SC, (2)The Wolfson Department of Chemical Engineering, Technion - Israel Institute of Technology, Haifa, Israel, (3)Department of Desalination and Water Treatment, Zuckerberg Institute for Water Research, Jacob Blaustein Institutes for Desert Research, Ben Gurion University, Midreshet Ben Gurion, Israel

MEXSY MemBrain in Water Treatment.

Alex Seregin[1] *and* ***Jeff Zhuk***[2]
(1)MetaFormula Inc, Palo Alto, CA, (2)ITS Inc, Englewood, CO

MetaExpert Systems (MEXSY) – is a new class of expert systems for universal applications [1-2]. MEXSY MemBrain project studing MEXSY solutions in the field of membrane science and technology [3] is moving to real applications. So it's necessary to discuss the actual MEXSY MemBrain knowledge base content with professional community (North American Membrane Society).

This presentation considers MEXSY MemBrain knowledge base in whole and in part of water treatment [4] to develop related part of MEXSY MemBrain TechnOntology knowledge portal [1]. For another examples see [5-7].

REFERENCES:

[1] Alex Seregin, Jeff Zhuk. MEXSY MemBrain TechnOntology / NAMS 2016 (poster session)

[2] Alex Seregin, Jeff Zhuk. MetaExpert Systems for Enterprises / Smart Data Conference, August 18-20, San Jose, CA, 2015; http://smartdata2015.dataversity.net/sessionPop.cfm?confid=91&proposalid=7707

[3] Alex Serein et al. NAMS Meetings 1995, 1996, 1997, 2002, 2003, 2005, 2006, 2007, 2009, 2011

[4] Isabel Escobar, Dibakar Bhattacharyya, Ben Weaver. Workshop 2: Measurement Methods for Membranes. NAMS 20016

http://www.membranes.org/2016/program.htm

[5] Alex Seregin and Jeff Zhuk. MEXSY MemBrain at Stanford, Exaple: Membrane Gas Separation. NAMS 2016 (session "Gas separation")

[6] Alex Seregin and Jeff Zhuk. MEXSY MemBrain in Membrane Characterization. NAMS 2016 (session "Design and processing of polymeric and organic membranes")

[7] Alex Seregin. Knowledge Based Data Integration and Knowledge Management in Fuel Cell Technology. NAMS 2007, Orlando, FL

http://toc.proceedings.com/17876webtoc.pdf (p. 112)

169 **Highly Permeable Artificial Water Channel in Block Copolymer Membranes.**

Yuexiao Shen[1], *Tingwei Ren*[2], *Daniel Tsai*[2], *Ian Sines*[2], *and Manish Kumar*[3]
(1)Department of Chemical Engineering, The Pennsylvania State University, University Park, PA, (2)Pennsylvania State University, (3)Chemical Engineering, The Pennsylvania State University, University Park, PA

Commercial Nanofiber Nonwoven As a Support for Thin Film Composite Membrane for Forward Osmosis.

Maqsud R. Chowdhury[1], *Liwei Huang*[2], *and Jeffrey R. McCutcheon*[1]
(1)Department of Chemical and Biomolecular Engineering, University of Connecticut, Storrs, CT, (2)Department of Chemical & Biomolecular Engineering, University of Connecticut, Storrs, CT

Product scale-up for any new emerging technology is key to seeing that technology succeed. Forward osmosis (FO) is one such technology that has been struggling to be a commercially viable technology. A major impediment to this is associated with the challenges that are there due to membrane manufacturing. Those companies that have seen some success with FO scale-up have developed FO membranes that resemble conventional RO membranes, though having certain chemical or structural differences. Many of the highly novel membranes, such as those widely discussed in the academic literature, will likely never see commercial reality because of the challenges associated with making those membranes in large quantity. Electrospun nanofiber supported thin film composite membranes are generally thought to belong to this group. Nanofibers have the properties sought after in FO TFC membrane support materials (high porosity, low tortuosity), and these membranes have shown great promise in the academic literature. However, the production of nanofiber mats at scale has been criticized. Only recently has industrial scale production of nanofibers from companies like DuPont, Donaldson, and Hollingsworth & Vose brought nanofiber membranes into the commercial space. In this work, we demonstrate the use of a commercial nanofiber nonwoven based support media to manufacture thin film composite (TFC) membranes for forward osmosis. We evaluate the viability of this material by comparing it to laboratory grade nanofiber based TFC membranes and commercially available solution casted TFC membranes. The fiber diameter of the nanofiber mat ranges from 445± 135 μm. ATR-FTIR spectrum confirms the successful formation of a polyamide selective layer by in-situ polymerization process and SEM imaging demonstrates the uniformity of the polyamide layer. The TFC membrane has a very high Young's modulus, equivalent tensile strength, and low ductility as compared to a laboratory grade nanofiber based TFC membrane. The TFC membrane generates two times more water flux and ten times less reverse solute flux as compared to a commercial HTI solution cast TFC membrane. These results prove that the commercial nanofiber support is capable of producing better membranes than commercial HTI TFC membrane while maintaining similar standards as the laboratory grade nanofiber supported TFC membrane.

171 **Modeling Nanofiltration: The Fundamentals and Its Application in Softening and Pretreatment.**

Omar Labban[1], *Chang Liu*[2], *Tzyy Haur Chong*[2], *and John Lienhard V*[1]
(1)Department of Mechanical Engineering, Massachusetts Institute of Technology, Cambridge, MA, (2)School of Civil and Environmental Engineering, Nanyang Technological University, Singapore, Singapore

Effect of Feed Spacer Thickness on the Fouling Behavior in Reverse Osmosis Process - a Pilot Scale Study.

Hyung-Gyu Park[1], *Sung-Geun Cho*[2], *Kang-Jin Kim*[2], *and Young-Nam Kwon*[3]

(1)Toray Chemical Korea Inc., Gumi, Korea, Republic of (South), (2)Toray Chemical Korea Inc., (3)Ulsan National Institute of Science and Technology (UNIST)

The performance deterioration of RO membrane processes caused by the accumulation of rejected solutes on membrane surface is an inevitable phenomenon in membrane processes. The feed spacer in spiral wound reverse osmosis (RO) membrane modules can provide the structural support to keep feed channel open and also allow turbulent flow to mitigate solute concentration build-up at the vicinity of membrane surface. The objective of this studywas to investigate the effect of feed spacer thickness on both membrane fouling behavior and cleaning efficiency in a pilot test during a 659 h operation. Furthermore, fouling load distribution was studied by measuring normalized differential pressure of individual elements in pressure vessels. Foulant analysis according to feed spacer thickness was also conducted to compare fouling propensities. This study showed that a thicker feed spacer could reduce membrane fouling and subsequently decrease membrane cleaning frequency and allow an even fouling load distribution along the modules installed in a pressure vessel.

172 **Investigation of Environmentally Friendly Solvents for Casting of Water Treatment Membranes.**

Xiaobo Dong[1], *Sneha Chede*[2], *and Isabel Escobar*[1]
(1)Department of Chemical and Materials Engineering, University of Kentucky, Lexington, KY, (2)Chemical and Environmental Engineering, The University of Toledo, Toledo, OH

Functionalized Cellulose Nanofiber Membranes for Water Purification.

Nithinart Chitpong[1] *and Scott M. Husson*[2]
(1)Department of Chemical and Biomolecular Engineering, Clemson University, Clemson, SC, (2)Chemical and Biomolecular Engineering, Clemson University, Clemson, SC

In this project, high-performance, nanofiber-based ion-exchange membranes were prepared for removal of heavy metals from impaired waters. Our hypothesis was that such membranes would have high metal ion capacities resulting from large surface area-to-volume ratios and high water permeability relative to barrier-type membranes such as reverse osmosis and nanofiltration. The nanofiber membranes were produced by electrospinning cellulose acetate nanofiber mats, which were strengthened by thermal-mechanical annealing and then converted to regenerated cellulose nanofiber mats. Subsequent surface modification transformed the membranes into ion-exchange membranes for removal of trace metal ions in water.

Cellulose acetate was selected as the starting material because it has been used widely to electrospin fibers due to simple processing. The electrospinning formulation and conditions yielded fiber mats having a fiber diameter in submicron range and a thickness range of 160 to 330 μm. Following annealing, the nanofiber mats were converted to regenerated cellulose, which is a low-fouling material that can be modified using convenient surface chemistry to introduce the ion-exchange ligands.

Ion exchange using carboxylate anion ligands is a potential strategy to capture cadmium and nickel ions found in industrial wastewaters. Thus, the regenerated cellulose nanofiber mats were modified by grafting poly(itaconic acid) (dicarboxylic acid side groups) and poly(acrylic acid) (monocarboxylic acid side groups) to the surfaces of individual nanofibers that had been modified with poly(glycidyl methacrylate).The success of poly(itaconic acid) and poly(acrylic acid) grafting was confirmed by attenuated total reflectance Fourier-transform infrared spectroscopy.

Direct-flow filtration measurements showed that the nanofiber membranes have water permeabilities of 400-1600 $L/m^2/bar/h$, much higher than reverse osmosis or nanofiltration membranes. The single-component ion-exchange isotherms were measured at constant pH for cadmium, nickel, and calcium ions. Competitive sorption measurements were made using environmentally relevant concentrations of these ions to determine the selectivity of the membranes for cadmium ions. Experimental isotherms for Cd-Ca and Cd-Ni were compared to model predictions. The multi-component isotherms for Cd-Ca and Cd-Ni systems were modeled using the competitive Langmuir model with Langmuir parameters obtained by fitting single-component isotherms.

Results show that the polymer-modified nanofiber membranes have outstanding permeability and high selectivity for heavy metals over a common competing anion present at much higher concentration. The nanofiber membrane adsorbers bind cadmium ions with static capacities that are similar to and, in some cases, superior to commercial ion-exchange resins and activated carbon.

173 **Advancements in Switchable Polarity Solvent Forward Osmosis (SPS FO) Water Treatment Technology.**

Birendra Adhikari[1], *Christopher J. Orme*[1], *Michael G. Jones*[1], *Catherine M. Hrbac*[1], *Daniel S. Wendt*[2], *and Aaron D. Wilson*[1]
(1)Biological and Chemical Processing, Idaho National Laboratory, Idaho Falls, ID, (2)Energy Resource Recovery and Sustainability, Idaho National Laboratory, Idaho Falls, ID

Development of Novel Low-Fouling RO Membrane.

Takafumi Ogawa[1], *Jun Okabe*[1], *Koji Nakatsuji*[2], *Hiroki Tomioka*[1], *and Masahiro Kimura*[1]
(1)Global Environment Research Laboratories, Toray Industries, Inc., Shiga, Japan, (2)RO Membrane Products Dept., Toray Industries, Inc., Ehime, Japan

Membrane fouling is one of the most significant problems for the stable operation of reverse osmosis (RO) membranes. To improve low-fouling property of RO membranes, a study of membrane surface was carried out. With the control of hydrated water on membrane surface, a new membrane performed adhesion inhibition of various foulant, such as non-ionic surfactant, anionic surfactant, and cationic surfactant. In cooperation with Toray Singapore Water Research Center, the remarkable reduction of foulants on the membrane surface was observed through optical coherence tomography analysis, and the new membrane performed a superior flux retention than a conventional membrane under a secondary treated sewage.

174 **The Water Flux Properties of Negative and Positive Electric Charged Monodisperse Particles Using Water Treatment Membrane.**

Sang Yong Nam[1], ***Seung Moon Woo***[2], *and dong Hee Choi*[3]
(1)Department of Polymer Science and Engineering, Gyeongsang National University, Jinju, Korea, Republic of (South), (2)Gyeongsang National University, Jinju, Korea, Republic of (South), (3)gyeongsang national university, jinju-si/South Korea, Korea, Republic of (South)

175 **Exploring Water Reuse with a Novel FO-DCMD Hybrid Membrane Process.**

Gavin Zirkel[1], *Jairo Luque Villanueva*[1], *and Andrea Achilli*[2]
(1)Humboldt State University, Arcata, CA, (2)Environmental Resources Engineering, Humboldt State University, Arcata, CA

177 **Optimizing Design Parameters for Thin Film Composite Hollow Fiber Membranes and Module for Forward Osmosis.**

Jian Ren *and Jeffrey R. McCutcheon*
Department of Chemical and Biomolecular Engineering, University of Connecticut, Storrs, CT

Numerical Analysis and Experimental Investigation of Nano-Composite Ultrafiltration Membranes for Water Treatment.

Milad R.Esfahani[1] *and Ehsan Languri*[2]
(1)Center for the Management, Utilization and Protection of Water Resources, Tennessee Tech University, Cookeville, TN, (2)Mechanical Engineering, Tennessee Technological University, Cookeville, TN

Nanocomposite membranes have been used as a promising technology in water treatment. A variety of nanoparticles have been reported to increase the membrane permeability, selectivity and antifouling properties. Nanoparticles with different physicochemical properties have different effects on the membrane structural and operational behavior. There is a need of predicting nanocomposite membranes for design and optimization purposes to control the membrane flux, fouling, module position and to consider energy consumption. A three-dimensional computational fluid dynamics (CFD) model was used to simulate cross-flow membrane process and fluid flow by solving Navier-Stocks equations coupled with the Darcy's law. Nanocomposite membrane flux was modeled at different cross-flow velocity and different operational pressure. Also, effect of membrane structural properties such as porosity and thickness was studied. This CFD model can be used to better apprehend fluid flows in different geometries and to test the impact of operational parameters on nanocomposite membranes flux and finally could help to effectively increase the permeate flux, reduce the fouling on the membrane. The modeling data is validated with experimental data regarding to nanocomposite membrane flux.

Development of Thin Film Composite Polyamide Membranes with Polydopamine/Molecular Sieve Layers for Water Desalination through Reverse Osmosis.

Pinar Cay Durgun *and Mary Laura Lind*
School for Engineering of Matter, Transport, and Energy, Arizona State University, Tempe, AZ

We developed a method to increase membrane performance and particle stability of molecular sieves in traditional polyamide thin film nanocomposite (TFN) membranes. Compared to common particle addition into polyamide casting solution, we stabilized particles by thin bio-inspired adhesion polydopamine layer, on microporous polysulfone support, and then we interfacially synthesized polyamide on top of this layer. In this work, we present our results for development of thin film composite (TFC) polyamide membranes on polydopamine layer with molecular sieves particles on polysulfone support membrane. First, we synthesized dopamine layer with different particle content and different particle size. Then, we synthesized a series of polyamide membranes via interfacial polymerization. For comparison, we also synthesized regular TFC and TFN polyamide membranes. Then we measured the permeability and selectivity performance using dead-end filtration and cross-flow testing system. We report the synthesized membrane surface and morphology result with contact angle, Fourier transform infrared spectroscopy, scanning electron microscopy, and atomic force microscopy.

The Impact of Slickwater Fracturing Fluid Composition / Shale Interactions on Membrane Fouling of Flowback Water.

Boya Xiong[1], *Selina Roman-White*[2], *Travis Tasker*[1], *Benjamin Farina*[2], *William Burgos*[1], *Manish Kumar*[2], *and Andrew L. Zydney*[2]
(1)Department of Civil and Environmental Engineering, Pennsylvania State University, University Park, PA, (2)Department of Chemical Engineering, Pennsylvania State University, University Park, PA

Although there is broad interest in the development of membrane systems for treatment of flowback water from natural gas production, there is currently little information on the critical components governing the fouling behavior of these complex fluids. Synthetic flowback water was generated by subjecting mixtures containing friction reducer, breaker, crosslinker and surfactants to the high pressures and temperatures associated with fracking operations in a 1 L Parr reactor that contained pulverized shale outcrops. Dead-end filtration experiments were performed to evaluate the impact of fluid composition and fracturing conditions on the fouling characteristics of flowback water during microfiltration (MF). Polyacrylamide (PAM) based friction reducer (FR) was found to be the most significant membrane foulant. Dynamic light scattering results suggested that the presence of breaker (ammonium persulfate) reduced the mean hydrodynamic diameter of PAM from more than 500 to 80 nm after activation with sodium tetraborate pentahydrate at high temperatures (80). This decrease in size led to a significant reduction in fouling during MF. The adsorption of organic additives within the shale formation during treatment in the Parr reactor was enhanced at high pressures (1200 psi), further reducing fouling by the recovered water. The addition of surfactants to the fracturing fluid led to the formation of a stable emulsion, which eliminated almost all of the fouling during subsequent MF treatment. These results demonstrate the significant interactions between various components used in hydraulic fracturing operations and its impact on membrane fouling, providing valuable suggestions on the development of efficient treatment strategies for flowback waters.

178 **Fabrication and Characterizations of Hollow Fiber Membranes for Removing Virus and Bacteria.**

Seung Moon Woo[1], *Sang Yong Nam*[2], *and Youn Suk Chung*[3]
(1)Gyeongsang National University, Jinju, Korea, Republic of (South), (2)Department of Polymer Science and Engineering, Gyeongsang National University, Jinju, Korea, Republic of (South), (3)Puremem Co. Ltd., Hwasung, Korea, Republic of (South)

179 **The Validation of a Stepwise Model of Direct Contact Membrane Distillation for Large-Scale Applications.**

***Joanna R. Murphy**, Jacob Rowe, and Andrea Achilli*
Environmental Resources Engineering, Humboldt State University, Arcata, CA

POSTER PRESENTATION III- GAS AND MOLECULAR SEPARATION

Carbon Molecular Sieve Membranes for Mixed Olefins-Paraffins Separations.

***Yu-Han Chu**[1] and William J. Koros[2]*
(1)School of Chemical and Biomolecular Engineering, Georgia Institute of Technology, Atlanta, GA, (2)Chemical & Biomolecular Engineering, Georgia Institute of Technology, Atlanta, GA

Light olefins, such as ethylene (C_2H_4) and propylene (C_3H_6), are key building blocks for the petrochemical industry. The separation of olefins from paraffins is typically carried out by energy-intensive distillation. It is, therefore, appealing to develop olefins-selective membranes for use in a hybrid membrane-distillation process to retrofit the current distillation separation processes to lower the overall energy usage.

Carbon molecular sieve (CMS) membranes are able to exceed the limitation of the trade-off between productivity and efficiency of traditional polymeric membranes. In this work, starting materials and processing conditions were investigated in order to explore the impact on CMS membranes separation properties for mixed olefins/paraffins. CMS membranes were made having useful olefins/paraffins separation performance, which will be reported.

Physical Aging of Perfluoropolymers in Ultrathin Film Composite Membranes.

***Milad Yavari**[1], Sajjad Maruf[2], Yifu Ding[3], and Haiqing Lin[4]*
(1)Chemical and Biological Engineering, University at Buffalo, Buffalo, NY, (2)Mechanical Engineering, University of Colorado at Boulder, Boulder, CO, (3)Department of Mechanical Engineering, University of Colorado at Boulder, Boulder, CO, (4)Department of Chemical and Biological Engineering, University at Buffalo, The State University of New York, Buffalo, NY

Thin film composite (TFC) membranes for gas separation are comprised of a selective layer of glass polymers, which are usually subjected to physical aging, i.e. permeability decreases with time. The aging behavior has been ascribed to the rearrangement of polymer chains in the thin film, resulting in changes in free volume (f) and glass transition temperature (T_g). Our objective is to investigate the effect of physical aging on

gas permeance, selectivity and T_g of the selective layer in the TFC membranes.
Two-layer TFC membranes consisting of perfluoropolymers of Teflon® AF or Hyflon® AD at various film thicknesses (50 – 400 nm) on polyethersulfone ultrafiltration support were prepared. Pure-gas permeances of CH_4, N_2, H_2 and CO_2 at 35 °C were measured and monitored for aging over 1000 hr. Apparent T_g of the thin selective layer was measured over time by a new thermo-mechanical method, Nano Thermal Analysis. In general, aging decreases gas permeances and increases T_g, and the rate of changes is more significant for thinner films. For example, after aging for 2000 hr, N_2 permeance decreases from 1000 gpu to 550 gpu by 45% and T_g increases from 160 °C to 172 °C in a TFC membrane with 100-nm thick Teflon® AF1600. In this talk, we will compare the aging behavior and gas permeance reduction of the TFC membranes with those of freestanding thin films. More importantly, for the first time we have developed a correlation between the gas permeance reduction and T_g increase using the free volume model to explain the physical aging behavior.

182 **Polyamide Thin-Film Composite Membranes for Potential Raw Biogas Purification: Experiments, Modelling and Scale-up.**

Pavel Izák[1]*, Mirek Simcik*[2]*, Zuzana Sedlakova*[3]*, Magda Karaszova*[4]*, Karel Friess*[5]*, Jiri Vejrazka*[3]*, and Marek Ruzicka*[1]
(1)Institute of Chemical Process Fundamentals Academy of Sciences of the Czech Republic, Prague 6, Czech Republic, (2)Institute of Chemical Process Fundamentals Academy of Sciences of the Czech Republic, Prague, Czech Republic, (3)Institute of Chemical Process Fundamentals, Prague, Czech Republic, (4)Institute of Chemical Process Fundamentals, Prague 6, Czech Republic, (5)Institute of Chemical Technology, Prague 6, Czech Republic

Colloidal ZIF-8 PBI/6FDA-DAM Immiscible Polymer Blend Membranes for H2/CO2 Separation.

Yu Huang[1]*, Nimanka Panapitiya*[2]*, Edson Vladimir Perez Jimenez*[1]*, John P. Ferraris*[2]*, Kenneth J. Balkus Jr.*[2]*, and Inga H. Musselman*[2]
(1)Department of Chemistry, The University of Texas at Dallas, Richardson, TX, (2)Department of Chemistry and Biochemistry, The University of Texas at Dallas, Richardson, TX

Colloidal ZIF-8 PBI/6FDA-DAM Immiscible Polymer Blend Membranes for H_2/CO_2 Separation

Yu Huang, Nimanka P. Panapitiya, Edson V. Perez, John P. Ferraris, Kenneth J. Balkus, Jr., Inga H. Musselman

ABSTRACT

High pressure and high temperature separation of H_2/CO_2 is of great interest for clean energy generation in processes such as coal gasification. Flat and tubular membranes of the polybenzimidazole (PBI)/6FDA-DAM immiscible blend containing colloidal ZIF-8 were prepared to test the gas separation performance of single gases and the H_2/CO_2 gas mixture at various pressures and temperatures. It was found that the amount of colloidal ZIF-8 in the blend has a remarkable effect on membrane morphology and gas separation. An increase of the loading for the colloidal ZIF-8 in the blend leads to a decrease in the size for the dispersed 6FDA-DAM domains. A higher experimental temperature resulted in higher gas permeance through the membrane, while the gas selectivity remained constant. At 30 atm and 300 °C, a 5% colloidal ZIF-8 PBI/6FDA-DAM membrane exhibited a H_2 permeance of 29.74 GPU (approximately 476 Barrer), an ideal H_2/CO_2 selectivity of 16.1, and a H_2/CO_2 gas mixture selectivity of 11.4. In comparison to other membranes, this polymer blend membrane shows high gas permeance and high gas selectivity, and it surpasses the upper bound of the 2008 Robeson plot for H_2/CO_2.

Membrane Separation of CO2 from Syngas Using PDMS and Mixed Matrix Membranes.

F. Handan Tezel[1] *and* ***Lauren Rose***[2]
(1)Department of Chemical and Biological Engineering, University of Ottawa, Ottawa, ON, Canada, (2)Chemical and Biological Engineering, University of Ottawa, Ottawa, ON, Canada

Membrane Separation of CO_2 from Syngas Using PDMS and Mixed Matrix Membranes

Carbon dioxide (CO_2), being one of the leading global greenhouse gas emissions, can be important to remove from other gaseous components in downstream processing. In processes such as gasification or dry CH_4 reforming, CO_2 along with syngas (CO and H_2) are the primary products. In order for syngas to be used in other processes, such conversion into methanol or ammonia, these components must be separated. Polymeric membranes have been proven to selectively separate CO_2 over syngas, due to the favoured transport mechanism. Along with these materials, mixed matrix membranes are also proving to be novelty separation methods, combining the adsorption and membrane separation mechanisms to further improve their performances. In this study, commercial pure PDMS membranes, along with PDMS mixed matrix membranes including activated carbon and zeolites were investigated for CO_2 removal from syngas.

Mixed Matrix Membrane Development for CO2 Capture.

Surendar Venna[1], *Anne Marti*[1], *Ali Sekizkardes*[1], *Victor A. Kusuma*[1], *Shan Wickramanayake*[1], *Ganpat J. Dahe*[1], *Jie Feng*[1], *and David Hopkinson*[2]
(1)National Energy Technology Laboratory, Pittsburgh, PA, (2)National Energy Technology Laboratory, Morgantown, WV

Mixed matrix membranes (MMM) have the potential to achieve high performance to capture CO_2 from flue gas. The most difficult challenge is fabricating defect free and optimized MMMs. Defects are generally due to incompatible interfaces between the polymer and the filler particle. Here, we present the fabrication of defect-free MMM by engineering the polymers and metal organic frameworks. Thermal, mechanical and gas transport properties confirmed the improved interaction between the polymer and the MOFs. Also, an attempt was made to fundametally understand the gas transport through the MMMs. MMMs showed significantly enhanced CO_2 permeability and CO_2/N_2 selectivity compared to the pure polymer membranes. These results confirm the success of the proposed technique to mitigate defective Polymer/MOF interfaces. The potential polymers and MMMs were coated on the porous hollow fibers supports using dip coating technique and gas separation performance testing of these thin membranes under flue gas conditions showed the stable CO_2 permeability and CO_2/N_2 selectivity.

184 **Modeling Gas Sorption in Thermally Rearranged Polymers:.**

Michele Galizia[1], *Kevin A. Stevens*[1], *Zachary Smith*[2], *Donald Paul*[3], *and Benny D. Freeman*[4]
(1)McKetta Department of Chemical Engineering, The University of Texas at Austin, Austin, TX, (2)Department of Chemistry, University of California, Berkeley, Berkeley, CA, (3)Department of Chemical Engineering, The University of Texas at Austin, Austin, TX, (4)Mcketta Department of Chemical Engineering, Center for Energy and Environmental Resources, and Texas Materials Institute, The University of Texas at Austin, Austin, TX

Bulky Diamines Modified Crosslinkable Polyimides Hollow Fiber Membranes for Natural Gas Purification.

Gongping Liu[1], *Nanwen Li*[2,3], *Ying Labreche*[4], *Stephen Miller*[5], *and William J. Koros*[6]
(1)School of Chemical & Biomolecular Engineering, Georgia Institute of Technology, Atlanta, GA, (2)Georgia Institute of Technology, Atlanta, GA, (3)State Key Laboratory of Coal Conversion, Chinese Academy of Sciences, Taiyuan, China, (4)School of Chemical and Biomolecular Engineering, Georgia Institute of Technology, Atlanta, GA, (5)Chevron, Richmond, CA, (6)Chemical & Biomolecular Engineering, Georgia Institute of Technology, Atlanta, GA

6FDA-based polyimide membranes have received particular attention in natural gas separation. Covalent ester crosslinking is considered as an effective approach to stabilize the membranes against CO_2 plasticization. In the past, it was found that during the crosslinking, collapse of the thin transition layer of asymmetric hollow fibers can occur, thereby causing large losses in permeance. The current work explores partial replacement $-CH_3$ groups of PDMC (propanediol monoester crosslinkable 6FDA-DAM:DABA 3:2

polyimide) by more bulky –CF_3 groups to inhibit the transition layer collapse responsible for loss in permeance of the standard PDMC fibers during crosslinking. It was found that the introduction of bulky and rigid diamine increases the β-transition temperature of PDMC by increasing the rotational barrier for side group mobility. Low molar fractions (5%) of CF_3-diamine modified PDMC polymer was synthesized and engineered to defect-free hollow fiber membranes by optimizing the dope composition and spinning conditions. The crosslinked PDMC-CF_3 asymmetric hollow fiber membranes, without sacrifice of permeance, exhibited an excellent CO_2 separation performance under aggressive feeds (high pressure, impurities of hydrocarbons), showing great potential in natural gas purification.

186 **Synthesis and Characterization of Tetraamino Diphenyl Sulfone (TADPS) Based Polybenzimidazoles and Post-Modification for Gas Separation Membranes.**

Ran Liu[1], *Ronald M. Joseph*[1], *Joshua D. Moon*[2], *Shreya Roy Choudhury*[1], *Hailun Borjigin*[1], *Kevin A. Stevens*[2], *Benny D. Freeman*[2], *and J.S. Riffle*[1]
(1)Macromolecules and Interfaces Institute, Virginia Tech, Blacksburg, VA, (2)McKetta Department of Chemical Engineering, The University of Texas at Austin, Austin, TX

Cellulose Triacetate Doped with Ionic Liquids for CO2/CH4 Separation.

Haiqing Lin[1], *Benjamin Lam*[2], *Lingxiang Zhu*[1], *Shuangjiang Luo*[3], *Ruilan Guo*[3], *Atsushi Morisato*[4], *and Paschalis Alexandridis*[5]
(1)Department of Chemical and Biological Engineering, University at Buffalo, The State University of New York, Buffalo, NY, (2)Dept. of Chemical and Biological Eng., University at Buffalo, The State University of New York, Buffalo, NY, (3)Department of Chemical and Biomolecular Engineering, University of Notre Dame, Notre Dame, IN, (4)Cameron-Cynara Membrane Systems, (5)Department of Chemical and Biological Engineering, University at Buffalo - The State University of New York (SUNY), Buffalo, NY

Polymeric membranes have been widely used for the removal of CO2 from natural gas. It is a leading technology for offshore platform applications and is gaining more competitive advantages over conventional amine absorption for onshore applications. The work horse membrane material for industrial CO2/CH4 separation is cellulose acetate/triacetate (CTA). The more widely use of CTA is, however, limited by its high crystallinity, leading to low CO_2 permeability. The goal of this work is to reduce the crystallinity in CTA by doping with ionic liquids such as [emim](BF_4) and [emim](DCA), to improve the CO2/CH4 separation properties. Both ionic liquids have CO_2/CH_4 solubility selectivity higher than CTA. In this project, we have prepared series of CTAs doped with various contents of [emim](BF_4) or [emim](DCA). The effect of ionic liquids on the crystal formation in CTAs was characterized using DSC and

WAXD. Doping of ionic liquids can reduce or completely inhibit the crystal formation in CTA, resulting in an increase in gas solubility and permeability. The presentation will discuss the effect of the doping of ionic liquids on the physical properties of the blends, including morphology and pure-gas solubility, diffusivity and permeability. The blends exhibit the highest CO_2/CH_4 solubility selectivity among the polymers known. The addition of ILs also moves the separation performance towards the upper bound, which makes them interesting platform for CO_2/CH_4 separation.

Membrane Process Optimization for Carbon Capture.

Norfamila Che Mat *and Glenn Lipscomb*
Department of Chemical and Environmental Engineering, University of Toledo, Toledo, OH

Concerns over the effects of anthropogenic carbon dioxide (CO_2) emissions from fossil-fuel electric power plants has led to significant efforts in the development of processes for CO_2 capture from flue gas. Options under consideration include absorption, adsorption, membrane, and hybrid processes.

The US Department of Energy (DOE) has set goals of 90% CO_2 capture at 95% purity followed by compression to 140 bar for transport and storage. Ideally, the levelized cost of electricity (LCOE) would increase by no more than 35%.

Membrane Technology and Research, Inc. (MTR) has proposed a promising design that combines membrane capture with cryogenic condensation [JMS 359 (2010) 126-239]. The process utilizes a two-stage membrane process in which the feed air to the furnace sweeps the flue gas in the second stage to reduce the flow of CO_2 in the effluent to 10% of that leaving the furnace. Such a design significantly reduces capture costs but leads to a detrimental reduction in the oxygen concentration of the feed air to the boiler.

The transport properties and operating pressures of the membrane stages are optimized in this study. Membrane selectivity is varied over a broad range encompassing the values considered by MTR. Permeability is varied with selectivity according to the variation anticipated by the upper bound of the Robeson plot for CO_2 and nitrogen. Membrane permeance is calculated assuming membranes can be fabricated with an effective thickness of 0.1 micron. Additionally the two stages may utilize different membrane materials. The feed and permeate pressures also are varied over ranges encompassing the values proposed by MTR.

The optimization space of membrane properties and operating conditions is scanned globally to determine the process design that minimizes LCOE. The oxygen concentration to the boiler is evaluated during the optimization process and can be used to constrain viable alternatives. The results indicate a fairly broad range of membrane properties can yield comparable LCOE near the minimum. The optimal operating pressure range is somewhat narrower. The minimum allowable oxygen concentration can constrain viable designs significantly and is critical to process economics.

N/A **Towards a Truly Hybrid Membrane for Organic Solvent Filtration.**

F.Petrus Cuperus *and Ingrid M. Wienk*
Solsep BV, Apeldoorn, Netherlands

MEXSY MemBrain at Stanford, Exaple: Membrane Gas Separation.

Alex Seregin[1] *and* ***Jeff Zhuk***[2]
(1)MetaFormula Inc, Palo Alto, CA, (2)ITS Inc, Englewood, CO

MetaExpert Systems (MEXSY) – is a new class of expert systems for universal applications [1-2]. MEXSY MemBrain project studing MEXSY solutions in the field of membrane science and technology [3] is moving to real applications. So it's necessary to discuss the actual MEXSY MemBrain knowledge base content with professional community (North American Membrane Society).

This presentation considers MEXSY MemBrain knowledge base in whole (membrane science and technology) and in part of membrane gas separation, carbon capture, and some other problems studing at Stanford University [4-5] to set connection between MEXSY MemBrain project (developing since 1999) and MEXSY StanfordBrain project (developing since 2012). Two parts of StanfordBrain (universal knowledge base of Stanford University), the LabBrain (laboratories, centers, projects, publications, etc) and the OntoEducation (schools, departments, programs, courses, etc), need to connect with MEXSY MemBrain. First of all, to connect with Stanford related part of the MEXSY MemBrain knowledge base. Then the experience of LabBrain and OntoEducation connections will transfer to another parts of the MEXSY MemBrain system (Membraine Characterization, Water Treatment, Fuel Cells [6], etc).

REFERENCES:

[1] Alex Seregin, Jeff Zhuk. MEXSY MemBrain TechnOntology / NAMS 2016 (poster session)

[2] Alex Seregin, Jeff Zhuk. MetaExpert Systems for Enterprises / Smart Data Conference, August 18-20, San Jose, CA, 2015;
http://smartdata2015.dataversity.net/sessionPop.cfm?confid=91&proposalid=7707

[3] Alex Serein et al. NAMS Meetings 1995, 1996, 1997, 2002, 2003, 2005, 2006, 2007, 2009, 2011

[4] Jennifer Wilcox. Carbon Capture, Springer, 2012

http://www.amazon.com/Carbon-Capture-Jennifer-Wilcox/dp/1461422140

[5] Jennifer Wilcox et al. NAMS 2011, 2013, 2015

[6] Alex Seregin. Knowledge Based Data Integration and Knowledge Management in Fuel Cell Technology. NAMS 2007, Orlando, FL

http://toc.proceedings.com/17876webtoc.pdf (p.112)

High Temperature and High Pressure Gas Separation Properties of Polybenzimidazole Membranes Dried at Different Temperatures.

Edson V. Perez, *John P. Ferraris, Kenneth J. Balkus Jr., and Inga H. Musselman*
Department of Chemistry and Biochemistry, The University of Texas at Dallas, Richardson, TX

Polybenzimidazole (PBI) membranes are mechanically and chemically resilient making them suitable for the separation of gases at high pressure and high temperature. Hydrogen/carbon dioxide gas selectivities ranging from 5 to 24 have been reported for PBI membranes measured at temperatures up to 250 °C, illustrating significant temperature-dependent permselectivity properties. We investigated the effect of drying temperature and testing conditions on the gas permeation properties of PBI membranes. Four sets of PBI membranes were cast and then subjected to a final drying temperature of 250, 300, 350, or 400 °C. The gas separation properties of the membranes dried at 250 °C and at 300 °C were measured at 35, 100, 200, and 300 °C with upstream pressures of 5, 15, 30, and 45 atm. The PBI membranes measured at 35 °C exhibited H_2/CO_2 selectivities that ranged from 18 to 50 when tested with pure gases and up to 72 when tested with gas mixtures over the pressure range studied. PBI membranes dried at 250 °C and tested at 35 °C demonstrated H_2/CO_2 ideal selectivities that ranged from 18 at 5 atm to 33 at 30 atm. When the same membranes were tested at 300 °C, the H_2/CO_2 ideal selectivity remained almost unchanged (28 at 5 atm and 30 at 30 atm). Membranes dried at 300 °C exhibited improved separation properties when tested at 35 °C, showing a H_2/CO_2 ideal selectivity of 49 at 5 atm, 53 at 30 atm, and 47 at 45 atm. Using a gas mixture, the H_2/CO_2 selectivity at 35 °C and 30 atm was even higher at 72. These results suggest that PBI membranes need to be dried at high temperatures to fully maximize their separation properties.

Field Trials of Spiral-Wound Proteus Membrane Modules for H2/CO2 Separation from Shifted Coal-Derived Syngas.

Trevor Carlisle, *Jay Kniep, Zhen Sun, Randa Itani, Tim Merkel, Ken Chan, and Vincent Nguyen*
Membrane Technology and Research, Inc., Newark, CA

A membrane test skid designed to treat 50 pounds per hour of coal-derived syngas has been installed and operated at the Power Systems Development Facility (PSDF) located at the National Carbon Capture Center (NCCC) in Wilsonville, Alabama, USA. This membrane system has been used for several 500 to 1000-hour long operating campaigns to evaluate the separation performance and stability of the hydrogen-selective Proteus membrane developed by MTR. Syngas from the air-blown PSDF gasifier is subjected to a water-gas-shift prior to being delivered to the membrane skid at 12 barg and ambient temperature with a typical (dry-basis) composition of 70% N_2, 12% CO_2, 12% H_2, 3% CO and balance CH_4. The H_2S content of the feed stream ranges from 100-400 ppm. The skid is equipped to heat the syngas feed and operate membrane modules continuously at elevated temperatures (100-140°C). This work summarizes the results from Proteus field test campaigns that were completed during the past year at the NCCC. Parametric tests performed on bench-scale (2.5-inch diameter) Proteus modules at 120°C revealed a strong dependence of module H_2 permeance and H_2/CO_2 selectivity on feed flowrate. Enhancement in module H_2 permeance was observed with increased temperature from 100 to 135°C, with little change in H_2/CO_2 selectivity. Furthermore, full-length 4-inch diameter modules operated at 120°C for the entirety of the campaign (~600 hours) showed stable performance and excellent selectivity (H_2/CO_2 = 15-20).

Highly-Efficiently Enhancing PEO Membrane Properties Toward Sustainable CO2 Separations.

Lu Shao *and Shuai Quan*
Harbin Inst Technol, Harbin, China

Herein, we first report an extremely versatile strategy to significantly improve gas transport properties of PEO-containing cross-linked membranes based on their specific swelling properties. Low-molecular-weight poly (ethylene glycol) (PEG) was deliberately incorporated into synthesized swellable pure PEO cross-linked membrane via immersing PEO membrane into PEG-water solution and membranes with various PEG loadings are obtained. The imbedded PEG molecules can be located between polymer chains to improve chain mobility and increase chain distance, resulting in the increase of fractional free volume (FFV), which is in favor of gas diffusion. On the other hand, the increment of CO_2-philic building units resulting from the imbedded PEG molecules leads to higher CO_2 solubility. Finally, the simple treated membranes exhibit significant increments in both CO_2 permeability (up to 762%) and CO_2/H_2 selectivity (up to 86%). The separation performance was correlated nicely with the microstructure of the membranes. This study may provide useful insights about the formation and mass transport behaviour of highly-efficient polymeric membrane applicable to clean energy purification and CO_2 capture, and possibly bridge the material-induced technology gap between academia and industry. Most importantly, our discovered strategy has great potential for treating composite membranes and regenerating "aged" membranes to realize clean energy purification and CO_2 capture by membrane technology.

Novel Thermally Cross-Linked Polyimide Membranes for Ethanol Dehydration Via Pervaporation.

Sheng Xu[1] *and Yan Wang*[2]
(1)School of Chemistry and Chemical Engineering, Huazhong University of Science and Technology, Wuhan, China, (2)School of Chemistry and Chemical Engineering, Huazhong University of Science & Technology, Wuhan, China

In this work, two novel carboxyl-containing polyimides, 2,2′-bis(3,4-dicarboxyphenyl) hexafluoropropane dianhydride-4,4′-diaminodiphenylmethane/3,5-diaminobenzoic acid (6FDA-MDA/DABA, FMD) and 3,3′,4,4′-benzophenone tetracarboxylic dianhydride-4,4′-diaminodiphenylmethane/3,5-diaminobenzoic acid (BTDA-MDA/DABA, BMD), are synthesized via chemical and thermal imidization methods, respectively, and employed as pervaporation membranes for ethanol dehydration. Chemical structures of the two polyimides are examined by FTIR and TGA to confirm the successful synthesis. A post thermal treatment of the polyimide membranes with the temperature range of 250 to 400 °C is applied, and its effects on the membrane morphology and separation performance are studied and characterized by FTIR, TGA, WXRD, solubility and sorption test. It is believed that the thermal treatment of the carboxyl-containing polyimide membrane at a relative low temperature only leads to the physical annealing, while it may cause the decarboxylation-induced cross-linking at a higher temperature. In addition, the operation temperature in pervaporation is also varied and shown to be an important factor to affect the final membrane performance. Performance benchmarking shows that the developed polyimide membranes both have superior pervaporation performance to most other flat-sheet dense membranes. This work is believed to shed useful insights on polyimide membranes for pervaporation applications.

187 **CO_2-Selective Membranes Containing Sterically Hindered Polyvinylamine.**

Zi Tong[1] *and W.S. Winston Ho*[1,2]
(1)William G. Lowrie Department of Chemical and Biomolecular Engineering, The Ohio State University, Columbus, OH, (2)Department of Materials Science and Engineering, The Ohio State University, Columbus, OH

Light-Responsive Membranes for Gas Separation.

Bradley P. Ladewig *and Nicholaus Prasetya*
Department of Chemical Engineering, Imperial College London, London, United Kingdom

Recent developments in the field of coordinated porous materials (including metal organic frameworks and zeolitic imidazolate frameworks) have demonstrated that it is possible to tailor CO2 sorbents that respond dynamically to illumination with UV and/or broadband light, such that they cyclically adsorb and desorb CO2 according to the illumination state. This has been achieved via several routes, including through the use of light-responsive ligands, light-responsive guest molecules located within the pores, and via adsorption of a light-responsive dye.

Since the use of these materials as light-responsive sorbents is challenging (due to engineering constraints in providing unhindered gas and light access to as much of the sorbent as possible), it is highly desirable to incorporate them into thin polymeric membranes to create mixed-matrix membranes with light responsive properties. This could lead to membranes with switchable separation characteristics, or other desirable properties such as some form of self-cleaning or self-regeneration after pore blocking.

This presentation will unveil the recent developments in our lab as we develop these novel mixed-matrix membrane materials, including an overview of the light-responsive porous materials we have synthesised and characterised, the design and performance of our newest illuminated test cells for both flat sheet and hollow-fibre membranes (using a unique in-situ LED lighting arrangement to provide maximum illumination and simplify gas sealing).

Carbon Molecular Sieve Membranes from Compatibilized Polyimide-Polybenzimidazole Immiscible Polymer Blends for Gas Separations.

Chamaal Karunaweera*, Nimanka P. Panapitiya, Inga H. Musselman, Kenneth J. Balkus Jr., and John P. Ferraris*
Department of Chemistry and Biochemistry, The University of Texas at Dallas, Richardson, TX

Separation of gases from mixtures is an important industrial process. High gas selectivity of one polymer and high gas permeability of another could be combined by the use of polymer blend membranes to improve gas pair separation. Compatibilization of immiscible polymers must be done in order to further improve the properties of these membranes. Although, better selectivities for gas pairs could be obtained from polymer precursor membranes, the permeability values remain low. Carbonization of the precursor membranes at high temperatures under inert environment is employed to improve the separation properties of these membranes. Especially, high H_2 permeability values are obtained together with high gas pair selectivities, which are two basic requirements for the industrial applications.

189 **CFD Numerical Simulations for Gas Separation Membrane Module Designs.**

Jiefu Ma[1]*, David Hasse*[2]*, and Edgar S. Sanders*[3]
(1)Research and Development, Air Liquide, Newark, DE, (2)Research & Development, Air Liquide, Newark, DE, (3)Air Liquide Advanced Separation, Newport, DE

190 **Metallic Membranes and N2 Separation.**

Simona Liguori*, Kyoungjin Lee, and Jennifer Wilcox*
Department of Energy Resources Engineering, Stanford University, Stanford, CA

Development of Sorption Controlled Hollow Fiber Membrane for CO2 Separation.

SangYong Nam[1]*, TaeHyun Kim*[2]*, ChiHoon Park*[3]*, DeukJu Kim*[1]*, and JeongHo Park*[1]
(1)Department of Materials Engineering and Convergence Technology, Gyeongsang National University, JinJu-si, Korea, Republic of (South), (2)Department of chemistry, Incheon national university, Incheon, Korea, Republic of (South), (3)Department of Energy Engineering, Gyeongnam National University of Science and Technology, Jinju-si, Korea, Republic of (South)

The object of this study is to develop hollow fiber membrane with high CO_2 sorption property using chemical synthesis and modification method. Novel polyimide based material with excellent permeation property is systhesized using chemical imidazation method. High CO_2 soluble materials such as polyethylene glycol and ionic liquid are used as additive in main polymer for increase of CO_2 selectivity. Flat sheet type high CO_2 permeable polymer was fabricated and measured the gas separation property. And then the hollow fiber membrane is prepared to confirm potential for commercialization. The Ionic liquid mainchain polymer is also developed to investigate the gas permeability and potential for utilization to coating materials of supported membrane. Polyimide with pendant ionic liquid (Im-PpC) membrane showed the high $\alpha(CO_2/N_2)$ value and the main chain polymer prepared by UV crosslinking with PEG & ILMP crosslinker also showed high $\alpha(CO_2/N_2)$, especially, ILM[3:1] polymer has the highest value in the world because of high solubility of PEG moiety.

191 **ZIF-67 Filled PDMS Mixed Matrix Membranes for Comprising of Recovery of Bio Ethanol Via Pervaporation.**

Amin Khan
Chemical Engineering, COMSATS Institute of Information Technology, Lahore, Pakistan

193 **Sorption Enhanced Mixed Matrix Membranes for H2/CO2 Separation at Elevated Temperature.**

Lingxiang Zhu*, Deqiang Yin, Shailesh Konda, Mark T. Swihart, and Haiqing Lin*
Department of Chemical and Biological Engineering, University at Buffalo, The State University of New York, Buffalo, NY

Gas Separation Properties of Unsupported Carbon Molecular Sieve Membranes Obtained from Blends of High Temperature Rigid Aromatic Polymers.

José Manuel Pérez-Francisco, *José Luis Santiago-García, María Isabel Loria-Bastarrachea, and Manuel de Jesús Aguilar-Vega*
Materials Unit, Yucatan Center for Scientific Research, Mérida, Mexico

Carbon molecular sieve membranes (CMSM) is an emerging technology in gas separations that could offer an outstanding relationship of permeability and selectivity for different gas pairs. In the present work, CMSM's from polyimide PI DPPTD-MIMA, polybenzimidazole (PBI) and their blends are reported. CMSM's preparation was performed at a pyrolysis temperature of 600°C following a method reported by Ning and Koros (2014). Wide angle X-ray diffraction (WAXD) measurements were performed on CMSM using a SIEMENS 5000 X-ray diffractometer with CuKα radiation (wavelength 1.54 Å). Diffraction patterns show an amorphous halo with two maxima d-spacing obtained by Bragg's law ($n\lambda=2d \sin\theta$): the first between 6.17 - 6.37 Å and the second was located on ~4 Å. In CMSM`s obtained from PI DPPTD-MIMA/PBI blends, the intensity of each maxima d-spacing show a variation related with the presence of PBI in membrane precursors. Gas transport properties were determined by a variable-pressure constant-volume method. Gas permeability coefficients were measured at 35°C and 2 atm upstream pressure for He, O_2, N_2, CH_4 and CO_2. Permeability coefficients follow an order related to gas kinetic diameters ($P_{He} > P_{CO2} > P_{O2} > P_{N2} > P_{CH4}$). Higher permeability coefficient than in the PI DPPTD-MIMA and PBI polymer precursor was obtained for CO_2 (P_{CO2} = 435.79 Barrer) and O_2 (P_{O2} = 118.10 Barrer).CMSM from PI DPPTD-MIMA/PBI surpass separation performance trade-off for various industrial interesting gas pairs, CO_2/CH_4 (selectivity = 56.04) and O_2/N_2 (selectivity = 8.47) separation. It was found that PBI concentration in membrane precursors affects gas separation properties in CMS membranes by diminishing permeability coefficients for all gases tested and selectivity for the gas pairs studied. Gas permeability coefficients temperature dependence in CMS membranes between 35 and 65°C at 2 atm upstream pressure was examined. The results show that gas permeability coefficients for CMSM membranes increase with temperature following an Arrhenius type behavior.

194 **The Investigation of Perfluorocompound Gases Separation with Hollow Fiber Membrane.**

Joohwan Lim[1], *Sujung Jeong*[1], ***Hyung Chul Koh***[2], *and* ***Seong Yong Ha***[3]
(1)R&D, Airrane, Chung ju, Korea, Republic of (South), (2)Airrane Co., Ltd., Daejeon, Korea, Republic of (South), (3)Airrane Co., Ltd., Cheongju, Korea, Republic of (South)

195 **The Study of Solid Polymer Electrolyte Composite Hollow Fiber Membrane for Olefin Gas Separation.**

Joohwan Lim[1], *Sujung Jeong*[1], ***Hyung Chul Koh***[2], *and* ***Seong Yong Ha***[3]
(1)R&D, Airrane, Chung ju, Korea, Republic of (South), (2)Airrane, Cheongju, Korea, Republic of (South), (3)Airrane Co., Ltd., Cheongju, Korea, Republic of (South)

196 **Investigation of High Permeable Polyimide Hollow Fiber Membrane for Gas Separation.**

Joohwan Lim[1], *Sujung Jeong*[1], ***Hyung Chul Koh***[2], *and* ***Seong Yong Ha***[3]
(1)R&D, Airrane, Chung ju, Korea, Republic of (South), (2)Airrane, Cheongju, Korea, Republic of (South), (3)Airrane Co., Ltd., Cheongju, Korea, Republic of (South)

The Polymeric Upper Bound for N2/NF3 Separation and Beyond; ZIF-8 Containing Mixed Matrix Membranes.

Heseong An[1], *Sunghwan Park*[2], *Soon Man Hong*[3], *Hae-Kwon Jeong*[4], *and* ***Jong Suk Lee***[5]
(1)Chemical & Biomolecular Engineering, Sogang University, Seoul, Korea, Republic of (South), (2)Department of Chemical and Biological Engineering, Korea univ., Seoul, Korea, Republic of (South), (3)Hybrid Materials Research Center, Korea Institute of Science and Technology, Seoul, Korea, Republic of (South), (4)Chemical Engineering, Texas A & M University, College Station, TX, (5)Green City Technology Institute, Korea Institute of Science & Technology, Seoul, Korea, Republic of (South)

Global production for NF_3 is continuously increasing, especially due to its heavy consumption in the semiconductor industry. Even though the amount of its emission is relatively small compared to other greenhouse gases, especially CO_2, its relatively long atmospheric lifetime makes its emissions cumulative, possibly causing global climate change. Membrane-based separation technique can be an attractive alternative for the energy-efficient NF_3 recovery, and thereby, evaluating the N_2/NF_3 separation performance by using commercial polymeric membranes is crucial. For the first time, the empirical N_2/NF_3 upper bound relationship was established by using a wide variety of commercial polymeric membranes including both glassy and rubbery polymers based on their single gas (i.e., N_2 and NF_3) permeation characterization. Among them, 6FDA-DAM:DABA (3:2), Teflon® AF 2400 and PTMSP were found to show relatively high N_2/NF_3 separation performance. The theoretical N_2/NF_3 upper bound curve was also defined and it was comparable with our empirical upper bound limit. In addition, diverse zeolitic imidazolate frameworks (i.e., commercial ZIF-8 (Basolite® Z1200), house-made ZIF-8 or ICA-ZIF-8) were incorporated into Matrimid® 5218 as molecular sieves to improve the N_2/NF_3 separation performance. More specifically, the effect of solvent, particle size and ligands on the transport properties in mixed matrix membranes was investigated.

197 **Tuning PIM-PI/Z-MOF-Based Mixed-Matrix Membranes for Highly Efficient Propylene/Propane Separation.**

Ramy Swaidan[1], *Xiaohua Ma*[2], *and Ingo Pinnau*[1]
(1)Advanced Membranes and Porous Materials Center, King Abdullah University of Science and Technology, Thuwal, Saudi Arabia, (2)KAUST

Impact of Humidity on Sulfonyl-Containing Polybenzimidazoles for Gas Separations.

Joshua D. Moon[1], *Michele Galizia*[1], *Hailun Borjigin*[2], *Ran Liu*[2], *Pellegrino Musto*[3], *Pietro La Manna*[3], *Judy S. Riffle*[4], *Benny D. Freeman*[1], *and Donald R. Paul*[1]
(1)McKetta Department of Chemical Engineering, The University of Texas at Austin, Austin, TX, (2)Macromolecules and Interfaces Institute, Virginia Tech, Blacksburg, VA, (3)Institute of Polymers, Composites and Biomaterials, National Research Council of Italy, Pozzuoli, Italy, (4)Chemistry, Virginia Tech, Blacksburg, VA

Due to their high chemical and thermal stability, polybenzimidazoles (PBIs) are promising candidates for use as gas separation membranes for high temperature applications. A series of sulfonyl-containing PBIs based on the 3,3',4,4'-tetraaminodiphenylsulfone (TADPS) monomer were recently synthesized and shown to have enhanced solubilities in common organic solvents relative to commercial PBI (Celazole™), thus improving their processability. Celazole™ and TADPS-based PBIs show a strongly hydrophilic behavior, and can uptake up to 25% water by mass at room temperature. Due to the potential effects of water uptake on both mechanical and gas transport properties, this study seeks to understand the fundamentals of the interactions between water and PBIs. Water vapor gravimetric sorption and dilation measurements have been performed at 35 °C, and a significant hysteresis was observed during desorption experiments, due to swelling and plasticization. The specific volume of the water-polymer mixture as well as the partial molar volume of water in the glassy polymer have also been calculated. Furthermore, a separate FTIR analysis showed that strong hydrogen-bonding interactions occur between water molecules and the PBI backbone.

Development and Characterization of Hollow Fiber Membrane with High CO2 Selectivity.

SangYong Nam[1], *TaeHyun Kim*[2], *ChiHoon Park*[3], *DeukJu Kim*[1], *and JeongHo Park*[1]
(1)Department of Materials Engineering and Convergence Technology, Gyeongsang National University, JinJu-si, Korea, Republic of (South), (2)Department of chemistry, Incheon National university, Incheon, Korea, Republic of (South), (3)Department of Energy Engineering, Gyeongnam National University of Science and Technology, Jinju-si, Korea, Republic of (South)

In this study, we prepared the polyimide based hollow fiber membrane with High CO2 permeance property. In other to prepare high permeable gas separation membrane, we synthesized novel polyimide material using 6FDA, Durene and PEG monomers. And then general property of the polyimide membrane is characterized using flat sheet type of membrane. The membranes were prepared under various controlled conditions such as retention time and concentration of the polymer. And then the hollow fiber membrane is also prepared and then characterized for confirmation of their potential. The Ionic liquid mainchain polymer is also developed to investigate the gas permeability and potential for utilization to coating materials of hollow fiber membrane. Polyimide with pendant ionic liquid (Im-PpC) membrane showed the high $\alpha(CO_2/N_2)$ value and the main chain polymer prepared by UV crosslinking with PEG & ILMP crosslinker also showed high $\alpha(CO_2/N_2)$

198 **Performance Prediction of a Tubular Membrane Contactor for CO2 Absorption in Reactive Aail-Based Solvent.**

Muhammad Usman, *Liyuan Deng, and Magne Hillestad*
Chemical Engineering, Norwegian University of Science and Technology, Trondheim, Norway

Analysis of the Transport Properties of Thermally Rearranged (TR) Polymers and Polymers of Intrinsic Microporosity (PIM) Relative to Upper Bound Performance.

Lloyd M. Robeson[1], ***Michelle Dose***[2], *Benny D. Freeman*[2], *and D. R. Paul*[3]
(1)Lehigh University, Macungie, PA, (2)McKetta Department of Chemical Engineering, The University of Texas at Austin, Austin, TX, (3)Department of Chemical Engineering and Texas Materials Institute, The University of Texas at Austin, Austin, TX

The properties of polymers which lie along the upper bound are not necessarily determined purely by diffusion selectivity. For many materials, solubility selectivity helps push these limits, while diffusion selectivity is dominant for select gas pairs. A critical analysis of the comparison of diffusion selectivity, solubility selectivity, diffusivity, and solubility constants has been used to analyze the transport properties of thermally rearranged (TR) polymers and polymers of intrinsic microporosity (PIMs), two types of polymers which consistently preform at or beyond the polymer upper bound for select gas pairs (O_2/N_2 and CO_2/N_2). PIMs and TR polymers have higher diffusion coefficients than glassy polymers with similar size sieving abilities and have high O_2 and CO_2 solubility. Additionally, the previously determined diffusion gas kinetic diameters, d_g, provide the best correlation with diffusion coefficients of PIMs and TR polymers and serve to better analyze transport properties of upper bound materials. Based on this analysis of diffusion and solubility data for PIMs and TR polymers, upper bound performance can be achieved when diffusion selectivity, diffusion coefficients, and gas solubility are all maximized.

Poster-Membrane Process Optimization for Carbon Capture.

Norfamila Che Mat and Glenn Lipscomb
Department of Chemical and Environmental Engineering, University of Toledo, Toledo, OH

Membrane Process Optimization for Carbon Capture
Norfamila Che Mat and Glenn Lipscomb
Chemical and Environmental Engineering Department
University of Toledo

Key Words: carbon dioxide, Post-combustion CO_2 capture, hollow fiber membrane

Concerns over the effects of anthropogenic carbon dioxide (CO_2) emissions from fossil-fuel electric power plants has led to significant efforts in the development of processes for CO_2capture from flue gas. Options under consideration include absorption, adsorption, membrane, and hybrid processes.

The US Department of Energy (DOE) has set goals of 90% CO_2 capture at 95% purity followed by compression to 140 bar for transport and storage. Ideally, the levelized cost of electricity (LCOE) would increase by no more than 35%.

Membrane Technology and Research, Inc. (MTR) has proposed a promising design that combines membrane capture with cryogenic condensation [JMS 359 (2010) 126-239]. The process utilizes a two-stage membrane process in which the feed air to the furnace sweeps the flue gas in the second stage to reduce the flow of CO_2 in the effluent to 10% of that leaving the furnace. Such a design significantly reduces capture costs but leads to a detrimental reduction in the oxygen concentration of the feed air to the boiler.

The transport properties and operating pressures of the membrane stages are optimized in this study. Membrane selectivity is varied over a broad range encompassing the values considered by MTR. Permeability is varied with selectivity according to the variation anticipated by the upper bound of the Robeson plot for CO_2 and nitrogen. Membrane permeance is calculated assuming membranes can be fabricated with an effective thickness of 0.1 micron. Additionally the two stages may utilize different membrane materials. The feed and permeate pressures also are varied over ranges encompassing the values proposed by MTR.

The optimization space of membrane properties and operating conditions is scanned globally to determine the process design that minimizes LCOE. The oxygen concentration to the boiler is evaluated during the optimization process and can be used to constrain viable alternatives. The results indicate a fairly broad range of membrane properties can yield comparable LCOE near the minimum. The optimal operating pressure range is somewhat narrower. The minimum allowable oxygen concentration can constrain viable designs significantly and is critical to process economics.

200 **Dual CO2- and N2-Selective Intrinsically Microporous Hydroxyl-Functionalized Triptycene-Based Polyimide for Natural Gas Upgrading.**

Fahd I Alghunaimi[1], *Bader Ghanem*[1], *Nasser Alaslai*[1], *Mohsin Mukaddam*[1], *and Ingo Pinnau*[2]
(1)Advanced Membranes and Porous Materials Center, KAUST, Thuwal, Saudi Arabia, (2)Advanced Membranes and Porous Materials Center, King Abdullah University of Science and Technology, Thuwal, Saudi Arabia

POSTER PRESENTATION IV -LIFE SCIENCE, BIOMEDICAL, AND/OR SENSORS

Recovery of Organic Solvents from Vegetable Oil Streams.

F.Petrus Cuperus, *Ingrid M. Wienk, and Frank Bosman*
Solsep BV, Apeldoorn, Netherlands

Membrane separation is increasingly used for the processing of organic solvents systems. Apart from targeting at very specific separations, such membranes are suitable for upgrading of solvent recovery. In current industry, a combination of membrane technology and distillation/evaporation is the most relevant.

Currently used distillation/evaporation is expensive (energy) and sometimes introduces degradation of products (lowering quality). Less heat input in the products by using membranes should improve continuous quality at lower costs. But when solvent is recovered with membranes the throughput of the membrane is not constant-due to concentration polarization and fouling. This leads to a point where membranes are less effective. In a constellation of membrane and evaporation this means that there are optimal conditions –for membranes in technical sense and for the process for the overall economics.

We have modeled and experimented a few systems to explore the optimal regimes. In this area membrane permeation and separation as well as ease of evaporation and energy costs play a role. Solsep has also experimented with systems that can handle high viscosity feeds and promise relatively high and well-defined mass-transfer properties. This could shift the optimal point to less evaporation. For eg acetone-veggy oil streams it has been found that membranes can be extremely attractive giving well-defined product quality with decreased costs. Enviroment-friendliness and low space needed is also an attractive feature.

Advantage (over current membrane systems)

-robust in organic solvents (here proposed: methylene chloride, acetone, toluene)

-also designed for high viscosity

Advantage over currently used technology (evaporation):

-low temperature, low(er) energy costs

-better product quality

-adaptable to current technology (evaporation, "flashing")-hybrid

-less or other fouling that may be better controllable (also evaporation has fouling)

We will discuss work on the removal of solvents and possible use in processing of oils and fats industry. A special pilot is on acetone recovery from seed oil fractionation plan has been built. On this system extensive piloting is done and still in progress.

Mixed Matrix Membrane for Protein Separation.

Junfen Sun
Donghua University, Shanghai, China

In this study, strontium-substituted nano hydroxyapatite (Sr-HAP) was synthesized by ultrasonic method. The surfaces of Sr-HAP particles were grafted with acrylic acid (AAc) by low temperature plasma technology and Sr-HAP grafted with AAc (Sr-HAP-AAc) was obtained. The dual layer polyether sulfone (PES)/HAP mixed matrix membranes (MMMs) were prepared by an immersion phase separation process. A particle free PES membrane layer was co-cast with the active support layer, which led to formation of dual layer MMMs. The active support layer consisted of Sr-HAP-AAc particles embedded in macroporous PES. The dual layer MMMs showed separation of protein via diffusion as well as adsorption. The bovine serum albumin (BSA) was used as a model protein. The BSA adsorption capacity of MMMs increased with extending equilibrium time and almost reached stable at 20 h. The BSA adsorption capacity of dual layer MMMs was 48.6 mg/(g MMMs). The dual layer MMMs with top PES layer kept good filtration and adsorption properties.

201 **Adsorption Behavior and Self-Aggregation of 5,10,15,20-Tetrakis-(4-sulfonatophenyl)-Porphyrin on Quaternized Polysulfone Membrane.**

Yuzhong Zhang[1], *Lizhi Zhao*[2], *and Manman Liu*[3]
(1)School of Materials Science and Engineering, Tianjin Polytechnic University, Tianjin, China, (2)State Key Laboratory of Separation Membranes and Membrane Processes,, Tianjin Polytechnic University, Tianjin, China, (3)State Key Laboratory of Separation Membranes and Membrane Processes, Tianjin Polytechnic University, Tianjin 300387, PR China, Tianjin, China

202 **Free Diffusivity of Viruses: Experimental Measurements and Model Predictions.**

Ruth E. Baltus[1], *Appala Raju Badireddy*[2], *Armin Delavari*[3], *and Shankar Chellam*[4]
(1)Chemical and Biomolecular Engineering, Clarkson University, Potsdam, NY, (2)Department of Civil and Environmental Engineering, University of Vermont, Burlington, VT, (3)Chemical & Bimolecular Engineering, Clarkson University, Potsdam, NY, (4)Department of Civil Engineering, Department of Chemical Engineering, Texas A&M University, College Station, TX

204 **Functionalization of Silica Membranes with Lipid Bilayers for Sugar Recovery.**

Shanshan Zhou, *Daniel Schlipf, Stephen E. Rankin, and Barbara L. Knutson*
Department of Chemical and Materials Engineering, University of Kentucky, Lexington, KY

205 **Living Biofilm-Resistant Membranes.**

Manish Kumar[1], *Thammajun L. Wood*[1], *Rajarshi Guha*[1], *Li Tang*[2], *Michael Geitner*[2], *and Thomas Wood*[3]
(1)Department of Chemical Engineering, Pennsylvania State University, University Park, PA, (2)Department of Chemical Engineering, Penn State University, University Park, PA, (3)Chemical Engineering, Penn State University, State College, PA

206 **A Weak Cation Exchange Macroporous Hydrogel Membrane for Monoclonal Antibody Capture.**

Huayu Niu[1] *and Christine Moresoli*[2]
(1)Chemical Engineering, University of Waterloo, Waterloo, ON, Canada, (2)University of Waterloo

A Novel Method for the Covalent and Site Specific Immobilization of Proteins.

Xinyi Zhang[1]*, Priyesh Wagh*[2]*, Sebastián Hernández*[3]*, Dibakar Bhattacharyya*[3]*, Yinan Wei*[4]*, and Isabel Escobar*[3]
(1)Chemistry Department, University of Kentucky, Lexington, KY, (2)University of Kentucky, Lexington, KY, (3)Department of Chemical and Materials Engineering, University of Kentucky, Lexington, KY, (4)Department of Chemistry, University of Kentucky, Lexington, KY

Protein immobilization is an important step of bioengineering. The attachment of functional proteins onto a solid support or inside a polymer matrix is usually necessary to exploit the activity of the protein. However, proteins can easily denature and lose their function during these manipulation. Maintaining the protein structure, stability, and activity are the major challenges during the development of immobilization technology. Here we developed a method to covalently and site-specifically attach bioactive proteins through the metabolic incorporation of a non-natural amino acid azidohomoalaine (AHA). Due to its structural similarly to methonine, AHA can be used to replace methionine in protein synthesis. In the next step, AHA-containing protein can be covalently immobilized through its bio-orthogonal reaction to an alkyne group via the well-established click chemistry. Using a pyrophosphatase PpaC as the model protein, we demonstrated that the protein activity was not affected by the AHA incorporation. However, since the location of methonine residues are usually inside the folded protein structure, the limited accessibility prevented efficient immobilization. Therefore, we designed and introduced a novel methionine tag at the C-terminus of the protein and shown that the tag could be used to attach PpaC to a supporting matrix via AHA incorporation and subsequent click chemistry reaction. We expect this method and the methionine tag to be generally useful to a broad range of proteins including membrane proteins. Our future plan include the application of this method into the construction of functional biomimetic membranes.

207 **Concentrating Membrane Proteins without Concentrating Detergent Using Ultrafiltration.**

Hasin Feroz[1], *Craig Vandervelden*[2], *Bon Ikwuagwu*[3], *Bryan Ferlez*[4], *Carol S. Baker*[5], *Mariusz Grzelakowski*[6], *John H. Golbeck*[7], *Andrew L. Zydney*[8], *and Manish Kumar*[8]
(1)Department of Chemical Engineering, Pennsylvania State University, University Park, PA, (2)Chemical engineering, University of Kansas, Kansas, KS, (3)Chemical engineering, University of Pittsburgh, Pittsburgh, PA, (4)Department of Biochemistry and Molecular Biology, Pennsylvania State University, University Park, PA, (5)Chemical engineering, Pennsylvania State University, State college, PA, (6)Applied Biomimetic Inc., Cincinnati, OH, (7)Department of Chemistry, Department of Biochemistry and Molecular Biology, Pennsylvania State University, University Park, PA, (8)Chemical Engineering, The Pennsylvania State University, University Park, PA

Antibody Purification from Cell Culture Supernatant Using New Polymeric Multimodal Membranes.

Juan Wang[1] *and Scott M. Husson*[2]
(1)Chemical and Biomolecular Engineering Department, Clemson University, Clemson, SC, (2)Chemical and Biomolecular Engineering, Clemson University, Clemson, SC

This poster will describe our latest research efforts on devising improved methods to capture and purify antibodies from cell culture supernatant using overload and elute chromatography with new multimodal membranes (MMMs). The membranes incorporate functional groups that provide orthogonal modes of interactions, which dramatically increases the range of ionic strength for operation of these membranes relative to conventional ion-exchange and hydrophobic interaction chromatography media, and makes them attractive candidates to face emerging challenges in downstream processing of biologics...

Protein binding experiments with IgG show the MMM has a high equilibrium binding capacity (180 mg IgG/ml) and excellent salt tolerance. It maintains binding capacities in excess of 100 mg IgG/ml at salt concentrations that are typical for elution buffers used in multi-stage chromatographic bioseparation processes. Protein dynamic binding capacities in excess of 50 mg IgG/ml were determined from breakthrough curve analysis.

In one study, the MMMs were used as the stationary phase following a Protein A purification step in the purification human IgG1 from Chinese hamster ovary (CHO) cell culture supernatant. The MMM stationary phase greatly enhanced load productivity compared to resin-based ion-exchange media. Further, no buffer exchange step was required between the Protein A and MMM chromatography purification steps.

In a second study, the MMMs were used as the stationary phase for the primary capture step, in place of Protein A resin chromatography. By loading the same column volumes of feed solution, the MMMs demonstrated a high selectivity for capturing human IgG from CHO cell culture supernatant. A high quality protein product was obtained from the MMM chromatography purification step.

This work shows that the new MMM chromatography operation can effectively decrease the number of purification steps and shorten the purification time and/or increase load productivity in downstream processing of biologics. It further shows that the new MMMs may be suitable for capture step purification of non-mAb proteins or when Protein A cannot be used.

The Effects of Filtration Condition on Virus Clearance.

Xianghong Qian[1], *S. Ranil Wickramasinghe*[2], *and Rigen Te*[1]
(1)Department of Biomedical Engineering, University of Arkansas, Fayetteville, AR,
(2)Department of Chemical Engineering, University of Arkansas, Fayetteville, AR

Validation of adequate virus clearance is critical in order to obtain FDA approval for a manufacturing process. Clearance can be due to inactivation or removal from the process stream. During the purification, manufacturers of monoclonal antibodies must demonstrate reduction of 10^3 to 10^5 or more virus particles than is estimated in a single dose equivalent of the unprocessed bulk. Depending on the antibody titer, estimates of the number of virus particles in a single dose equivalent could be as high as 10^{10} to 10^{15} retrovirus-like particles per mL. In addition to the removal of adventitious virus, using model parvovirus is required. Today unit operations such as virus filtration are frequently used to validate virus clearance.

Virus filtration is conducted industrially in dead end mode. Membrane performance is shown to be affected by feed pH, ionic strength and feed buffer components. Since virus filtration experiments using model mammalian viruses are expenses, complicated to run and time consuming. The purpose of this work was to determine if model bacteriophages Phix174 and PR772 can be used to rapidly determine the effect of operating conditions on membrane performance. The clearance of two model bacteriophages was tested using 30, 100 and 300 kDa polyethersulfone (PES) membranes. Flux and virus clearance were measured through the dead-end filtration experiments. The presence and concentration of Bovine Serum Albumin (BSA) and Lysozyme on the virus removal were investigated. In addition, the effects of pH on virus removal and filtration flux were determined. Our results show that 100 kDa PES membrane performs the best with regard to filtration flux and virus clearance performance in the presence and absences of tested proteins. Feed concentration close to neutral leads to a higher flux while virus titer stays more or less the same compared to other pH conditions. Mechanisms and insights on the effects of proteins and pH on virus removal are obtained.

209 **Protein Adsorption and Desorption of pH-Responsive Membrane Based on Ethylene Vinyl Alcohol Copolymer.**

Yuzhong Zhang[1] *and Hui Ye*[2]

(1)School of Materials Science and Engineering, Tianjin Polytechnic University, Tianjin, China, (2)State Key Laboratory of Hollow Fiber Membrane Materials and Processes, Tianjin Polytechnic University, Tianjin, China

Development of Polybenzimidazole Nanofiltration Membranes with Biomimetic Surfaces.

Priyesh Wagh[1], *Isabel Escobar*[2], *Xinyi Zhang*[3], *and Yinan Wei*[4]
(1)University of Kentucky, Lexington, KY, (2)Department of Chemical and Materials Engineering, University of Kentucky, Lexington, KY, (3)Chemistry Department, University of Kentucky, Lexington, KY, (4)Department of Chemistry, University of Kentucky, Lexington, KY

Aquaporins are membrane protein water channels present in cells, and they restrict the passage of contaminants including bacteria, viruses, minerals, proteins, DNA, dissolved gases, salts, detergents, and even protons without preventing the passage of water. Small molecules such as urea and boric acid, the removal of which is inefficient by conventional membranes are also rejected. Biomimetic membranes, an innovation in water filtration technology attempts to replicate a natural process occurring at the cell level; specifically, the highly-selective and efficient transport of different molecules across a cell membrane. Therefore, aquaporins have received worldwide attention because of their potential to form biomimetic membranes with high flux and selectivity for water reuse and desalination. However, challenges involved in the incorporation of aquaporin proteins in membranes limit their applicability. One of them is to attach aquaporins to the membranes without chemically altering or damaging the aquaporins during the binding to the membrane. The second challenge is to design and prepare an assembly that allows biomimetic membranes with aquaporins to sustain hydraulic water pressure gradients without losing their integrity and performance. The overarching objective of this project is to form a biomimetic membrane made of unaltered aquaporins dispersed in a polymeric membrane selective layer and capable of operation under high hydraulic pressure. To this end, polybenzimidazole (PBI) membranes were surface modified with treated aquaporins in order to achieve higher water flux and selectivity. Membranes modified with aquaporins displayed lower flux declines and higher flux recovery values after backwash as compared to unmodified PBI membranes. Also, modified membranes showed improved rejection values for both protein and salt solutions of different concentrations. However, there was leakage of ions between the channels. Therefore, in order to improve the rejection of protons, ions and other impurities, aquaporin channels were aligned with the direction of water flow. Functional groups were installed on AqpZ for covalent attachment to the polymer matrix so that the proteins could be immobilized to the membranes and aligned in the direction of the flow.

210 **Measurement of Aquaporin Z Compatibility with Biomimetic Membranes Using a Developed Membrane Protein Compatibility Evaluation Method.**

Tingwei Ren[1], *Mustafa Erbakan*[2], *Mariusz Grzelakowski*[3], *Peter J. Butler*[4], *and Manish Kumar*[5]
(1)Chemical Engineering, the Pennsylvania State University, State College, PA, (2)Department of Biosystem Engineering, Bozok University, Yozgat, Turkey, (3)Applied Biomimetic Inc., Cincinnati, OH, (4)Bioengineering, Penn State University, University Park, PA, (5)Chemical Engineering, The Pennsylvania State University, University Park, PA

212 **Functionalization of Silver Nanoparticles on Membranes and Its Influence on Biofouling.**

Conor Sprick
Chemical and Materials Engineering, University of Kentucky, Lexington, KY

Functionalization of Membranes with Ompf Porins.

Cassandra Porter[1], *Sebastián Hernández*[1], *Xinyi Zhang*[2], *Yinan Wei*[2], *and Dibakar Bhattacharyya*[1]
(1)Department of Chemical and Materials Engineering, University of Kentucky, Lexington, KY, (2)Department of Chemistry, University of Kentucky, Lexington, KY

Even more impressive than artificial membranes are the biological membranes that nature offers, with higher selectivity, resistance to fouling, and regenerative capabilities. Thus, incorporating into membranes the structures responsible for high biological performance could potentially increase stimulus-responsive behavior and salt rejection while attaining larger fluxes than other highly selective membranes. In order to achieve proper alignment and stabilization of biomolecules, certain techniques are utilized. This research first explains the process and behavior of layer-by-layer assembly of polycations and polyanions. Multilayers of electrolytes have been shown to increase the number of immobilized biomolecules by 25-fold with insignificant distortion of biomolecules.

Porins, or cylindrical beta barrel protein pores, are the biomolecule of interest in this research. Membranes with solely porins acting as their channels could produce novel behaviors and transport only small metabolites while rejecting macromolecules. These beta barrel proteins are immobilized within the pores of polyvinylidene fluoride (PVDF) membranes using layers of poly(acrylic acid) (PAA) hydrogel intercalated through in situ free radical polymerization and poly(allylamine) hydrochloride (PAH) accumulated during convective flow. The major challenge with this study is achieving an orientation of the porins parallel to the length of membrane pores so fluids may flow through their channels. This poster examines methods of functionalizing the membranes with OmpF porins from *Escherichia coli* and characterization of the membrane during each stage of the process, including zeta potential analyses, pH responsiveness, changes in flux, and sugar rejection. Characterization of solutions containing OmpF porins is also expounded.

Acknowledgement: This research is made possible by NSF KY EPSCOR and by NIH-NIEHS-SRC.

213 **Nicotine Delivery Via Programmable Carbon Nanotube Membrane Transdermal Delivery Devices.**

Lily Berger
Materials Science & Engineering, University of Washington, Seattle, WA

214 **Dynamic Electrochemical Membranes for Cyclic Protein Transport.**

***Daniel Shea**, Susrut Akkineni, and Bruce J. Hinds*
Materials Science and Engineering, University of Washington, Seattle, WA

215 **Ultrafiltration to Concentrate Membrane Proteins without Concentrating Detergents.**

***Hasin Feroz**[1], Craig Vandervelden[2], Bon Ikwuagwu[3], Bryan Ferlez[4], Carol S. Baker[5], Mariusz Grzelakowski[6], John H. Golbeck[7], Andrew L. Zydney[8], and Manish Kumar[8]*
(1)Department of Chemical Engineering, Pennsylvania State University, University Park, PA, (2)Chemical engineering, University of Kansas, Kansas, KS, (3)Chemical engineering, University of Pittsburgh, Pittsburgh, PA, (4)Department of Biochemistry and Molecular Biology, Pennsylvania State University, University Park, PA, (5)Department of Biochemistry and Molecular Biology, Pennsylvania State University, State college, PA, (6)Applied Biomimetic Inc., Cincinnati, OH, (7)Department of Chemistry, Department of Biochemistry and Molecular Biology, Pennsylvania State

University, University Park, PA, (8)Chemical Engineering, The Pennsylvania State University, University Park, PA

POSTER SESSION V - MEMBRANE MATERIALS, MODELING, AND OTHER NOVEL APPLICATIONS

217 Fabrication and Characterization of Polybenzimidazole-Multiwall Carbon Nanotubes Composite Membrane for Gas Separation.

Moonki Jeong *and SangYong Nam*
Department of Materials Engineering and Convergence Technology, Gyeongsang National University, JinJu-si, Korea, Republic of (South)

Nanoporous Metallic Films As Components of Composite Membranes.

Michael Detisch[1]*, Dibakar Bhattacharyya*[2]*, and John Balk*[1]
(1)Dept. of Chemical and Materials Engineering, University of Kentucky, Lexington, KY, (2)Chemical & Materials Engineering, University of Kentucky, Lexington, KY

Magnetron sputtering is a physical vapor deposition method that allows the deposition of thin films of different materials on a variety of substrate materials. Sputtering allows fine control of the film thickness and composition through co-sputtering from multiple target materials. As part of this study thin films have been sputtered on top of microfiltration polysulfone membrane substrates. The resulting composite membranes have remained permeable under testing with deionized water. The base Nanostone PS35 membrane showed permeability of 1941 LMH at 1.7 bar, while the membrane-film composite had a permeability of 163 LMH at 1.7 bar.

Thin films of metallic alloys deposited in this way can be made nanoporous through a process called dealloying. The process involves the removal of the less noble component of an alloy by an etchant creating an open nanoporous structure. The pores created by this method commonly vary from a few nanometers to a few hundred nanometers.

This research focuses on using magnetron sputtering to deposit precursor metallic alloy films from 100 to 250nm thick on top of porous membrane substrates. These dense precursor films are then dealloyed to produce pore/ligament structures of approximately 10nm characteristic size. In these studies iron and palladium were chosen as a precursor alloy. A portion of the iron is etched away with sulfuric acid to generate a nanoporous structure. Fe/Pd nanoparticles have been used with success to dechlorinate various chlorinated organic compounds (COCs) for wastewater treatment purposes. Nanoporous Fe/Pd shows promise to be similarly reactive with COCs due to its high surface area and curved structure. Taken together this means the composite membrane produced by fabricating a high surface area, porous Fe/Pd film on top of a microfiltration or nanofiltration membrane substrate shows promise both as a catalyst and as a platform for separations.

This project is funded by NIH-NIEHS-SRC and by NSF KY EPSCOR at the University of Kentucky.

Novel Amphiphilic Zwitterionic Copolymer Membranes of ~ 1nm Pore Size for Industrial Wastewater Treatment and Reuse.

Prity Bengani-Lutz[1] *and Ayse Asatekin*[2]
(1)Department of Chemical & Biological Engineering, Tufts University, Medford, MA,
(2)Department of Chemical Engineering, Massachusetts Institute of Technology, Cambridge, MA

Membranes with ~1 nanometer effective pore size that allow removal of high oil/organic content (dyes, industrial chemicals, surfactants etc.) have numerous applications in the treatment of wastewater from textile plants, oil refineries, fracking and oil extraction. The small effective pore size of such membranes can potentially provide better effluent quality and prevent the environmental release of a wider range of pollutants. However, there are very few commercial membranes in this pore size range, and those that are available exhibit poor selectivity, charged surfaces which lead to complex separation mechanisms, and fouling due to the accumulation of feed components on the membrane surface. In addition to high flux, sharp selectivity and exceptional fouling resistance to common foulants in wastewaters, membranes for these applications should also have high chemical resistance and be easily fabricated into large-area membranes. Zwitterionic groups have been shown to strongly resist biomacromolecular fouling due to their high degree of hydration, which makes them promising for membrane applications. We have introduced a new class of membranes whose selective layers are made of self-assembling zwitterionic amphiphilic random copolymers. These membranes derive not only fouling resistance but also their permeability and selectivity from this self-assembled nanostructure. In this study, we evaluate the potential and performance of these new membrane materials in wastewater treatment. We synthesized random copolymers of a hydrophobic monomer (trifluoroethyl methacrylate) and two zwitterionic monomers (sulfobetaine methacrylate, phosphobetaine methacrylate) by free radical polymerization. These copolymers self-assemble into bicontinuous networks of ~1 nm nanochannels, driven by the strong electrostatic interactions between the zwitterionic groups, and documented by transmission electron microscopy (TEM). Membranes manufactured by coating these copolymers on commercially available support membranes exhibit size-based selectivity with a ~1 nm size cut-off demonstrated by filtering negatively charged dyes. We have studied how feed composition such as ionic strength and presence of foulants in varying wastewater streams affects membrane performance. Specifically, we have tested the performance of these membranes by filtering single foulants, and simulated wastewater feeds. These membranes are extremely resistant to fouling by proteins and oil emulsions, fully retaining their permeability even during the filtration of the foulant solution in dead-end mode for up to 7 days. They are also stable in a wide range of chemical solutions: salt solutions (upto 1M NaCl), acidic and basic buffers, as well as chlorine. They exhibit a sharp, size-based cut-off for organic molecules such as dyes found in textile wastewater. Their low salt rejection allows operation at lower pressures. These results indicate the promise of this new family of membrane materials for industrial wastewater treatment, reclamation and reuse.

Preperation and Characterization of Cation/Anion Exchange Membrane for Electroadsorption-Deionization Process.

TaeYang Son, *DongHee Choi, Deuk Ju Kim, and SangYong Nam*
Department of Materials Engineering and Convergence Technology, Gyeongsang National University, JinJu-si, Korea, Republic of (South)

The electrodeionization process is a mixture of ion exchange resin method and electrodialysis. There are a cation exchange membrane and an anion exchange membranes for electrodialysis, and cation and anion exchange resin filling in between two films to prepare a module. And Homogenous size distribution and the equivalent ratio of the ion exchange resin are also very important factor in determining the performance of the module. On the basis of these point, EAD (Electroadsorption-deionization) process module using the bipolar membrane with the ion exchange resin is produced thesedays. We prepared on the rubber polymer to improve the ion exchange capacity than the module currently produced. So, Sulfonated and aminated SEBS polymer (Poly styrene-ethylene-butylene-styrene triblock copolymer) was used to prepare the cation and anion exchange membrane in the module. Thus, this study is information on the cation and anion exchange membrane prepared blending the particles introduced into the ion exchange capacity and the high ion exchange capacity polymer to improve the performance of the module.

218 **Fabrication of a Superhydrophilic PVDF Based Membrane Using a Facile Prefabricated TiO2 Nano-Particle Surface".**

Hassan Younas *and Jiahui Shao*
School of Environmental Science and Engineering, Shanghai Jiao Tong University, Shanghai, China

A Sequential Use of UV/H2O2 – (PSF/TiO2/MWCNT) Mixed Matrix Membranes for Removal of Dye from Aqueous Solution.

Negin Koutahzadeh[1], *Milad R.Esfahani*[2], *and Pedro Arce*[3]
(1)Department of Chemical Engineering, Tennessee Technological University, Cookeville, TN, (2)Center for the Management, Utilization and Protection of Water Resources, Tennessee Tech University, Cookeville, TN, (3)chemical engineering, Tennessee Technological University, cookeville, TN

Performance of a hybrid UV/H_2O_2 – Mixed matrix membrane system for an azo dye, Acid Black 1 [AB1], removal in a water purification process was studied. Different mixed matrix membranes embedded with titanium dioxide (TiO_2) nanoparticles, multi-wall carbon nanotubes (MWCNT), and a mixture of them were fabricated by the phase inversion method. Mixed matrix membranes embedded with MWCNTs resulted in higher pure water flux, and mixed matrix membranes embedded with TiO_2 showed lower flux declines in the presence of AB1. However, all the membranes exhibited very low Total Organic Carbon (TOC) rejection and none of the mixed matrix membranes could decolorize the AB1 solution. UV/H_2O_2 pretreatment of the AB1 solution resulted in enhanced TOC rejection, decolorization and enhanced anti-fouling membrane behavior. Combining UV/H_2O_2 with each type of polysulfone (PSF) mixed matrix membranes (PSF/TiO_2, PSF/MWCNT and PSF/TiO_2/MWCNT) resulted in optimal performance in terms of permeation, flux decline, antifouling, rejection and decolorization. The hybrid process of UV/H_2O_2-PSF/TiO_2/MWCNT mixed matrix membrane resulted in 270 ($L.m^{-2}.h^{-1}$) permeation, 29% flux decline, 90% TOC rejection, 99% decolorization and 99% Flux Recovery Ratio (FRR %).

219 **The Potential to Enhance Membrane Module Design with 3D Printing Technology.**

Jian-Yuan Lee
Singapore Membrane Technology Centre, Nanyang Environment & Water Research Institute, Singapore, Singapore

220 **Theoretical Study of Fluid Flow through Screened Channel in a Tangential Flow Filtration Process.**

Armin Delavari[1] *and* ***Yu Zou***[2]
(1)Chemical & Bimolecular Engineering, Clarkson University, Potsdam, NY, (2)EMD Millipore Corp., Bedford, MA

221 **Weak Polyelectrolyte Multilayers As Tunable Membranes for Solvent Resistant Nanofiltration (SRNF).**

Shazia Ilyas[1,2], *Nithya Joseph*[2], *Kitty Nijmeijer*[1], *Ivo Vankelecom*[2], *and Wiebe M. de Vos*[1]
(1)Membrane Science and Technology, University of Twente, Enschede, Netherlands,
(2)Centre for Surface Chemistry and Catalysis, K.U. Leuven, Heverlee, Belgium

Boldly Going Where No UV Has Gone before: Producing UVC inside Membrane Modules Via X-Ray Radioluminescence for Biofouling Prevention.

Ezra L. Cates[1], *Timothy A. Johnson*[1], *and David A. Ladner*[2]
(1)Environmental Engineering and Earth Sciences, Clemson University, Anderson, SC,
(2)Environmental Engineering & Earth Sciences, Clemson University, Anderson, SC

We are exploring methods for producing germicidal UV radiation inside spiral-wound RO modules to directly inactivate bacteria and potentially provide a more robust method for reducing, or even eliminating biofouling in water treatment applications. The technology employs UVC-radioluminescent (RL) materials incorporated onto the feed spacers which can be activated by irradiation through the module with hard X-rays via an external X-ray source. We have demonstrated 2-log bacterial inactivation using UVC-RL materials at modest X-ray doses, and additionally studied the impact of X-ray and UVC irradiation on membrane integrity. Our ongoing work involves studying the impact of X-ray photon energy on module penetration and phosphor excitation. Progress towards prototype development and achieving an effective and affordable antifouling technology will be analyzed and discussed.

The Effect of Sulfonated Polysulfone on the Compatibility and Structure of Polyethersulfone-Based Blend Membranes.

Jianxin Li Sr.
School of Materials Science and Engineering, Tianjin Polytechnic University, Tianjin, China

Polymer blend is the most common method to improve membrane hydrophilicity, in which the compatibility between the polymers is an important factor affecting the structures and properties of blend membranes. Polyethersulfone (PES)-based blend membranes were prepared from PES/polysulfone (PSf)/N, N-dimethylacetamide (DMAc) and PES/sulfonated polysulfone (SPSf)/DMAc casting solutions with water as coagulant via immersion precipitation phase inversion, respectively. The compatibility of blend systems and membrane formation process were investigated through thermodynamics and kinetics analyses. Ultrasonic through-transmission measurements were performed to quantify the rate of membrane formation during the phase inversion process. The effect of SPSf on the compatibility and morphologies of PES-based blend membranes were characterized by scanning electron microscopy (SEM), pure water flux and BSA rejection. Results indicated that the PES/PSf blend system was partially compatible, leading to instantaneous demixing and the obtained membrane with finger-like structure, whereas the PES/SPSf blend system was entirely miscible, resulting in delayed demixing and the obtained membrane with sponge-like asymmetrical gradient structure. Simultaneously, the surface segregation behavior of SPSf with hydrophilic sulfonic acid groups during mutual diffusion process resulted to the enhancement of PES/SPSf membrane hydrophilicity. Consequently, PES/SPSf blend membrane showed excellent water permeability and better antifouling ability.

222 **Study on Insoluble Polyimide Solvent-Resistant Nanofiltration Membranes Fabricated By a Two-Step Method.**

Pei Li, *Yuan Li, Jingjing Xue, and Bing Cao*
Beijing University of Chemical Technology, Beijing, China

223 **Permeation-Deformation Relationships for Porous Polymer Membranes.**

Yifu Ding[1], *Sajjad Maruf*[2], *Masoud Aghajani*[2], *and Alan R. Greenberg*[3]
(1)Department of Mechanical Engineering, University of Colorado at Boulder, Boulder, CO, (2)Mechanical Engineering, University of Colorado at Boulder, Boulder, CO, (3)Department of Mechanical Engineering, University of Colorado Boulder, Boulder, CO

224 **Phase Inversion Kinetics of Polymer- Ionic Liquid Solutions in Relationship with Membrane Morphology and Macrovoid Formation.**

Elif Nur Durmaz *and Zeynep Culfaz-Emecen*
Middle East Technical University, Ankara, Turkey

228 **Cellulose Membranes for Organic Solvent Nanofiltration.**

Faqih Muhamad Sukma[1] *and Zeynep Culfaz-Emecen*[2]
(1)Chemical Engineering, Middle East Technical University, Ankara, Turkey, (2)Middle East Technical University, Ankara, Turkey

Concentration and Recovery of Ionic Liquids from Aqueous Solutions By Nanofiltration.

Yinhua Wan[1], *Junfeng Wang*[2], *Jianquan Luo*[2], *and* ***Xiangping Zhang***[3]
(1)State Key Laboratory of Biochemical Engineering, Institute of Process Engineering, Chinese Academy Of Sciences, Beijing, China, (2)State Key Laboratory of Biochemical Engineering, Institute of Process Engineering, Chinese Academy of Sciences, Beijing, China, (3)Institute of Process Engineering, Chinese Academy of Sciences, Beijing, China

Ionic liquids (ILs) have been considered as a class of promising solvents for dissolution of natural polymer such as cellulose, silk, and wool chitin as well as raw biomass like lignocellulose and crustacean shells due to their high dissolving capacity of biopolymer. Given the fact that ILs are much more expensive than conventional solvents, it is of significance to recycle ILs during the dissolution of natural polymer and raw biomass.

In this work, the filtration behavior of three ILs ([AMIM]Cl, [BMIM]Cl and [BMIM]BF_4) in aqueous solutions by two nanofiltration (NF) membranes (NF90 and NF270) were investigated. The experimental results showed that the permeate flux increased linearly with the pressured applied at a constant IL concentration and decreased with the IL concentration at a constant pressure. Compared to NF270 membrane, NF90 showed the higher rejection because of its smaller pore size. For both membranes examined, the retention of IL was in the order of [BMIM]Cl > [BMIM]BF_4 > [AMIM]Cl, and was inversely proportion to the IL diffusion coefficients in water. The experimental results indicated that the organic cation in IL had an important effect on the IL rejection because of its larger size compared to anion. The permeate fluxes of ILs were calculated using the solution-diffusion transport model coupled with film theory. The results showed that this approach was able to calculate the permeate fluxes of ILs for IL/H_2O mixtures studied. Under the optimal experimental conditions determined in the present work, [BMIM]Cl in the aqueous solution could be concentrated to 18.85 wt% from its initial concentration of 5 wt% using the NF90 membrane, and a recovery of around 96% was obtained. The outcome of the present work will provide necessary guide for the application of concentrating ILs in aqueous solutions by nanofiltration.

Direct Observation of Gel Deposit Dynamics on Nanofiltration Membranes.

Elina Borisover, Naama Segev, Uta Cheruti, and ***Guy Z. Ramon***
Civil & Environmental Engineering, Technion - Israel Institute of Technology, Haifa, Israel

Membrane-based water production can be severely limited by organic and biological fouling, the prevention and management of which is a long standing theme of research.

Such fouling materials are often characterized by gel structures comprised of long chain polymers such as polysaccharides. Here, we report a direct observation method that was used to monitor alginate fouling and cleaning with the aid of fluorescent marker beads. These markers are trapped in the alginate layer and their motion is hindered due the visco-elastic medium. Observation of these marking particles trapped at various distances from the membrane, enables to follow the micro-rheology and formation kinetics of the deposited layer.

Fouling and cleaning experiments were conducted with and without the presence of calcium, enabling controlled manipulation, albeit qualitatively, of gel rigidity – the presence of calcium ions forms a more rigid alginate layer, which is manifested through the near immobilization of the marker beads; this is contrasted by the more fluid-like layer formed in the absence of calcium. Interestingly, rigidity of the deposited 'gel' correlated with a faster permeate flux decline. Finally, direct observation experiments suggest that detachment occurs more rapidly and completely when the applied pressure is shut off, while significantly less detachment was observed when only the permeation flux was shut off during a cleaning stage. This insight into the dynamics of 'soft', deformable deposits reveals interesting mechanisms of fouling and important implications for the development of efficient cleaning protocols.

Influence of Imogolite Nanotubes on the Structure and Performance of PVA and Psf Nanocomposite Membranes.

Ming Li
University of Wyoming, Laramie, WY

Presentation Title: Influence of imogolite nanotubes on the structure and performance of PVA and PSf nanocomposite membranes.

Authors: Ming Li and Jonathan Brant

Abstract: Engineered nanomaterials, imogolite nanotubes, have been used to improve the mechanical strength and performance (flux and selectivity) of membranes used in desalination processes. Enhanced flux is purported to result from water transport through nanotubes because of their unique geometry and chemical properties. Though promising results have been obtained with carbon nanotubes, these materials are not without their drawbacks, namely cost, clumping and poor loading in polymer films. The structures and chemical properties of imogolite nanotubes are highly desirable for to further the application of nanocomposite membranes. Our presentation will cover the relationships between imogolite nanotube properties, namely length, throat diameter, and surface (inner and outer) functionalities, on water flux and ion rejection. Membrane performance is evaluated using pressure-driven filtration and thermaly-driven configurations with an emphasis on treating high-salinity waters (TDS > 50,000 ppm), representative of oil and gas produced waters. Membranes are synthesized using the following base polymers: poly (vinyl alcohol) (PVA) and polysulfone (PSf). Nanotube dispersion will be evaluated in these two polymers and the resulting impacts of these variables on membrane performance. The change of membrane robustness and surface chemistry by adding imogolite nanotubes will also be discussed.

231 **Carbon Nanotubes/Polyethersulfone (CNT/PES) Nanocomposite Membranes Fabrication and Characterization.**

Vedat Uyak[1] *and Ismail Koyuncu*[2]
(1)Environmental Eng., Pamukkale University, Denizli, Turkey, (2)Environmental Eng., ITU, Istanbul, Turkey

232 **Charged Nanopore Membranes By Amphiphilic Random Copolymer Micelle Assembly.**

Ilin Sadeghi[1], *Jacob Kronenberg*[2], *and Ayse Asatekin*[3]
(1)Chemical and Biological Engineering, Tufts university, Medford, MA, (2)Chemical and Biological Engineering, Tufts University, Medford, MA, (3)Department of Chemical and Biological Engineering, Tufts University, Medford, MA

233 **The Effect of Pore Penetration on Composite Membranes.**

Pingjiao Hao[1], *J. G. Wijmans*[2], *Tim Merkel*[3], *Jenny He*[4], *and L. S. White*[5]
(1)Membrane Technology and Research,Inc, Newark, CA, (2)Membrane Technology and Research, Inc, Newark, CA, (3)Membrane Technology and Research, Inc., Newark, CA, (4)Membrane Technology and Research, Inc., Menlo Park, CA, (5)Membrane Technology and Research, Inc (MTR), Menlo Park, CA

234 **The Effect of Pore Entrance on the Particles Concentration Profile in Porous Membranes.**

Armin Delavari[1] *and Ruth E. Baltus*[2]
(1)Chemical & Bimolecular Engineering, Clarkson University, Potsdam, NY, (2)Clarkson University

Abstracts:

237 **Nanoparticle Deposition with Controlled Density and Placement.**

Audra DeStefano
Chemical Engineering, University of Wyoming, Laramie, WY

Transport of Deformable Particles in Porous Media.

Eduard Benet*, Aly Badran, Franck Vernerey, and John Pellegrino*
Department of Mechanical Engineering, University of Colorado, Boulder, CO

Filtration is critical to the purification in food and residual water in fossil fuels industries. In the pharmaceutical industry, new drug delivery techniques to transport and deliver drugs in a more efficient way use a filtration process in the fabrication and purification of liposomes—small, soft vesicles formed by a lipid bilayer that encapsulate a water solution where the drugs are stored. And in a similar vein, new designs of microfiltration devices are used to capture circulating tumor cells, and then be able to design better treatments. In all these cases, the transport of soft colloids, soft vesicles and other small deformable particles through the tortuous network is a key figure-of-merit. To further the ability to design membranes for fractionation of deformable particles, we have developed a (2-d) computational multiscale approach. We use a combination of the extended finite element method to describe the creeping fluid flow ($Re \sim 0$) inside a portion of a filtration membrane with an embedded interface, coupled with a particle method that interpolates the particle's membrane (interfacial) position and corresponding fields using least square fitting.

We have now modeled (in 2-d) how the combination of porous network geometry (porosity, pore size, and connectivity) and soft particle deformation-related properties (elasticity and surface tension) influence the particles' sieving coefficients. The focus in this paper is on the relationship between the elasticity of the particle and the porosity of the system, and its effects on the overall permeability of the system (and the sieving of the particles). First we present the base case of two immiscible fluids (such as an oil droplet in water), and show that the transport of such particles consistently scale with the concepts of a critical pressure gradient across the pore being determined by the ratio of the particle and pore diameter and the interfacial tension, as has bee analytically derived for the case of an axisymmetric spherical droplet entering a capillary. The base case of a single pore and droplet is then extended to include arrays of obstacles (creating a porous network) and multiple droplets. In this case, the applied trans-network prerssure drop is not the same presssure gradient each droplet experiences when it needs to deform to pass between obstacles. These cases provide a separate set of scaling rules that can guide membrane design for fractionation purposes.

Finally, and most importantly we examine the difference between the simple immiscible droplet and an elastic particle. In the former case, the "penalty" for deformation is in creating curvature, while in the latter case the deformation requires creation of surface area. Thus, we find that

- elastically deformable particles have different modes of deformations, and

- the pressure landscapes on their permeation are drastically different than surface tension based particles.

238 **Alkali Metal Chloride Sorption in Cross-Linked Polymer Membranes.**

***Eui-Soung Jang*[1]**, *Don R. Paul*[2], *and Benny D. Freeman*[1]
(1)McKetta Department of Chemical Engineering, The University of Texas at Austin, Austin, TX, (2)Department of Chemical Engineering, The University of Texas at Austin, Austin, TX

Preparation of Attapulgite-Silver-Nanoparticles and Its Application on Fabricating Antimicrobial Thin-Film Composite Membranes.

Jian Xu *and Yiqun Liu*
SINOPEC Beijing Research Institute of Chemical Industry, Beijing, China

In this study, a novel way of preparation of attapulgite (AT)-Silver-nanoparticles (AT-AgNP) composite powder via a non-chemical-reduction method was achieved by blending attapulgite suspension and silver nitrate solution, followed by heat-stirring treatment, spray drying and calcination process. The prepared AT-AgNP was used for fabricating antimicrobial NF and RO membranes.

It was firstly studied the effect of preparation process on the formation of AT-AgNP. XRD was used to analyze the structure and chemical composition of the composite powder, SEM and TEM were used to characterize the nano-silver in the composite powder, and TGA and FTIR were used to analyze the formation mechanism of nano-silver.

The results showed that attapulgite-nano-silver can only be obtained by first stirring the AT suspension and $AgNO_3$ solution at elevated temperature, then spray drying the mixture and finally calcining the sprayed powder at a temperature above 200 . The formation mechanism of nano-silver is correlated with the formation of Ag_2CO_3, which was formed by the reaction of CO_3^{2-} in the AT with the Ag^+ in $AgNO_3$ and then decomposed and covered on the surface of AT during calcination. The release of silver ions of AT-AgNP and its antimicrobial properties were measured with GC-ICP-MS and disk test method. The result showed the content of silver ions increased with increasing soaking time. AT-AgNP showd a bactericidal rate of 99.7% at the concentration of 0.5ppm.

To utilize AT-AgNP's antimicrobial properties, AT-AgNP was incorporated in the functional layer of NF membrane through interfacial polymerization. The prepared NF membrane showed a bactericidal rate of 90% at a loading of AT-AgNP 0.005% (wt/v) in the organic phase. With increasing the loading of AT-AgNP in the organic phase, the salt rejection of the prepared NF membranes decreased, and the water flux increased, indicating the incorporation of AT-AgNP may lead to defects in the polyamide separation layer.

To avoid the formation of defects in the separation layer, AT-AgNP was also added in PVA solution first, and then coated onto the surface of composite RO membranes to fabricate the antimicrobial membrane. The coated homemade RO membrane showed a bactericidal rate of 99.5%, at loading of 0.075% (wt/v) AT-AgNP in PVA coating solution and the membrane exhibited a water flux of 30.3L/m^2·h, a salt rejection of 99.4%.

Ultrafiltration Membranes Produced from Polymerization of Water/Oil/Block Copolymer Self-Assembly.

Sahar Qavi[1], *Aaron Lindsay*[1], *Millicent A. Firestone*[2], *and Reza Foudazi*[1]
(1)Chemical and Materials Engineering, New Mexico State University, Las Cruces, NM,
(2)Materials Physics & Applications Division, Center for Integrated Nanotechnologies, Los Alamos National Laboratory, Los Alamos, NM

As human population continues to grow, demands for clean water will increase, necessitating further development of water purification technologies. Ultrafiltration is a necessary step in the treatment of both drinking water and wastewater in which suspending particles, viruses, and bacteria are removed. Conventional ultrafiltration (UF) membranes are produced through non-solvent induced phase separation (NIPS). However, the NIPS technique is not ecofriendly and resultant membranes are anisotropic with low surface area and porosity. The self-assembly of block copolymers at an oil/water interface has been widely studied for applications ranging from drug delivery to energy storage. The purpose of this research is to utilize self-assembly as a template for producing mesoporous polymeric materials to be used as UF membranes. A tri-block copolymer of poly(ethylene oxide), PEO, and poly(propylene oxide), PPO, with $(PEO)_{13}$-$(PPO)_{30}$-$(PEO)_{13}$ composition is used, and different water/oil/block copolymer concentrations and compositions are studied. The oil phase consists of butyl acrylate as well as ethylene glycol dimethacrylate, a cross-linker, and can be polymerized via photo- and thermal initiators. Upon self-assembly of the water/oil/block copolymer mixture into a desired mesostructure, the oil phase is polymerized and a mesoporous polymer is obtained. Resultant materials are characterized through polarized light microscopy (PLM), rheology, BET, and small angle X-ray scattering (SAXS). The produced membranes are found to have mesopores in the range of 70 nm with improved permeability over conventional UF membranes, as measured in a dead-end flow using an oil-in-water emulsion as feed stream.

Hydrophilization of Hydrophobic Membrane Surfaces By the Adsorption of Water-Soluble Polymers.

Do Hyiung Shin, Yi Seul Jeon, and ***Ji Won Rhim***
Advanced Materials and Chemical Engineering, Hannam University, Daejeon, Korea, Republic of (South)

The adsorption of biological compounds such as proteins on membrane surfaces which is driven by different protein-surface forces, including van der Waals, hydrophobic and electrostatic forces, is important in the fields of separation science. The physical adsorption of natural polymers such as proteins on hydrophobic surfaces is referred to as a part of 'fouling phenomena', which is a major drawback of the continuous flux decline due to the adsorption in question, pore blocking, precipitation and cake formation in membrane-based water purification processes. The low surface tension energies of hydrophobic membrane surfaces give rise to several problems of easy wetting, ease of initial flux, adsorption or spreading. Therefore, it is required that the surface energy should be increased through the introduction of polar functionality using chemical modification, gas-discharge treatment (plasma, corona, etc., g-radiation, UV irradiation, and coating. Since, however, these chemical approaches are limited to lab scale to date because of the complicated production equipments, strict operation requirements and high costs, the coating or impregnating with hydrophilization agents, such as alcohols, surfactants, or amphiphilic polymers is more favorable method to endow the hydrophilicity with hydrophobic membranes.

In this study, to improve the water flux of porous hydrophobic membranes, various water-soluble polymers including neutral, cationic and anionic polymers were adsorbed using 'salting-out' method. The adsorbed hydrophobic membrane surfaces were characterized mainly via the measurements of contact angles and scanning electron microscopy (SEM) images. To enhance the durability of the modified membranes, the water-soluble polymers such poly(vinyl alcohol) (PVA) were crosslinked with glutaraldehyde (GA) and found to be resistant for more than 2 months in vigorously stirred water. The water flux was much more increased when the ionic polymers used as the coating materials rather than the neutral polymer and in this case, about 70% of 0.31 $L/m^2 \cdot h$ (LMH) to 0.50 LMH was increased when 300 mg/L of polyacrylamide (PAAm) was used as the coating agents. Among the cationic coating polymers such as poly(styrene sulfonic acid-co-maleic acid) (PSSA_MA), poly(acrylic acid-co-maleic acid) (PAM) and poly(acrylic acid) (PAA), PSSA_MA was found to be the best in terms of contact angle and water flux. In the case of PSSA_MA, the water flux was enhanced about 80%. The low concentration of the coating solution was better to hydrophilize while the high concentration inclined to block the pores on the membrane surfaces. The best coating condition was found: (1) coating concentration 150 to 300 mg/L, (2) ionic strength 0.15, (3) coating time 20 min.

Fabricaiton of Hierarchically Structured Helical Micro/Nanofibrous Nonwovens and Its Mechanism Study.

Yongchun Zeng
College of Textiles, Donghua University, Shanghai, China

Electrospinning and melt blowing are the most commonly used technologies to produce micro/nanofibrous nonwovens. Helical fibers resembling plant tendrils in micro/nanoscale have been of increasing interest because of their unique characteristics. The introduction of helical structures into the micro/nanofibrous nonwovens finds potential applications in fields such as micro/nanoscale sensors, filtration materials, and oil sorbents. The three-dimensional structure of the helical fibers can provide the nonwovens larger specific surface area, larger porosity, and better mechanical properties in terms of resiliency and flexibility. In this study, we report the fabrication of micro/nanoscale helical fibers by electrospinning and melt blowing. Two kinds of polymer, an elastomeric component and a rigid component, are combined in the electrospinning and melt-blowing processes to produce helical fibers. A Core-shell spinneret is used in electrospinning. While during melt blowing, a swirl-die is adopted. Due to the different shrinkages displayed by the two components, helical fibers with three-dimensional spiral shape are formed after electrospinning and melt blowing. This work can provide a promising technique for producing micro/nanofibrous materials with helical fiber morphology.

239 **Antifouling Modification of Ultrafiltration Membranes By Grafting Polyzwitterionic Brushes.**

Douglas Davenport*, Jongho Lee, and Menachem Elimelech*
Department of Chemical and Environmental Engineering, Yale University, New Haven, CT

240 **Mechanical Enhancement of Thin-Film Composite Membranes Using Helically-Coiled Carbon Nanotubes for Engineered Osmosis.**

Jaime Idarraga-Mora[1]*, Jinxiang Zhou*[1]*, Apparao M. Rao*[2]*, and Scott M. Husson*[3]
(1)Department of Chemical and Biomolecular Engineering, Clemson University, Clemson, SC, (2)Physics and Astronomy, Clemson University, Clemson, SC, (3)Chemical and Biomolecular Engineering, Clemson University, Clemson, SC

Special Applications of Nanofiltration Membranes.

Yong-Doo Jung
RO Development Team, Toray Chemical Korea Inc., Gumi, Gyeongsangbuk-do, Korea, Republic of (South)

Nanofiltration membranes have established itself as a mainstay in many applications involving drinking water and beverage manufacturing. However, their use in other special applications is not often discussed in literature. This study takes on the challenge of applying commercial nanofiltration membranes for various separation/concentration processes with growing commercial interest. NE70 (Toray Chemical CSM) nanofiltration membrane reduced COD and BOD found in landfill leachate treatment water from 5,400ppm to 180ppm and 3,400ppm to 50ppm respectively. Polysaccharides with low conductivity were concentrated to over 80% by using NE40 and 70 membranes. Raw materials used in the cosmetics industry were concentrated using NE70 membranes with significant energy savings as compared to traditional thermal fractionation techniques. In order to recover valuable ions found in the chlor-alkali wastewater, NE70 membranes were used to selectively separate out sulfate (SO42-) ions. NE70 membranes were also used to effectively separate lithium ions from other multivalent species for battery manufacturing processes. And last, nanofiltration membranes were applied to concentrate glyphosates, most often used in pesticides, from 0.1% up to 2.5% via a membrane de-watering process.

241 **Modification of UF Membranes with a Polyampholyte Hydrogel for Fouling Mitigation.**

Roy Bernstein *and Zhe Young*
The Department of Desalination & Water Treatment, Ben-Gurion University, Sde Boker, Israel

N/A **Zwitterionic Modification of Ultrafiltration PVDF Membranes for Improved Fouling Resistance.**

242 **One-Step Synthesis of Hybrid Zeolite Membranes Containing Catalytic Metal Nanoclusters.**

Seok-Jhin Kim[1], *Shuai Tan*[2], *Micaela Taborga Claure*[2], *David Sholl*[2], *Christopher W. Jones*[2], *Sankar Nair*[2], *Yujun Liu*[3], *Jason S. Moore*[3], *Ravindra S. Dixit*[3], *John Pendergast*[3], *and Karren L. More*[4]
(1)School of Chemical Engineering, Oklahoma State University, Stillwater, OK,
(2)School of Chemical & Biomolecular Engineering, Georgia Institute of Technology,

Atlanta, GA, (3)Engineering & Process Sciences, The Dow Chemical Company, Freeport, TX, (4)Metals and Ceramics Division, Oak Ridge National Laboratory, Oak Ridge, TN

243 **Preparation of PVDF UF Membrane Using Lithium Chloride and Amphiphilic Block Copolymer Via Non-Solvent Induced Phase Separation.**

Sang-Hee Park[1], *Jinhan Cho*[2], ***Soon Man Hong***[3], *and Kyung-Youl Baek*[4]
(1)Center for Materials Architecturing, Korea Institute of Science and Technology, Seoul, Korea, Republic of (South), (2)Department of Chemical & Biological Engineering, Korea university, seoul, Korea, Republic of (South), (3)Hybrid Materials Research Center, Korea Institute of Science and Technology, Seoul, Korea, Republic of (South), (4)Center for Materials Architecturing, Korea Institute of Science and Technology, seoul, Korea, Republic of (South)

244 **Preperation and Characterization of Novel Polyetheretherketone Based Polymer Cation Exchange Materials.**

Moonki Jeong, *JeongHo Park, and SangYong Nam*
Department of Materials Engineering and Convergence Technology, Gyeongsang National University, JinJu-si, Korea, Republic of (South)

N/A **Preperation and Morphological Control of Porous Polybenzimidazole Membrane.**

TaeYang Son, *Moonki Jeong, Deuk Ju Kim, and SangYong Nam*
Department of Materials Engineering and Convergence Technology, Gyeongsang National University, JinJu-si, Korea, Republic of (South)

.

245 **Pulsed Low Energy Positron System (PLEPS): A Promising Tool to Investigate Dispersion in Mixed Matrix Membrane.**

Ahmad Choudhary[1,2], *Tönjes Koschine*[1], *Werner Egger*[3], *Peter Van Kessel*[2], *Yujiro*

Itami[2], and Ivo F. J. Vankelecom[1]
(1)Center for Surface Chemistry and Catalysis (COK), K.U. Leuven, Leuven, Belgium, (2)Fujifilm Manufacturing Europe B.V., Tilburg, Netherlands, (3)Institut für Angewandte Physik und Messtechnik, Universität der Bundeswehr München, Neubiberg, Germany

N/A **Fabrication of Hierarchically Structured Helical Micro/Nanofibrous Nonwovens.**

Conditioning Membranes for Improved PRO Performance.

Lingling Xia *and Jeffrey R. McCutcheon*
Department of Chemical and Biomolecular Engineering, University of Connecticut, Storrs, CT

Maximizing energy production from PRO starts with designing membranes that are able to withstand PRO conditions. An appropriately designed PRO membrane exhibit high water permeance to limit resistance to water flux, excellent salt rejection to maintain the osmotic driving force, low structure parameter to minimize internal concentration polarization (ICP), and robust strength to withstand the required hydraulic pressure.

There are multiple trade-offs for making PRO membranes, however. As is typical with many membrane processes, increasing selectivity to reduce ICP will generally result in lower water permeance. Making low structural parameter membranes (thin, highly porous support layers) make them less tolerant to pressure. A low structural parameter may also make the membrane susceptible to severe compaction (which would ultimately enhance ICP). These tradeoffs make approaching membrane performance from a membrane design standpoint challenging.

We have recently identified a method of conditioning an existing membrane prior to PRO testing that resulted in improved overall PRO performance (higher water flux, lower salt flux, and higher power density). By conditioning the membrane under 150 psi pressure (essentially in RO) for 2 hours prior to testing under relevant PRO conditions, water flux increases by 17% and salt flux drops by an astonishing 85%. We believe this low pressure RO treatment may "heal" defects in the membrane through modest fouling caused by particles and organics present in any membrane system. The fouling results in a slightly lower permeance, which mitigates some of the water flux improvements in PRO, the lower salt flux lessens ICP. The results suggests that better utilization of osmotic pressure is possible in PRO systems when the membranes can be conditioned under pressure.

247 **Composite Self-Assembled Block Polymer Hollow Fiber Nanofiltration Membranes with Chemically-Tailored Pore Walls.**

Yizhou Zhang[1], *Jessica L. Sargent*[2], *Bryan W. Boudouris*[2], *and William Phillip*[3]
(1)Department of Chemical and Biomolecular Engineering, University of Notre Dame, Notre Dame, IN, (2)School of Chemical Engineering, Purdue University, West Lafayette, IN, (3)Chemical and Biomolecular Engineering, University of Notre Dame, Notre Dame, IN

Cellulose Membranes and Their Derivatives: Modeling of Casting Methods.

John Moore II[1], *Kevin Roberts*[2], *and Jamie A. Hestekin*[3]
(1)Chemical Engineering, University of Arkansas, Fayetteville, AR, (2)Ralph E. Martin Department of Chemical Engineering, University of Arkansas, Fayetteville, AR, (3)Chemical Engineering, University of Arkansas, Fayettville, AR

This project seeks to characterize the properties of experimental cellulose derivatives and their effect on membrane casting. Properties linked to several categories of derivatives ranging from cellulosic ester to nanocellulose have been investigated. As well as their corresponding casting methods including dry wet spinning and phase inversion by means of evapoporametry, tangential flux filtration, tensile strength, and FTIR. Additionally, data of physical and chemical alterations to the casting technique of cellulosic membranes have been examined and summarized.

248 **Rapid Uranium Isotopic Analysis Using Ultrafiltration Membranes and Alpha Spectroscopy.**

Christine Duval[1], *Timothy DeVol*[2], *and Scott M. Husson*[3]
(1)Department of Chemical and Biomolecular Engineering, Clemson University, Clemson, SC, (2)Department of Environmental Engineering and Earth Science, Clemson University, Anderson, SC, (3)Chemical and Biomolecular Engineering, Clemson University, Clemson, SC

Growth-Front Monolayer Plating of Precious Metals through Insulating Mesoporous Supports.

Alex Peek[1] *and Nicholas Linck*[2]
(1)Materials Science and Engineering, University of Washington, Seattle, WA, (2)University of Kentucky, lexington, KY

Monolayers of precious metals are deposited within the pores of insulating mesoporous membranes via a new electrochemical underpotential deposition growth front mechanism. A gold electrode is evaporation deposited onto one membrane face and a monolayer of copper is underpotentially deposited through the membrane and to the top surface, followed by spontaneous galvanic replacement of copper by platinum or iridium. Detection of precious metal on the opposite membrane face via XPS surface analysis and quantification of areal density by ICP-IES demonstrates a novel copper monolayer *growth front* plating mechanism through the pores. Catalytic water splitting for hydrogen generation is demonstrated for applications including polymer electrolyte membrane fuel cells, hydrogen generation, flow battery energy storage, and electrocatalytic conversions.

N/A **Low Fouling Membrane.**

Surface Energy Enhancement of Poly(butylene terephthalate) By Addition of TiO2 and Polyhedral Oligomeric Silsesquioxane (POSS) Nanoparticles.

Negin Koutahzadeh[1], *Holly A. Stretz*[1], *Mark Wieczorek*[2], *and Pedro Arce*[3]
(1)Department of Chemical Engineering, Tennessee Technological University, Cookeville, TN, (2)Cummins Filtration, Cookeville, TN, (3)chemical engineering, Tennessee Technological University, cookeville, TN

The presence of water droplets in fuels is a major factor affecting the performance of diesel engines, contributing to fuel system corrosion, fouling and performance degradation. Coalescing filters are efficient and effectual for the removal of water drops from fuel. One disadvantage of the filtering approaches is that the hydrophobicity of fibers can cause increased liquid holdup in the filter which in turn reduces the filter's permeability. Due to its excellent chemical and mechanical properties, poly(butylene terephthalate) (PBT) largely used to produce fibers for coalescing filters, for the removal of water drops from fuel. One disadvantage of the filtering approach is the hydrophobicity of PBT fibers which causes liquid holdup in the filter, which in turn reduces the permeability. Surface energy enhancement of PBT by addition of hydrophilic fillers could decrease the hydrophobicity and modify the filters performance.

In this research, poly(butylene terephthalate) (PBT) based nanocomposites containing nano- Titanium dioxide and Polyhedral Oligomeric Silsesquioxane (POSS) nanomaterial were prepared by melt compounding. The influence of these nano materials on surface energy of the resultant nanocomposites was investigated by the use of contact angle measurement (CAM). Contact angle measurements of the resultant PBT nanocomposites showed that the wettability of such composites depends on the particular nanoparticles used. PBT/TiO2 (5 wt%) and PBT/Trisilanolethyl POSS (5 wt%) showed the lowest water contact angle and the highest wettability.

Effect of Chemical Structure on Permeability and Fouling Resistance of Self-Assembling Zwitterionic Copolymer Membranes.

Prity Bengani-Lutz[1]*, Elisha Converse*[2]*, and Ayse Asatekin*[3]
(1)Department of Chemical & Biological Engineering, Tufts University, Medford, MA, (2)Tufts University, Medford, (3)Department of Chemical Engineering, Massachusetts Institute of Technology, Cambridge, MA

Polymer self-assembly is a promising tool for scalable manufacture of membranes while maintaining high permeability and controlled pore size. Tuning copolymer composition (chemical structure, monomer ratios) and processing methods can change copolymer behavior which can dramatically affect the self-assembly and hence the membrane performance (permeability, selectivity, fouling resistance). Membranes with ~1 nanometer effective pore size have numerous applications in the biochemical and pharmaceutical industries, as well as wastewater treatment processes. However, there are very few commercial membranes in this size range, and those that are available tend to show poor selectivity due to broad pore size distributions, and charged surfaces which lead to the rejections being affected by solute size, charge and chemical interactions between the membrane and solute. There is a need for new membrane materials that enable separation of small organic molecules while exhibiting ease of fabrication into large-area membranes. Zwitterionic groups are shown to strongly resist biomacromolecular fouling due to their high degree of hydration, which makes them promising materials for membrane applications. Studies based on zwitterionic surfactants and simulations have reported zwitterionic groups have different degrees of interaction with water. But studies on the influence of the chemical structure of zwitterionic groups on membrane performance, especially fouling resistance are rather limited. Zwitterions are also documented to self-assemble into channel-type clusters 0.6-2 nm in size. We introduce a new class of membranes whose selective layers are made of self-assembling zwitterionic amphiphilic random copolymers. These membranes derive not only fouling resistance but also their permeability and selectivity from this self-assembled nanostructure. We synthesized random copolymers of three hydrophobic monomers (acrylonitrile, methyl methacrylate, trifluoro ethyl methacrylate) and four zwitterionic monomers (sulfobetaine methacrylate, sulfopropylbetaine vinylpyridine, sulfobutylbetaine vinylpyridine and phosphobetaine methacrylate) by free radical polymerization. These copolymers self-assemble into bicontinuous networks of ~1 nm nanochannels, driven by the strong dipoles of the zwitterionic groups, and documented by transmission electron microscopy (TEM) imaging. Membranes prepared by coating these polymers on porous supports exhibit size-based selectivity with a ~1 nm size cut-off demonstrated by filtering negatively charged dyes. Performance of membranes whose selective layers are made of zwitterionic amphiphilic random copolymers depends on the zwitterionic chemical structure. By systematically varying the chemical structure of the zwitterionic monomer (anionic group, spacer length between the charges) and hence the copolymer, we have shown that for similar zwitterionic content in the copolymers, membranes formed from these copolymers have very different permeability, and fouling resistance behavior. This is attributed to the different hydrophobicities, and solubilities of the different zwitterionic groups. These structure-property relationships serve as a guideline to develop new materials for various applications such as textile wastewater treatment, pharmaceutical purification and bioseparation applications.

249 **Thin Film Nano-Composite (TFN) Membrane with Metal Organic Framework (MOF) for Antimicrobial Property.**

Sang-Hee Park[1], Jinhan Cho[2], Soon Man Hong[3], and ***Kyung-Youl Baek[4]***
(1)Center for Materials Architecturing, Korea Institute of Science and Technology, Seoul, Korea, Republic of (South), (2)Department of Chemical & Biological Engineering, Korea university, seoul, Korea, Republic of (South), (3)Hybrid Materials Research Center, Korea Institute of Science and Technology, Seoul, Korea, Republic of (South), (4)Center for Materials Architecturing, Korea Institute of Science and Technology, seoul, Korea, Republic of (South)

Understanding the Roles That Patterning and Chemistry Play on Membrane Fouling.

Anna Malakian[1]*, Steven Weinman[1], Ryan Monk[1], David A. Ladner[2], Linkel Boateng[1], Sapna Sarupria[1], and Scott M. Husson[1]*
(1)Chemical and Biomolecular Engineering, Clemson University, Clemson, SC, (2)Environmental Engineering & Earth Sciences, Clemson University, Anderson, SC

The fouling propensity of a membrane depends greatly on its surface properties such as chemistry and roughness. Literature and experience show that modification of membrane surfaces with appropriate chemical coatings can be effective but not sufficient for controlling membrane fouling. The discovery that sub-micron patterning of a membrane surface can improve its fouling resistance provides an orthogonal membrane design parameter.

In this poster presentation, we will describe the initial results of a collaborative experimental-modeling project aimed at developing fouling-resistant membranes. The research *hypotheses* are that (i) certain combinations of geometric patterns and chemical coatings will significantly reduce membrane fouling and (ii) experimentally-trained multiscale computational models will accelerate the discovery of new surface patterns with appropriate chemistries to improve membrane fouling resistance.

We will introduce the methodology used to pattern the membrane surfaces and report the preliminary results of a systematic study to understand how physical patterning and chemistry impact fouling propensity for colloidal foulants. Performance measurements and microscopy images will be analyzed to compare the fouling behavior of the new membranes to appropriate controls. Adsorption energies will be measured and compared to those from computational simulations.

Antimicrobial Activity of Zinc Oxide Nanoparticles in Nexar® Polymer Films.

Pietro Malky *and Lauren F. Greenlee*
Ralph E. Martin Dept. of Chemical Engineering, University of Arkansas, Fayetteville, AR

Bio-fouling on membranes, as well as in other applications including piping, container and transportation systems, continues to be a challenging problem faced by scientists and engineers. Industries face great costs trying to repair and maintain biofouled equipment, including membranes for water filtration and polymeric films used for coating applications in devices such as heat exchangers, piping systems and chemical reactors. To develop a method of controlling and potentially eliminating the possibility of bio-fouling would allow both economical and engineering relief for a wide range of industries. The process of embedding metal oxide nanoparticles in a polymer film matrix has demonstrated promising antimicrobial activity and may be used as an efficient and cost-effective method of protection against fouling and corrosion. This research focuses on developing a method of creating antimicrobial Nexar® polymer films using ZnO nanoparticles as the main antimicrobial agent. As a comparison material, TiO_2 nanoparticles will also be evaluated.

The focus of this poster presentation will be on the initial materials synthesis and development of nanoparticle-Nexar composite films. Nanoparticles will be tested at a concentration range of 1 – 20 g/L, where the nanoparticles are added into Nexar-solvent solutions before the films are cast. The Nexar polymer will be cast from two different solvents to evaluate the effect of polymer-solvent interactions on nanoparticle-polymer film formation. To test the cast films for antimicrobial activity, we will measure and report the growth of E. Coli for a 24-hr period. We will present on our results thus far regarding ZnO-Nexar and TiO_2-Nexar films. Results will include casting setup and procedure, initial film characterization, and initial bacterial testing with the cast films.

Modelling of Feed Flow in Ultrafiltration Channels Using Computational Fluids Dynamics – Effect of Spacer Geometry Design.

Amira Abdelrasoul*, Huu Doan, and Ali Lohi*
Chemical Engineering, Ryerson University, Toronto, ON, Canada

In this study a laminar crossflow filtration in a rectangular channel was simulated using the computational fluid dynamics (CFD) software ANSYSFLUENT. CFD was used to solve Reynolds-averaged Navier-Stokes equations and to assess the impact of change in the surface shear, boundaries, and shear strain, thus effectively studying their influence on the membrane fouling phenomenon. A three-dimensional numerical solution of the Navier–Stokes equations were developed using the Finite volume method. Varied ultrafiltration channel spacer sizes of 0.1, 0.2, 0.4 micrometers were simulated. SAMPLE of solution control was used as a pressure-velocity coupling and QUICK for momentum discretization. Scaled residual monitors were decreased up to 10^{-5} for convergence satisfaction conditions. The larger spacers of 0.4 microns led to increased pressure drop across the membrane, as well as, increasing the hydrodynamics shear stress. However, the larger spacers of 0.4 microns led to the formation of localized stagnation or dead zones, where the membrane fouling phenomena may be increased. The influence of the feed spacer size on flow patterns inside the unltrafiltration unit will be discussed for each geometry at different feed flow rates in the range of 1 LPM to 7 LPM. The study demonstrated that the thickness of the ultrafiltration channel of 0.2 microns had the optimum results for the velocity and pressure profiles at different flow rates.

250 **High Fouling Resistance Thin-Film Composite Membranes with Copolyazole Substrates for Osmotically Driven Processes.**

Phuoc H.H. Duong[1], *Stefan Chisca*[2], *Peiying Hong*[3], *Hong Cheng*[4], *Suzana Nunes*[5], *and Chung Tai-Shung*[6]
(1)King Abdullah University of Science and Technology (KAUST), Thuwal, Saudi Arabia, (2)Biological and Environmental Sciences & Engineering Division (BESE), King Abdullah University of Science and Technology (KAUST), Thuwal, Saudi Arabia, (3)Water Desalination and Reuse Center, Biological and Environmental Science and Engineering Division, King Abdullah University of Science and Technology (KAUST), Thuwal, Saudi Arabia, (4)Water Desalination and Reuse Center, Biological and Environmental Science and Engineering Division, King Abdullah University of Science and Technology (KAUST), Thuwal, Saudi Arabia, (5)Biological and Environmental Science and Engineering Division, King Abdullah University of Science and Technology (KAUST), Thuwal, Saudi Arabia, (6)Chemical and Biomolecular Engineering, National University of Singapore, Singapore, Singapore

MEXSY MemBrain TechnOntology: Portal and Analytic.

Alex Seregin[1] *and* ***Jeff Zhuk***[2]
(1)MetaFormula Inc, Palo Alto, CA, (2)ITS Inc, Englewood, CO

MetaExpert Systems (MEXSY) – is a new class of expert systems for universal applications, including corporate information systems [1]. Starting since original meta-theoretical approach to formal synthesis of theories in physics and chemistry (1995-1997) [2-4] and having brilliant experience of the Invention Machine project (Boston, 1998-2001) [5], MEXSY technology opens a new opportunities for knowledge based applications in modern information space (master data, big data, social networks, semantic web, etc). Many large companies (including Fortune Global 500 list participants) use the MEXSY CorpBrain technology benefits in their real practice of data management (solutions: MEXSY MDM, MEXSY Expert, etc) [5-7]. Another (parallel) project of the MEXSY solutions development is the MEXSY MemBrain [8], which now also ready to go from research mode (2002-2011) [8] to real applications.

Experience of Invention Machine, Ontologic Classifier, and MEXSY projects allows to develop a new approach to accelerate development of technologies – MEXSY TechnOntology, or the MEXSY MemBrain TechnOntology in the case of membrane science and technology, which (the approach) may be organized as a memrane technology knowledge portal. This is a way to accumalate and use all knowledge about semipermeable membranes and membrane technologies. First version of the knowledge portal needs a knowledge base with simple navigation on the domaign information (books, journals, conferences, web-sites, etc). Next version of the portal should empowered by "domaign-oriented search engine" MEXSY Analytics, a part of MEXSY CorpBrain 2.0 (for enterprises) [1] or MEXSY WebBrain solution (for the industry, the MemBrain TechnOntology option). It's necessary to discuss some special questions about the membrane technology and portal with the professional community (for example North American Membrane Society) for better understanding (and formal presentation in the knowledge base) of real current problems of the industry. The milestone of the MEXSY TechnOntology project is the MEXSY MemBrain TechnOntology portal, as real accelerator for the membrane science and technology development. Then the experience of acceleration may be used in another industries (environment engineering, materials engineering, alternative energy, etc)

REFERENCES:

[1] Alex Seregin, Jeff Zhuk. MetaExpert Systems for Enterprises / Smart Data Conference, August 18-20, San Jose, CA, 2015;
http://smartdata2015.dataversity.net/sessionPop.cfm?confid=91&proposalid=7707

[2] Alex Seregin. Development of the Unified System of Sorption Isotherm Models for Gases in Polymers. PhD Thesis, Russian Academy of Sciences, Moscow, 1996, in Russian

http://www.dissercat.com/content/razrabotka-edinoi-sistemy-modelei-izoterm-sorbtsii-gazov-v-polimerakh, in Russian

[3] Alex Seregin, Valery Bondar, Ben Mattes, Yuri Yampolsky, Vladimir Volkov. **Dual Mechanism Of Equilibrium Sorption In Polymer Systems - A Generalized Theory /** Polymer Science, 1996, #3:

http://serials.unibo.it/cgi-ser/start/en/spogli/df-s.tcl?prog_art=1661642&language=ENGLISH&view=articoli

[4] Alex Seregin et al. NAMS Meetings 1995, 1996, 1997

[5] TechOptimizer & Cobrain (with "pre-MEXSY" knowledge bases inside), products of Invention Machine Corporation:

http://www10.mcadcafe.com/nbc/articles/view_article.php?articleid=51406

Since 2010 Invention Machine is a part of IHS, and TechOptimizer & Cobrain are inside of IHS Goldfire solution, product of IHS:

https://www.ihs.com/products/design-standards-software-goldfire.html

[6] Manual for "MEXSY MDM" users in Gazprom (see sections 4.4 and 4.5), in Russian: http://komplektatsiya.gazprom.ru/d/textpage/72/114/manual_v2.09.pdf

[7] "MEXSY Expert" in TVEL Corporation (a division of Rosatom) / IBM magazine "Innovations in Technology & Business", 2009, isure 1, page 20 (in Russian) http://download.boulder.ibm.com/ibmdl/pub/software/dw/ru/magazines/2009/ibm1_09.pdf

[8] NAMS Meetings 2002, 2003, 2005, 2006, 2007, 2009, 2011

MEXSY MemBrain in Membrane Characterization.

***Alex Seregin*[1] *and* *Jeff Zhuk*[2]**
(1)MetaFormula Inc, Palo Alto, CA, (2)ITS Inc, Englewood, CO

MetaExpert Systems (MEXSY) – is a new class of expert systems for universal applications [1-2]. MEXSY MemBrain project studing MEXSY solutions in the field of membrane science and technology [3] is moving to real applications. So it's necessary to discuss the actual MEXSY MemBrain knowledge base content with professional community (North American Membrane Society) for better understanding (and formal presentation in the knowledge base) of real current problems of the industry.

This presentation considers MEXSY MemBrain knowledge base in whole (membrane science and technology) and in part of membrane characterization [4] to develop related part of MEXSY MemBrain TechnOntology knowledge portal [1]. For another examples see [5-7].

REFERENCES:

[1] Alex Seregin, Jeff Zhuk. MEXSY MemBrain TechnOntology / NAMS 2016 (poster session)

[2] Alex Seregin, Jeff Zhuk. MetaExpert Systems for Enterprises / Smart Data Conference, August 18-20, San Jose, CA, 2015; http://smartdata2015.dataversity.net/sessionPop.cfm?confid=91&proposalid=7707

[3] Alex Serein et al. NAMS Meetings 1995, 1996, 1997, 2002, 2003, 2005, 2006, 2007, 2009, 2011

[4] John Pellegrino, Ruth Baltus. Workshop 1: Measurement Methods for Membranes. NAMS 20016

http://www.membranes.org/2016/program.htm

[5] Alex Seregin and Jeff Zhuk. MEXSY MemBrain at Stanford, Exaple: Membrane Gas Separation. NAMS 2016 (session "Gas separation")

[6] Alex Seregin and Jeff Zhuk. MEXSY MemBrain in Water Treatment. NAMS 2016 (session "Water treatment, reclamation & reuse")

[7] Alex Seregin. Knowledge Based Data Integration and Knowledge Management in Fuel Cell Technology. NAMS 2007, Orlando, FL

http://toc.proceedings.com/17876webtoc.pdf (p. 112)

N/A **Comparative Analysis of Fouling Mechanisms in CTA and TFC Membranes.**

Farrukh Arsalan Siddiqui *and Robert Field*
Department of Engineering Science, University of Oxford, Oxford, United Kingdom

WATER TREATMENT, RECLAMATION & REUSE

251 **Optimized Ultrafiltration Backwash with Real-Time Control of Coagulant Dosing.**

Larry Gao[1], *Anditya Rahardianto*[2], *Han Gu*[3], *Panagiotis D. Christofides*[4], *and Yoram Cohen*[2]
(1)Chemical & Biomolecular Engineering, UCLA, Los Angeles, CA, (2)Chemical and Biomolecular Engineering, University of California, Los Angeles, Los Angeles, CA, (3)Chemical & Biomolecular Engineering, University of California, Los Angeles, Los Angeles, CA, (4)Department of Chemical and Biomolecular Engineering and Department of Electrical Engineering, University of California, Los Angeles, Los Angeles, CA

253 **Removal of Organic Micro-Pollutants (phenol, aniline and nitrobenzene) Via Forward Osmosis (FO) Process: Evaluation of FO As an Alternative Method to Reverse Osmosis (RO).**

Yue Cui[1], *Xiang-Yang Liu*[2], *Tai-Shung Chung*[3], *Martin Weber*[4], *Claudia Staudt*[5], *and Christian Maletzko*[6]
(1)Department of Chemical and Biomolecular Engineering, National University of Singapore, Singapore, Singapore, (2)Department of Chemistry and Department of Physics, National University of Singapore, Singapore, Singapore, (3)Chemical and Biomolecular Engineering, National University of Singapore, Singapore, Singapore, (4)Advanced Materials & Systems Research, BASF, Ludwigshafen, Germany, (5)Advanced Materials & Systems Research GM-B001, BASF SE, Ludwigshafen, Germany, (6)BASF SE, Ludwigshafen, Germany

Onsite Treatment of Raw Produced Water through the Use of Forward and Reverse Osmosis.

Rudy Maltos[1], *Shalom Fox*[1], *Julia Regnery*[1], *Bryan D. Coday*[2], *and Tzahi Cath*[1]
(1)Civil and Environmental Engineering, Colorado School of Mines, Golden, CO,
(2)Carollo Engineers, Broomfield, CO

The U.S. Department of Energy (DOE) projects the production of oil and gas (O&G) to increase 48 and 56 percent by the year 2040, respectively, compared to 2012 production levels. Once a well has been developed, every barrel of oil produced will also generate roughly 9.2 barrels of produced water. Produced water is known to be saline, high in chemical oxygen demand (COD), and contain large concentrations of minerals, factors that pose a danger to human and environmental health if not disposed of properly. For these reasons the handling of produced water is a costly logistical challenge to transport, treat, or dispose of. The most common disposal practice is deep well injection of produced water; however, tighter regulations in the eastern U.S restricting deep well injection and unpredictable water resources create a need for a new onsite treatment process.

Forward osmosis (FO) coupled with reverse osmosis (RO) could provide high quality permeate capable for industrial and agricultural reuse, alleviating water tensions in arid regions and disposal logistic of raw produced water. Compared to current onsite treatment trains, the FO/RO system has exceptional product water quality due to a dual semipermeable membrane system, which has a 97 percent removal of a broad range of contaminants. This study focused on the long term performance of a pilot scale FO/RO system treating raw produced water from the O&G industry. The system consists of a 200 gallon FO Feed tank, 4.93 of FO membrane and 8.4 of RO membrane. The FO/RO system was operated at water recoveries of 0, 25, and 50 percent. The system performance was evaluated through the following parameters: specific flux, specific reverse salt flux, trace organic removal, and metal removal. The FO/RO system was operated continuously for periods of two weeks, during which the FO water flux decreased from 3.1 LMH to 1.8 LMH and produced 4,200 L of high quality permeate.

There are three potential factors that drive this decrease in water flux over time 1) The reverse salt flux increases the salinity of the feed water, therefore decreasing the osmotic driving force across the membrane 2) over time a fouling layer is developed on the FO membrane by the raw produced water, reducing water flux across the membrane, and 3) a decreasing water permeability coefficient due to chemical changes of the polymeric FO membrane. This value corresponds to the membranes physical composition, which should remain constant after the system stabilizes. Because this value never remains constant, it could be hypothesized that the produced water is interacting with the membrane surface, possibly altering the surface chemistry that facilitates water diffusion through the membrane. From these findings a standard operating procedures was developed, which optimizes backwashing intervals, minimizes decrease in water flux, and maintains a constant water permeability coefficient. These improvements will increase the viability of the FO/RO technology, possibly resulting in a commercial implementation.

Direct Contact Membrane Distillation (DCMD) for Industrial Wastewater Reuse: Managing the Accumulation of Organic Compounds in DCMD Product Water.

Judith Winglee[1], *Nathan Bossa*[1], *Zachary Hendren*[2], *Lora Toy*[3], *and Mark R. Wiesner*[4]
(1)Duke University, Durham, NC, (2)Sustainability and Resource Management, RTI International, Research Triangle Park, NC, (3)Energy Technology Division, RTI International, RTP, NC, (4)Department of Civil and Environmental Engineering, Duke University, Durham, NC

Treating complex, highly saline industrial wastewaters, such as produced water from oil and gas extraction, is very costly, but direct contact membrane distillation (DCMD) is a novel process that can decrease water treatment costs. In DCMD, a heated feed stream is distilled at low–temperature (40-80°C) inside a membrane module and the distillate is collected in the permeate collection stream. While many researchers have shown that DCMD systems have high salt rejection, few studies have analyzed the rejection of organic compounds (OCs), which are present in many industrial wastewaters. The research outlined in this presentation indicates that due to the temperature gradient between the feed and permeate collection streams, volatile and semi-volatile compounds can accumulate in the permeate collection stream at higher concentrations than in the feed. As a result, the OC concentration in DCMD product water should be carefully monitored especially for applications with human exposure. To evaluate the accumulation of OCs in the DCMD product water, flux tests of synthetic wastewater (brine solutions containing industrially relevant OCs of varying volatility) were tested in a batch DCMD system. The water flux and concentration of OCs in the feed and permeate collection reservoirs were monitored over 5 hours. A Fickian mass transfer model was developed to fit the data and describe the change in reservoir concentration over time. This model can be used to predict whether the DCMD product water will meet the OC discharge requirements and determine if post-treatment is necessary.

Ss.

Volatile and Semi-Volatile Contaminant Transport in Direct Contact Membrane Distillation.

Kevin A. Salls[1], *Ryan D. Gustafson*[2], *Danbi Won*[3], *Edward P. Kolodziej*[4], *Amy E. Childress*[5], *and Sage R. Hiibel*[6]
(1)Civil and Environmental Engineering, University of Nevada, Reno, Reno, NV, (2)University of Southern California, Los Angeles, CA, (3)Civil & Environmental Engineering, University of Washington, Tacoma, WA, (4)Civil and Environmental Engineering, University of Washington, Tacoma, WA, (5)Sonny Astani Department of

Civil and Environmental Engineering, University of Southern California, Los Angeles, CA, (6)Civil and Environmental Engineering, University of Nevada Reno, Reno, NV

Membrane distillation (MD) is a thermally driven separation process based on mass transport of water across a microporous, hydrophobic membrane. Transport is caused by a vapor pressure differential across the membrane that is generated by a thermal gradient. Recent advances in new membrane technologies and waste heat capture have made MD systems the subject of renewed interest as an alternative membrane technology for water treatment. Theory indicates that MD has high rejection rates for non-volatile compounds, such as ions and metals, and high permeability of volatile and semi-volatile compounds. The general assumption has been that this broad classification of volatile versus non-volatile is sufficient for determining the applicability of MD to a given source water, but little research has been done to examine the mechanics and physical phenomena of volatile contaminant transport in MD systems. Recent experimental data confirms that the earlier assumption was too broad, and that rejection of volatile and semi-volatile compounds vary significantly.

A novel, gas-tight direct contact membrane distillation system (DCMD) was designed and constructed to evaluate factors influencing transport of volatile and semi-volatile compounds in MD. The system retains volatile components, allowing for greater contaminant recovery and closure of the mass balance. Various contaminant classes, including metals, ionic salts, nitrogenous disinfection byproducts, volatile organic compounds (including phthalates, phenols, polyaromatic hydrocarbons, anilines, chlorinated aromatics, and amines), and pharmaceutical and personal care products were tested for transport in the system. These contaminant mixtures include diverse compounds with a very wide range of chemical properties, and spanned 10 orders of magnitude in air-water equilibrium partitioning (K_H). Rejection of the non-volatiles was generally above 99%, as expected, while the rejection of volatile compounds varied significantly. Membrane passage was compared to the physicochemical characteristics of each compound. Volatility, as assessed by K_H, weakly correlated with rejection, although there exist a number of compounds whose observed transport was inconsistent with their Henry's law constant. It was also found that many hydrophobic compounds (e.g., PAHs, large chlorinated aromatics, and phthalates) exhibited extensive sorption and hydrophobic partitioning to system components such as tubing, containers, and the membrane itself, suggesting that DCMD is not an appropriate treatment for these types of compounds. The results of this study are a major step towards our mechanistic understanding of the transport of compounds in DCMD based on their physicochemical properties.

Membrane Distillation for Treating Hydraulic Fracturing Flow Back Waters.

Kamyar Sardari[1] *and* ***S. Ranil Wickramasinghe***[2]
(1)Ralph E. Martin department of chemical engineering, University of arkansas, fayetteville, AR, (2)Department of Chemical Engineering, University of Arkansas, Fayetteville, AR

Hydraulic fracturing flow back waters often exhibit very high TDS (in excess of 100,000 ppm). Treatment of these highly impaired wastewaters is a major challenge due to the presence of high TDS and organic contaminants. Pressure driven membrane processes such as reverse osmosis are impractical for treating very high salinity wastewaters due to the high osmotic back pressure that must be overcome. One of the technologies that is being investigated for treating these produced waters is membrane distillation.

Here we have screened a number of commercially available microporous hydrophobic membranes. We have characterized membrane surface as well as bulk properties. Using bulk membrane properties, we calculate a structural parameter that indicates membranes that should display the highest permeate flux. Permeate fluxes were determined for all membranes using a model feed stream containing a 20,000 ppm (0.34 M) NaCl. We have compared the observed permeate fluxes with values obtained for the structural parameter. Next membranes that displayed the highest permeate fluxes were challenged with feed streams containing 100,000 ppm (1.7 M) NaCl. The feeds stream was concentrated until the permeate flux rose quickly while the conductivity of the permeate increased rapidly above 50 mS cm^{-1} indicating the passage of the feed through the membrane pores. Finally, these membranes were tested with real produced waters. Our results indicate that a staged process using membranes with different bulk and surface properties is necessary to concentrate the feed to over 350,000 ppm. Further our results indicate the importance of membrane surface properties in order to prevent breakthrough of the feed.

POSTER SUBMISSIONS

POSTER PRESENTATION WITH STUDENT & POSTDOCTORAL AWARDS

N/A **Overcoming Challenges in Fabrication of Thin Metallic Membranes for Nitrogen Separation.**

Kyoungjin Lee
Department of Energy Resources Engineering, Stanford University, Stanford, CA

Biomimetic and Bioinspired Membranes: Opportunities, Challenges and New Directions

Manish Kumar*, Yue-xiao Shen, Tingwei Ren, Patrick Saboe
Department of Chemical Engineering, Pennsylvania State University, University Park, PA, 16802, USA
*Corresponding author: 155 Fenske Laboratory, Pennsylvania State University, University Park, PA, 16802, USA, (814) 865-7519, manish.kumar@psu.edu

Biomimetic and bioinspired membranes incorporate elements or paradigms that are inspired by biological membranes or incorporate biological elements themselves (Figure). Interest in these membranes in the membrane community has grown rapidly in recent years, from being a science-based enquiry to larger commercialization efforts. This talk will present an overview of biomimetic and bioinspired approaches to development of membranes for separations. It will focus on a few examples including several well-known classes of biomimetic and bioinspired membranes: bioinspired antifouling membrane materials, bioinspired antifouling strategies, membrane protein enhanced membranes, synthetic channel based membranes, and selective transport membranes.

This talk will also present a summary of the fundamental research needs for biomimetic and bioinspired membranes and the benefits that may accrue from such investigations. It will provide a contrasting discussion on the challenges to scale-up and use of biomimetic and bioinspired membranes in more industrial contexts.

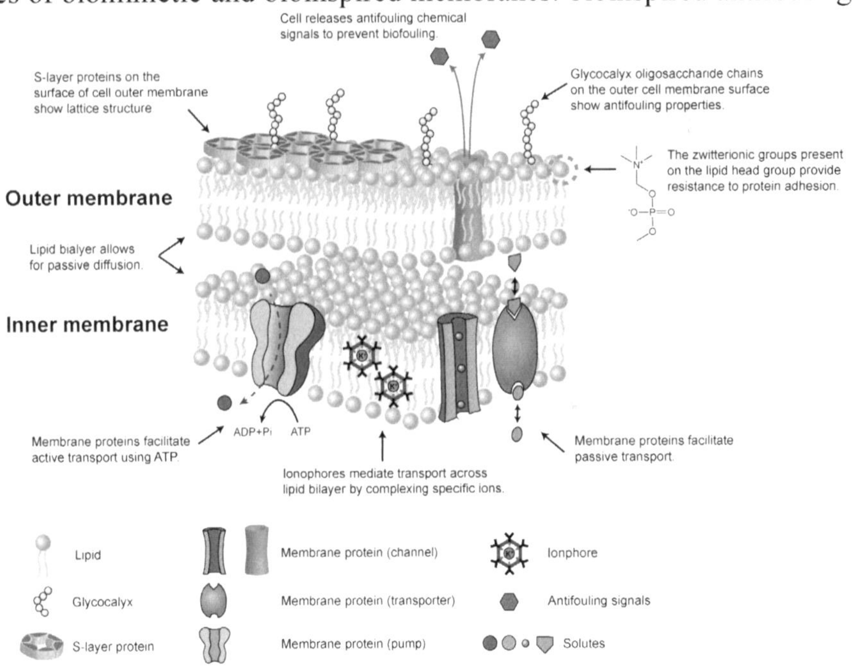

FUNCTIONAL DNA AS VERSATILE BUILDING BLOCKS FOR BIOMIMICKING MEMBRANES

Thomas Schäfer, POLYMAT University of the Basque Country, Donostia-San Sebastián, Spain and Ikerbasque Basque Foundation for Science, Bilbao, Spain
thomas.schafer@ehu.es

Iván Revilla, Abel de Cózar, Fernando Cossío, University of the Basque Country, Donostia-San Sebastián, Spain

Key Words: DNA-aptamers, gating membranes, controlled release, stimuli-responsive membranes

DNA-aptamers are oligonucleic acids that can be selected to specifically interact with in principle any kind of target molecule. An important asset of aptamer conjugates is the fact that upon specific binding, their spatial conformation may change drastically, depending on a fine equilibrium between mainly electrostatic and hydrophobic interactions. Recently, if has been shown that this specific conformational change can be exploited for creating biomimicking gating membranes, where aptamers serve as a "nanovalve" which selectively triggers the opening or closing of a nanopore depending on the presence of a target molecule. Such systems add an important degree of freedom in the design of stimuli-responsive systems which conventionally respond to bulk stimuli such as pH, temperature, light or electrical and magnetic fields.

In this presentation a short overview will first be given of how DNA can be used as a gating element in nanodevices and membranes with particular emphasis on how to quantify target-DNA interactions and respective conformational changes by advanced surface sensitive characterization techniques. It will be shown that DNA-aptamers can indeed be employed for creating gating or controlled delivery devices responding to very different kinds of targets. Insights will be given into how these molecules possibly fulfil their gating function and what potential limitations need to be expected despite of the versatility of this DNA-based platform.

Subsequently, an outlook will be presented on how far this versatility can be pushed while maintaining the function of such DNA-building blocks in membranes. For this purpose, we investigated on one hand their performance in non-conventional aqueous media such as aqueous ionic liquid solutions, and on the other the possible functions DNA might fulfil other than recognizing target molecules. With regard to the former, we could prove that DNA-aptamers possess a remarkable capacity to remain functional even in strongly non-ideal aqueous media containing ionic liquids, depending on the particular physico-chemical properties of the ionic liquid environment and its specific interactions with the DNA which we quantified. As regards the latter, DNAzymes were being studied as functional elements to be integrated into nanoporous membrane structures. The examples given will illustrate that DNA is indeed a highly versatile building block for creating biomimicking membranes of different responsive functions.

Salt-excluding artificial water channels in lipidic systems

Istvan Kocsis,[1] Manish Kumar,[2] Mark Baaden,[3] Mihail Barboiu[1]

[1]Adaptive Supramolecular Nanosystems Group, Institut Europeen des Membranes, Montpellier, France

[2]Department of Chemical Engineering, Pennsylvania State Univerity, University Park, USA

[3]Laboratoire de Biochimie Théorique, Institut de Biologie Physico-Chimique, Paris, France

Key words: synthetic channels, water transport, salt rejection.

Playing an important role in both mechanistic understanding as well as advances in development of novel materials, biomimetic approaches for synthetic water channels have emerged in an effort to produce highly efficient materials for water purification(1). One phenomenon to be considered when designing these materials is the change in properties of water when shifting from bulk to the confined environment. Under the pore confinement of biological channels such as Aquaporins (AQPs) and Gramicidine A, water adopts a highly order structure which determines the water/ionic selectivity across lipidic bilayers.(2)

Here we report a set of synthetic supramolecular water channels self-assembled from alkylureido-imidazole building blocks. The formed channels show a pore size of 2.6 Å containing oriented water wires resembling the structure of water from AQPs pore (Fig. **1**). Transport studies across lipidic bilayers show a water permeability of only two orders of magnitude below that of natural water channels while retaining the selectivity towards water in saline solutions. Based on X-ray structures, molecular modeling of the supramolecular assemblies embedded in a lipid bilayer show stabilized water channels with varying degrees of structuring and water occupancy.

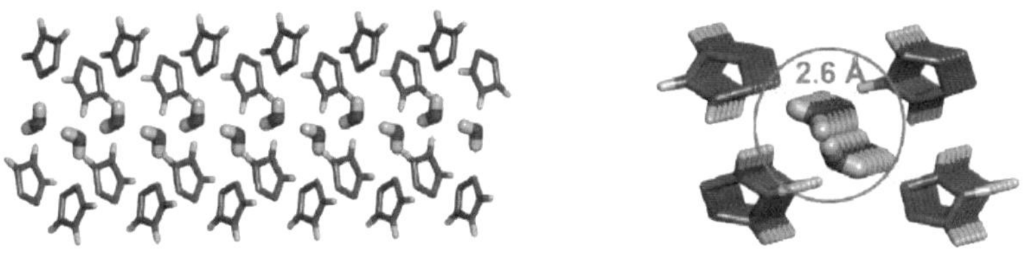

Figure 1. *Side and top view of imidazole quartet water channels*

References

(1) M. Barboiu, *Angew. Chem. Int. Ed.* **2012**, *51*, 11674; Y.-X. Shen *et al.*, *Proc. Natl. Acad. Sci. USA* **2015**, *112*, 9810.

(2) D. Zhong, S. K. Pal, A. H. Zewail, *Chem Phys. Lett.* **2011**, *503*, 1; T. W. Allen, O. S. Andersen, B. Roux, *Proc. Natl. Acad. Sci. USA* **2004**, *101*, 117; P. Agre, *Angew. Chem.-Int. Ed.* **2004**, *43*, 4278.

Modeling Low-Pressure Nanofiltration Membranes for Softening and Pretreatment

Omar Labban[a], Chang Liu[b;c], Tzyy Haur Chong[b;c], John H. Lienhard V[a];

[a]Rohsenow Kendall Heat Transfer Laboratory, Department of Mechanical Engineering, Massachusetts Institute of Technology, Cambridge MA 02139-4307, USA

[b]School of Civil and Environmental Engineering, Nanyang Technological University, Singapore 639798, Singapore

[c]Singapore Membrane Technology Centre, Nanyang Environment and Water Research Institute, Nanyang Technological University, Singapore 637141, Singapore

Abstract

Nanofiltration (NF) for softening and brine/seawater pretreatment has had surging interest lately in view of its ability to selectively reject scale-causing multivalent ions in feed waters. Two major limitations on the viability of NF for such applications has been the increased energy consumption associated with the pretreatment system, and a lack of "specialized" membranes for these applications. In this work, a recently developed cross-linked layer-by-layer (LbL) NF membrane for low-pressure applications will be modeled for the first time using the Donnan Steric Pore Model with Dielectric Exclusion (DSPME-DE). An overview of the modeling procedure along with validation experiments run on hard water solutions, including artificial seawater, will be presented. The results obtained by running parametric studies on different operating conditions, such as varying flow rates and applied pressure, will be addressed. The insights and implications of our modeling, and the reasons underlying the membrane's superior performance relative to other NF membranes will also be discussed.

A Module-scale Computational Fluid Dynamics Model to Evaluate Hybrid Osmotically-Driven Desalination Systems

Lori Jones and Dr. Andrea Achilli
Environmental Resources Engineering, Humboldt State University, Arcata, CA

Reverse osmosis (RO) is the leading technology employed in seawater desalination due to its proven effectiveness and relatively low energy consumption compared to thermal desalination [1]. However, it still consumes more energy than alternative technologies used to secure fresh water and brine discharge is also of environmental concern [2]. Osmotically driven membrane processes, combined with RO, can mitigate these issues. One system could use forward osmosis (FO) to dilute seawater with water from an impaired water source, reducing the osmotic pressure (and thus the energy requirement) of the RO feed water [3]. Another system could use pressure-retarded osmosis (PRO), which uses the brine from RO and the impaired water source to recover energy and pre-pressurize the RO feed water, reducing its energy requirement [4]. PRO can also be implemented before the RO step, where the seawater is initially diluted, and the recovered energy from this initial dilution step can be used to pre-pressurize the RO feed water [5]. In recent years, the three hybrid systems have been the topic of many different investigations [3-8]; however, only one direct comparison of the three systems has been previously conducted [9]. This comparison used a simplified, analytic model to evaluate the systems, and thus a comparison with a rigorous, validated model has not been completed. A preliminary investigation of these three systems performed with a numerical one-dimensional model showed potential energy reductions of 20% for FO-RO, 37% for RO-PRO, and 43% for PRO-RO.

In this investigation, a three-dimensional computational fluid dynamics (CFD) model of the PRO and FO processes, accounting for the geometry of a spiral wound module to accurately evaluate the system performance, was developed in the open source CFD package OpenFOAM. The novelty of the proposed CFD model, compared to existing FO and PRO models, lies in the fact that the baffled membrane leaves are modeled after the geometry of typical spiral wound membrane modules, which, to the author's knowledge, has not been reported in the literature. Particular effort was placed on accurately determining the pressure drops throughout the module due to the turbulence promoting baffles and other operating conditions. Models of the FO and PRO processes to date have been as flat-plate modules in either co-current, counter-current, or cross-current flow patterns; this approach does not capture the effect that the internal baffles have on flow velocities and pressures. By accurately determining these effects, the trade-off between overall pressure drops in the module and total water permeated across the membrane for varying operational conditions and geometrical parameters can be explored.

The proposed CFD model was integrated with the well known Dow Water and Process Solutions software Reverse Osmosis System Analysis (ROSA) to evaluate the three proposed hybrid treatment schemes with respect to the dilution of the discharged brine and specific energy expenditure. The advantages and drawbacks of each system, and scenarios in which one system may be favorable over the others, are discussed.

Additionally, using results from the CFD model, design considerations and recommendations for the operating conditions and geometry of the FO and PRO membrane modules are given.

References

[1] H. Cooley, M. Heberger, Key Issues in Seawater Desalination in California: Energy and Greenhouse Gas Emissions, Pacific Institute, (2013).
[2] H. Cooley, N. Ajami, M. Heberger, Key Issues in Seawater Desalination in California: Marine Impacts, Pacific Institute, (2013).
[3] G. Blandin, A.R.D. Verliefde, C.Y. Tang, P. Le-Clech, Opportunities to reach economic sustainability in forward osmosis–reverse osmosis hybrids for seawater desalination, Desalination, 363 (2015) 26-36.
[4] J. Prante, J. Ruskowitz, A. Childress, A. Achilli, RO-PRO desalination: an integrated low-energy approach to seawater desalination, Applied Energy, 120 (2014) 104-114.
[5] J. Kim, M. Park, S.A. Snyder, J.H. Kim, Reverse osmosis (RO) and pressure retarded osmosis (PRO) hybrid processes: Model-based scenario study, Desalination, 322 (2013) 121-130.
[6] A. Achilli, J. Prante, N. Hancock, E. Maxwell, A. Childress, Experimental Results from RO-PRO: A Next Generation System for Low-Energy Desalination, Environmental Science & Technology, 48 (2014) 6437-6443.
[7] D.L. Shaffer, N.Y. Yip, J. Gilron, M. Elimelech, Seawater desalination for agriculture by integrated forward and reverse osmosis: Improved product water quality for potentially less energy, Journal of Membrane Science, 415–416 (2012) 1-8.
[8] T.Y. Cath, N.T. Hancock, C.D. Lundin, C. Hoppe-Jones, J.E. Drewes, A multi-barrier osmotic dilution process for simultaneous desalination and purification of impaired water, Journal of Membrane Science, 362 (2010) 417-426.
[9] A. Altaee, A. Sharif, G. Zaragoza, A.F. Ismail, Evaluation of FO-RO and PRO-RO designs for power generation and seawater desalination using impaired water feeds, Desalination, 368 (2015) 27-35.

Hybrid Pressure Retarded Osmosis–Membrane Distillation (PRO–MD) Process for Osmotic Power and Clean Water Generation

Gang Han, Jian Zuo, Chunfeng Wan and Tai-Shung Chung*

Department of Chemical and Biomolecular Engineering, National University of Singapore, Singapore 117585, Singapore

Abstract

By integrating the PRO and MD processes, this study has proposed an advanced PRO–MD hybrid system for sustainable production of renewable osmotic power and clean water from various water effluents.

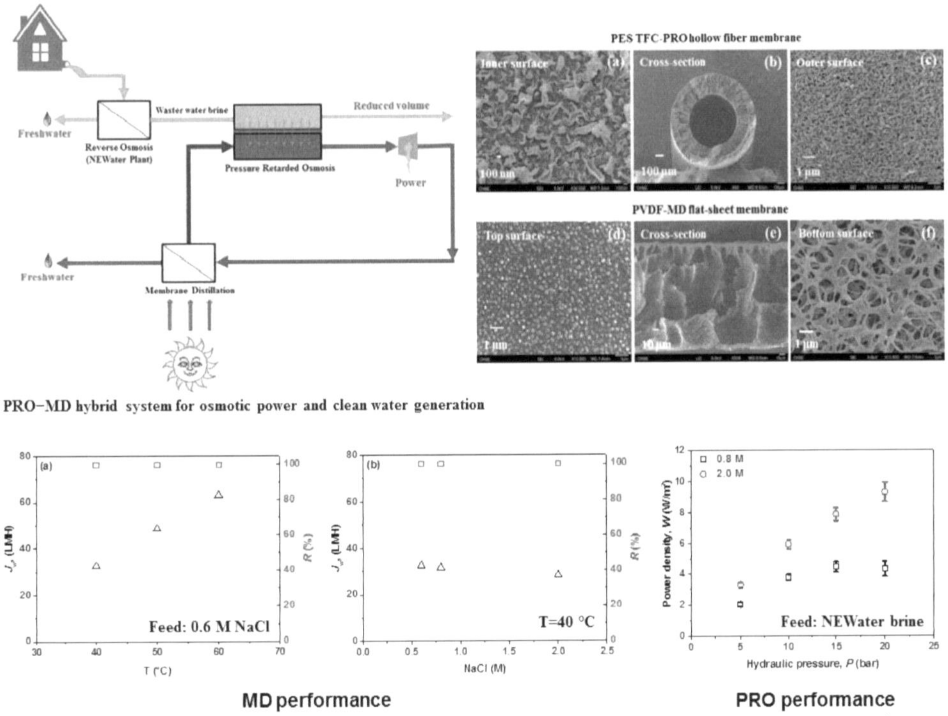

Fig. 1 PRO–MD hybrid system for osmotic power and clean water generation.

As illustrated in Fig. 1, the integrated system consists of one open-end loop using freshwater or wastewater as the feed for the PRO subsystem, and one closed loop including a PRO subsystem to produce osmotic power and a MD subsystem to re-concentrate the diluted draw solution and produce potable water using pretreated seawater or SWRO retentate. The PRO–MD hybrid system exhibits unique advantages of high water recovery rate, huge osmotic power generation, well controlled membrane fouling, and minimal environmental impacts. By employing state-of-the-art PRO and MD membranes, ultrahigh power densities of 31.0 W/m^2 and 9.3 W/m^2 have been demonstrated by the PRO subsystem using deionized water and real wastewater brine as the feeds, respectively, when employing a 2 M NaCl MD concentrate as the draw solution. Simultaneously, high purity potable water with a flux of 32.5–63.1 L/(m^2.h) can be produced by the MD subsystem at 40–60 °C without any detrimental effects of fouling. The energy consumption in the MD subsystem might be further reduced by applying a heat exchanger in the hybrid system and using low-grade heat or solar energy to heat up the feed solution. The newly developed PRO–MD hybrid process would provide insightful guidelines for the exploration of alternative green technologies for renewable osmotic energy and clean water production.

Novel Thermal Efficiency-Based Model for Geothermal Membrane Distillation

Johan Vanneste[1], John Bush[1], Christopher Marks[1], Tzahi Y. Cath[1, *]

[1]*Department of Civil and Environmental Engineering, Colorado School of Mines,*

1500 Illinois St., Golden, CO, USA

* tcath@mines.edu

A large amount of geothermal exploration wells exist in the US that don`t produce the quantity and quality of heat required for power generation with an Organic Rankine Cycle (ORC). At temperatures under 150 °C, power generation is currently not cost effective. Membrane distillation can desalinate water at temperatures as low as 50 °C and thus constitutes a possibility to generate revenue from these abandoned wells. Moreover, if the distillate can be applied for evaporative cooling of air-cooled geothermal plants, the efficiency and thus revenue of these plants can be augmented substantially. This narrative motivated the geothermal office of DOE to award a project between the National Renewable Energy Laboratory (NREL) and the Colorado School of Mines (CSM) to map the potential application in the US and build and install a pilot-scale membrane distillation unit at a test site of ORMAT.

First phase of this project is to select an appropriate membrane. The performance of 15 different membranes was analysed from 6 different manufacturers over a wide temperature range (40-70 °C). Most of these manufacturers don`t make dedicated membrane distillation membranes, but make microfiltration membranes for water and air purification with the appropriate pore size and contact angle. One manufacturer makes oleophobic membranes for humidity equilibration and oil rejection in automotive applications. Due to the different application of these membranes, some manufacturers don`t report characteristics or use different methods to determine essential parameters to model membrane distillation performance, like pore size and porosity for instance. Non of them report vapour fluxes or thermal conductivity.

Therefore a model was developed that does not require the knowledge of the pore size, the porosity or the thermal conductivity of the membrane. It was found that thermal efficiency is fairly constant for a wide range of salinities and temperature differences over the membrane. The Schofield method was adapted to incorporate the thermal efficiency instead of the thermal conductivity, pore size and porosity. Thermal efficiency is often not reported in MD literature, but it can have an important impact on the efficiency and hence the cost effectiveness of the process for geothermal desalination. The membranes tested exhibited thermal efficiencies from 15 to 55 %. In other words up to 85 % of the heat input was lost through conduction.

Development of highly permeable thin film composite polyamide membranes for water desalination through reverse osmosis with polydopamine/molecular sieve particles layer

Pinar Cay-Durgun and Dr. Mary Laura Lind

Current reverse osmosis technology widely uses ultrathin, crosslinked, aromatic polyamide (PA) thin film composite (TFC) membrane which is interfacially polymerized onto microporous polysulfone support membrane. The dense PA structure is highly selective and the thinness of this layer enables high water permeability. Recently, a class of mixed matrix membranes called thin film nanocomposite (TFN) membranes has been developed to tailor the water desalination performance by embedding particles into polyamide structure during interfacial polymerization. The addition of nanoparticles to TFC membranes is a promising technology to reduce energy costs and increase the performance in desalination processes. However, there may be some "non-ideal effects" in the membrane such as particle aggregation at higher zeolite loadings and void or rigidified polymer formation at the polymer/particle interface. Furthermore, there might be particle loss under high pressure during cross-flow desalination process which may reduce the performance of the membrane. To overcome these drawbacks, we developed a method to increase separation performance and particle stability of molecular sieves in TFN membranes. First, we stabilized particles on microporous polysulfone support by thin bio-inspired polydopamine layer, and then we interfacially synthesized polyamide on top of this layer. This method also minimizes the expensive particle usage compared with the common TFN synthesis procedure (particle addition into polyamide casting solution). In this work, we present our results for development of thin film composite polyamide membranes on polydopamine layer with molecular sieves particles on polysulfone support membrane. First, we synthesized dopamine layer with different particle content and different particle size. Then, we synthesized a series of polyamide membranes via interfacial polymerization. For comparison, we also synthesized regular TFC and TFN polyamide membranes. Then we measured the permeability and selectivity performance using dead-end filtration and cross-flow testing system. We report the synthesized membrane surface and morphology result with contact angle, Fourier transform infrared spectroscopy, scanning electron microscopy, and atomic force microscopy.

Ion and methanol sorption and transport in a cation exchange membrane based on crosslinked sulfonated polystyrene

Michele Galizia, Francesco M. Benedetti, Donald R. Paul, Benny D. Freeman

J.McKetta Department of Chemical Engineering, 200 E. Dean Keeton Street, 78712 Austin, and Center for Energy and Environmental Resources, 10100 Burnet Road, 78758 Austin, The University of Texas at Austin, Texas (USA)

Sorption and transport of alkali-chlorides (NaCl, KCl, LiCl), alkaline earth-chlorides ($CaCl_2$, $MgCl_2$) and liquid methanol in a commercial cation exchange membrane based on cross-linked polystyrene with fixed sulfonate ions were investigated.

Ion transport parameters were correlated with ion size and counter-ion valence. Divalent counter-ions establish stronger interaction with the polymer's sulfonate groups, which significantly reduces the actual polymer charge density. Due to this partial neutralization of fixed charges, Donnan exclusion [1] becomes less effective in the presence of alkaline-earth chlorides, which allows greater permeability and sorption coefficients relative to alkali-chlorides. For the same reason, the membrane in the divalent counter-ion form sorbs significantly less water than in the monovalent counter-ion form.

In highly charged polymers, electrostatic interactions between fixed ions and counter-ions sorbed from the external solution overwhelm the free volume effects related to the membrane's water content. Conversely, sorption and transport of neutral solutes, such as methanol, are significantly influenced by the membrane degree of swelling, with more swollen membranes exhibiting faster methanol transport rates.

The importance of electrostatic interactions was also demonstrated by measuring the solubility of mixtures of salts having the same counter-ion and different co-ions.

[1] F.G. Donnan, Chem. Rev. 1924 1 73-90

Controlling Water Content of Cross-linked Polymer Membranes and Its Effect on Ion Sorption Properties

Eui-Soung Jang, Donald R. Paul**, Benny D. Freeman*

McKetta Department of Chemical Engineering,The University of Texas at Austin, Austin, TX, 78758, U.S.A.

**freeman@che.utexas.edu, **drp@che.utexas.edu*

Cross-linking provides a good way to control polymer water content, which is a key property for ion/water separation membranes. In this study, two series of cross-linked polymer membranes (i.e., uncharged and charged membranes) were synthesized to study the effect of polymer water content on ion transport properties. Cross-linked poly(ethylene glycol)diacrylate hydrogels (XLPEGDA) were chosen as control uncharged polymers and negatively charged cross-linked poly(2-acrylamido-2-methylpropane sulfonic acid-r-diethylene glycol dimethacrylate) (XL(AMPS-DEGDMA)) was chosen as a charged counterpart. Cross-linking density of the polymers was controlled by varying the amount of solvent (i.e., de-ionized water for XLPEGDA and NMP for XL(AMPS-DEGDMA)) in the prepolymerization mixture. Salt sorption properties in the two series of membranes were characterized with three alkali metal chlorides (LiCl, NaCl, and KCl).

Salt sorption coefficient increased with increasing water content in both series of polymers. The order of salt sorption coefficients in XLPEGDA was: KCl > LiCl > NaCl, which does not correspond with the order of ion hydration size, i.e., $Li^+ > Na^+ > K^+$. These results suggest that ion sorption in uncharged XLPEGDA are influenced by both ion hydration and polymer-ion specific interactions. Fixed charge concentration is a crucial factor that influences ion sorption in ion exchange membranes. For a fair comparison, cross-linking density and monomer compositions were controlled to fix the fixed charged concentration while varying water content in a series of cation exchange membranes. Changes in water content did not greatly affect salt sorption coefficients, suggesting that salt sorption in ion exchange membrane is mainly influenced by the fixed charge concentration as Donnan theory predicts. Furthermore, based on the ion sorption measurements, electrolyte activity coefficients in the polymers were calculated to quantify the thermodynamic non-ideality of the ion-polymer-water mixing system. All of these findings assist in designing polymer membranes and understanding ion sorption properties of polymers as part of a long-term effort to improve fundamental understanding regarding ion uptake and ion transport in both charged and uncharged polymers.

Environmental and economic assessment of fertilizer drawn forward osmosis and nanofiltration hybrid system for desalination of mine impaired water for irrigation

Jungeun Kim, **Sherub Phuntsho***, Laura Chekli, Hokyong Shon**

School of Civil and Environmental Engineering, University of Technology, Sydney, Post Box 129, Broadway, NSW 2007, Australia.

Corresponding authors: Email: *sherub.phuntsho@uts.edu.au, **Hokyong.Shon-1@uts.edu.au

1. Background

As a driest continent on the earth, drought and water scarcity is a major issue in Australia. Like most parts of world, agriculture sector uses up to 60-70% of the total fresh water consumption in Australia. Coal mining is a globally significant industry in Australia and most coal mining activities involve massive volumes of produced water. With appropriate treatment, mine impaired water could be valuable source of irrigation water for growing food rather than simply dispose to the environment after treatment. Most produced water are however associated with high salinity, generally not fit for irrigation until salinity is lowered. Lately, forward osmosis (FO) process is seen as innovative low-energy technology for various applications including desalination for irrigation purpose [1]. One of the most promising applications of the FO process with potentially lower additional energy consumption is the fertilizer drawn FO (FDFO) process using fertilisers as DS in which the diluted DS containing fertiliser nutrients can be used as non-potable water for irrigation of crops [2, 3]. Recently a pilot-scale FDFO-NF desalination system with a capacity of 1000 - 4000 L/day was operated and tested at Centennial Coal Mine site at Newtan Colliary, Fassifern, New South Wales for the desalination of saline water [4]. Based on about six months pilot-scale operation in the field, it was observed that the FDFO-NF hybrid system could be a robust desalination technology in producing nutrient rich water for irrigation with the potential to support farming industry around the coal mining area.

FDFO-NF hybrid desalination technology is a relatively new technology and hence, the economic and environmental life cycle assessment (LCA) of the technology needs to be conducted for the FDFO-NF system to investigate not only the hot-spots of improving the whole process operation but also to understand the sustainability and economic viability of this new technology for commercial applications. In this study, we conducted an environmental impact assessment of the FDFO-NF hybrid system and compared with the two different conventional reverse osmosis (RO) hybrid scenarios for the desalination of mine impaired saline groundwater. The environmental impact assessment also aims to identify the major hot-spots that are responsible for the economic and environmental impacts including chemical consumptions for membrane cleaning, materials consumed for membrane replacement and energy consumption during process operation.

2. Methodology

A standard LCA was considered which generally consists of 4 stages; goal and scope definitions, life cycle inventory analysis (LCI), life cycle impact assessment, and interpretation. The three different hybrid desalination processes chosen for this goal includes MF-RO hybrid system for scenario A and UF-RO hybrid system for scenario B which was then compared to the hybrid FDFO-NF process for scenario C. To aggregate primary data for each scenario, the inventory analysis has been carried out using the Ecoinvent LCA database version 3.0 and the Australian LCA database in the software Simapro version 8.0 [5-7] with several common considerations for economic and environmental life cycle assessment of the three scenarios. Life-cycle impact assessment (LCIA) stage of LCA was performed based on the six Australian relevant impact assessment categories Australian indicator set v.3.01 [6, 8]. An economic assessment in this study considered annual operational expenditure (OPEX) that includes three cost components: electricity (energy) cost, membrane replacement cost, and cost of chemicals for membrane cleaning. The FO OPEX data such as FO membrane module cost were based on membrane manufacturers data while for NF, the existing OPEX data from the RO and NF were adapted from the published literatures.

3. Findings from the study

The following are some of the brief findings from this study:

- Environmental LCA results showed that the overall environmental impact of both energy consumption and cleaning chemical usage is lower for FDFO-NF hybrid system using thin film composite FO membrane while that of membrane requirement is higher for FDFO-NF hybrid system (using cellulose triacetate FO membrane) compared to conventional MF-RO and UF-RO hybrid systems.
- The sensitivity analysis of the FDFO process under different considerations such as average FO water flux and NF feed concentration showed that a minimum average FO flux of 6 $Lm^{-2}h^{-1}$ is required to compete with conventional desalination processes (MF-RO and UF-RO).
- If the dilution of fertilizer draw solution by the FO process is economically achieved osmotic equilibrium, potential exists for energy savings in the NF process by nearly 50% thus resulting in the reduction the OPEX costs in the FDFO-NF hybrid system.
- However, in order to better understand the performances of a commercial level full-scale FDFO-NF hybrid system, more comprehensive and detailed analysis is important through rigorous simulation studies or confirm through design and operation of a full-scale FDFO-NF hybrid system especially to assess the practicality of achieving osmotic equilibrium within the practical module settings.

References

1. Subramani, A., et al., Water Research, **45**(5): p. 1907-1920.
2. Phuntsho, S., et al., Journal of Membrane Science, **375**(1–2): p. 172-181.
3. Phuntsho, S., et al., Environmental Science & Technology, **46**(8): p. 4567-4575.
4. Phuntsho, S., et al., Journal of Membrane Science, **508**: p. 22-31.
5. Consultants, P., *SimaPro software*. 2008, SimaPro Version.
6. Grant, T., et al., The International Journal of Life Cycle Assessment, **6**(6): p. 323-324.
7. Frischknecht, R., et al., The International Journal of Life Cycle Assessment, **10**(1): p. 3-9.
8. Bengtsson, J., et al.

Abstract

For NAMS 2016 Annual Meeting

May 21-25, Seattle, Washington

Oral Session: Membrane Formation

Macro-porous α-Al_2O_3 support tubes for thin defect-free membranes, made by centrifugal colloidal casting

Ralph A. Bauer, John Phillips, and Hendrik Verweij

Department of Materials Science and Engineering, The Ohio State University

2041 College Rd, 291A Watts Hall, Columbus OH 43210, USA.

☎: +1 (614) 805-3015, +1 (510) 847-2372, +1 (614) 247-6987

E-mail: Bauer.348@osu.edu, and Verweij.1@osu.edu

Thin membranes are considered for applications, in water purification, gas separation and as components in thin supported fuel cells and oxygen pumps. The compositions of such membranes are typically inorganic to achieve high selectivities. For reasons of energy efficiency their thickness is preferably less than 1 μm and possibly as small as 50 nm. The membranes are made by colloidal deposition of homogeneous nano-particle dispersions on a porous support, followed by thermal consolidation. For applications that require high temperatures and pressures, the support must be fully ceramic. It is typically made by compaction and thermal consolidation of α-Al_2O_3. α-Al_2O_3 is chemically inert, hard and strong and has an excellent thermal conductivity. High quality powders to make ceramic porous α-Al_2O_3 are readily available and affordable. Since a membrane thickness of 50 nm…1 μm greatly increases the risk of defect formation, the support structure must have a surface roughness of <25 nm, and no surface defects with dimensions of >50 nm. These requirements cannot be met with state-of-the-art forming techniques for porous ceramic support structure.

Centrifugal colloidal casting (CCC) is a known ceramic forming method that we used to make supports that meet the requirements for thin membrane deposition. A purified, stable dispersion of 300 nm α-Al_2O_3 particles was introduced into a cylindrical mold. Centrifugal consolidation of homogeneous dense tubular structure took place by rotation the mold at 20,000 rpm. The structure was dried inside the mold, released and sintered to the target strength and gas permeance by heating at 950°C. By careful optimization of the dispersion formulation, the centrifuge's stability, the drying and release procedure, we have been able to make, for the first time by CCC, high quality membrane support tubes with an outer diameter of <10 mm. One first application of such tubes will be the deposition of fully selective thin dense O_2 membranes for operation at temperatures of <500°C.

ABSTRACT.

Large-Area fabrication of Graphene-based Membranes by Shear Alignment of Discotic Nematic Liquid Crystals of Graphene Oxide

Graphene-based membranes demonstrating ultra-fast water transport, precise molecular sieving of gas and solvated molecules, have been considered for a host of applications in desalination, gas separation and pervaporation; however scale-up of these membranes to large-areas remains an unresolved problem. Here we are going to present that the discotic nematic phase of graphene oxide (GO) can be shear aligned to form highly ordered, continuous, thin films of multi-layered GO on a support membrane by an industrially-adaptable method to produce large-area in less than 5 seconds. The pseudoplastic, shear thinning fluid of the nematic phase of GO promotes membrane formation by enhancing wetting and spreading at high shear rates, while resisting dewetting at low shear rates. Pressure driven transport data demonstrate high retention (> 90%) for charged and uncharged organic probe molecules with a hydrated radius above 5 Å as well as modest (30-40%) retention of monovalent and divalent salts. The highly ordered graphene sheets in the membrane plane make organized channels and enhance the permeability of the membrane (71 ± 5 l m^{-2} $hr^{-1}$$bar^{-1}$ for 150 ±15 nm thick membranes). Benchmarking experiments with a highly established commercial nanofiltration membrane demonstrate around 9 times higher water flux at comparable retention values and excellent flux recovery by chemical cleaning.

Nitrogen Selectivity of Molybdenum Metallic Membranes

Kyoungjin Lee[1], Simona Liguori[1], Ralph A. Bauer[2], Hendrik Verweij[2], and Jennifer Wilcox[1]

[1]Department of Energy Resources Engineering, Stanford University
[2]Department of Materials Science and Engineering, The Ohio State University

Nitrogen(N_2)-selective membranes are rare since nitrogen has low adsorption or dissolution affinity toward typical membrane materials. An efficient high-temperature N_2 separation using membranes would be beneficial if applied to post-combustion capture of CO_2 to increase the CO_2 concentration, or to the purification of natural gas, which contains N_2 as impurity. The ultimate goal of this study is to develop nitrogen-selective membranes that can be operated at high temperature.

A composite structure consisting of molybdenum (Mo) thin film on porous alumina was chosen as a model membrane system. Porous alumina substrates with pore sizes of approximately 100 nm were synthesized in the lab by colloidal casting and partial sintering. Subsequently, Mo membranes with thicknesses of 300 and 800 nm were deposited by physical vapor deposition techniques. Scanning electron microscope (SEM) images showed that the surface concentration of open pores along the triple boundaries decreased with the increase of thickness of the Mo layer. The electrical resistivities of the Mo layers were found to be 1.5×10^{-6} and 8.8×10^{-7} Ωm at 300 and 800 nm, respectively. The decreasing resistivity indicates that the structure becomes more dense, but the bulk resistivity of 5×10^{-8} Ωm was not yet attained. These results indicate that the membranes have defects but the thicker Mo membrane is more dense.

Gas permeation through the fabricated membranes was measured using pure N_2 and argon (Ar) to determine the ideal selectivity of the membranes at different temperature and pressure. The permeate pressure was maintained at 1 atm in the same gas as the feed in all experiments, and the reported pressure conditions in the following are Δp. The porous alumina showed an Ar permeance of 4.8×10^{-7} mol/(m^2·sec·Pa) at 22 °C and 4 bar. The permeance of Ar through the Mo membranes at the same operating conditions was significantly lowered to 38% and 0.9% of the support value after 300 and 800 nm deposition, respectively. For the 300 nm-thick membrane, a N_2 permeance of 1.0×10^{-7} mol/m^2·sec·Pa with an ideal permselectivity with respect to Ar of 1.4 was obtained at 800 °C and 4 bar. Because the open pores allowed Ar Knudsen diffusion, the selectivity was close to the Knudsen diffusion ratio of

N_2 to Ar (1.2), but the slight increment in selectivity indeed confirmed a solution-diffusion mechanism for N_2. For the 800 nm-thick membrane at 4 bar, the N_2 permeances were 6.5×10^{-10} and 7.2×10^{-10} mol/(m^2·sec·Pa), with the ideal permselectivities of 2.1 and 2.0 at 400 °C and 500 °C, respectively. Lower selectivity values of 2.0 (at 400 °C) and 1.8 (at 500 °C) were obtained at a higher Δp of 5 bar. The N_2 permeance, after the subtraction of the Knudsen diffusion permeance, showed a dependence on $p_{feed}^{0.5}-p_{permeate}^{0.5}$, which supports the hypothesis that transport of atomic nitrogen through the bulk Mo is possible.

The permeances of both Ar and N_2 at high temperature conditions were significantly lowered compared to what is expected from the temperature dependence of the Knudsen diffusion. This implies that defect pore sizes might have decreased at high temperature, possibly due to surface oxidation. Post-permeation analyses using X-ray photoelectron spectroscopy (XPS) and X-ray diffraction (XRD) revealed that the metallic Mo phase is still present after permeation tests, but also molybdenum oxide was formed. Nevertheless, the Mo membrane was stable for more than 90 hours at 400 °C with a trace amount of oxygen present in the flow system. After exposure to 600 °C, the Mo membrane appeared to have pinholes, observed by SEM. This is in agreement with the observation that no special selectivity persisted at 600°C due to substantial membrane degradation.

The results in this study suggest that Mo metal can be used for N_2-selective membranes at moderately high temperatures. The formation of some surface oxide may help in reducing the size of pinhole defects although heavy oxidation leads to destruction of membranes. In future studies, the selectivity may be increased by improved metal membrane layers with a more dense microstructure.

Process Design of Natural Gas Treatment using DDR-type Zeolite Membrane

Hiroaki Hasegawa[1], Keiichi Nishida[1], Shuichi Oguro[1], Yasushi Fujimura[1]
Kenji Yajima[2], Makiko Niino[2], Manabu Isomura[2], Toshihiro Tomita[2]
[1]JGC Corporation, Yokohama, Japan
[2]NGK Insulators, Ltd., Nagoya, Japan

Abstract

Natural gas is a widely-used energy source and its demand is increasing. Some of the natural gases, however, contain high levels of CO_2. Membrane separation is highly useful for CO_2 separation from such kind of natural gas. JGC Corporation (JGC) and NGK Insulators, Ltd. (NGK) have been jointly developing a separation process for natural gas treatment using the DDR-type zeolite membrane (DDR Membrane). NGK, a top class manufacturer of ceramic products, has been conducting the synthesis of the DDR Membrane and scaling up the DDR Membrane element. On the other hand, JGC, a world leading engineering company for oil and gas processing, has been conducting process design, membrane housing design for commercial facility and the performance evaluation of the DDR Membrane with high pressure condition which simulates natural gas treatment.

DDR is two-dimensional zeolite with 0.36 x 0.44 nm pores and the DDR Membrane is prepared using hydrothermal synthesis on the surface of porous ceramic substrate. The DDR Membrane has the following features.

(1) Large-Size DDR Membrane

NGK has succeeded in scaling up the DDR Membrane element into a large monolithic structure with a diameter and length of 180 mm and 1,000 mm, respectively (Large-Size DDR Membrane). The membrane area of the single DDR Membrane element is 12 m^2, which is the largest among the inorganic membrane elements in the world to the best of our knowledge(see Fig. 1). The membrane area corresponds to that of 400 tubes of tubular type membrane of 10 mm diameter with the same length. Also pressure resistance of the Large-Size DDR Membrane is 8 MPaG as maximum.

(2) High Separation Performance on CO_2/CH_4

DDR Membrane has high selectivity of CO_2/CH_4 comparing with other zeolite membranes and polymeric membrane[1, 2]. Based on the result of our laboratory testing, the CO_2/CH_4 separation factor of the Large-Size DDR Membrane for equimolar CO_2/CH_4 gas feed is higher than 160 at 303 K and 0.3 MPaG. This separation factor is as same as that of the much smaller size DDR Membrane whose diameter and length are 30 mm and 160 mm, respectively (see Table 1). The performance of the DDR Membrane was not lowered with scale-up.

(3) Applicable to high CO_2 partial pressure condition

The DDR Membrane shows no CO_2-induced plasticization which occurs in the conventional polymeric membrane at high CO_2 partial pressure[3]. Thus the DDR Membrane is applicable to gas fields with high CO_2 partial pressure.

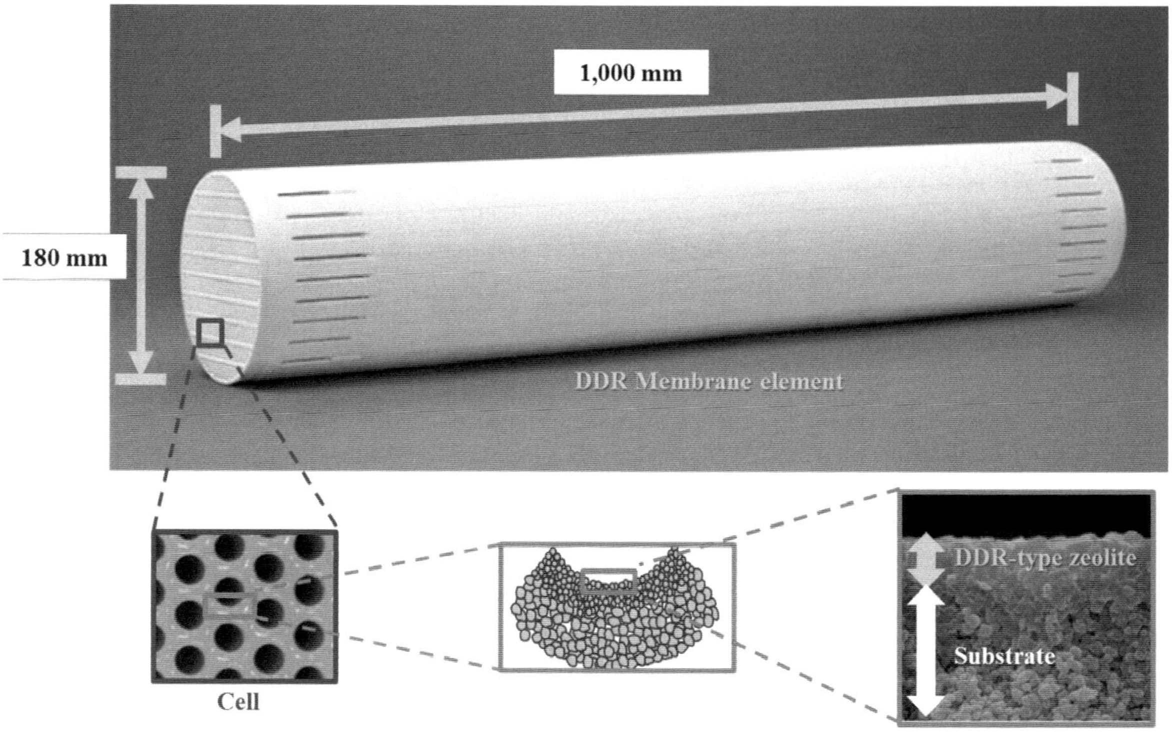

Fig. 1. Overview of DDR Membrane

Table 1. Separation factor of the DDR Membrane

Case	Feed gas composition [mol%]	Feed gas Pressure [MPaG]	Separation factor [-]	Permeate gas composition [%]
CO_2 & CH_4	CO_2 : CH_4 = 50 : 50	0.3	> 160	CO_2 > 99
H_2 & CH_4	H_2 : CH_4 = 60 : 40	0.4	> 100	H_2 > 99
CO_2 & H_2	CO_2 : H_2 = 40 : 60	0.4	12	CO_2 = 89
N_2 & CH_4	N_2 : CH_4 = 50 : 50	0.3	9	N_2 = 90

Membrane size : Diameter 30 mm – Length 160 mm

To evaluate the DDR Membrane for natural gas processing, we simulated the simple material balance based on our laboratory test results of the DDR Membrane. We have utilized the actual conditions of LaBarge gas field in USA as a practical example to evaluate the performance of the DDR Membrane. Gas composition of LaBarge is 65% CO_2, 21% methane, 7% nitrogen, 5% hydrogen sulfide (H_2S), 0.6% helium and others[4]. Separation of CO_2 in LaBarge is basically a solvent absorption and separated CO_2 is used for the enhanced oil recovery (EOR) in LaBarge field. The CO_2 concentration of the gas removed from the solvent absorption is more than 95 mol%. The DDR Membrane can be applied for bulk removal before the chemical absorption. Feed gas containing CH_4 and CO_2 is introduced to the membrane to be separated to the CH_4-rich treated gas by removing CO_2-rich permeate gas (see Fig. 2). The pressures of the feed and permeate gases are assumed as 8.0 MPaG and 0.1 MPaG, respectively. The specified CO_2 concentration of the treated gas is 15 mol% for bulk removal before chemical absorption. As the result of the calculation, the

CH_4 loss of the DDR Membrane is 2.8 % and CO_2 concentration in permeate gas is approximately 99 mol%. This low CH_4 loss improves the yield of CH_4-rich treated gas and high CO_2 concentration in permeate gas is suited for EOR. This means the probability that CO_2 separation technology can be partially replaced from current chemical absorption process to DDR Membrane. Energy requirement can be significantly reduced comparing with chemical absorption process by using the DDR Membrane. As mentioned previously, DDR Membrane can be a solution for natural gas treatment and also EOR application.

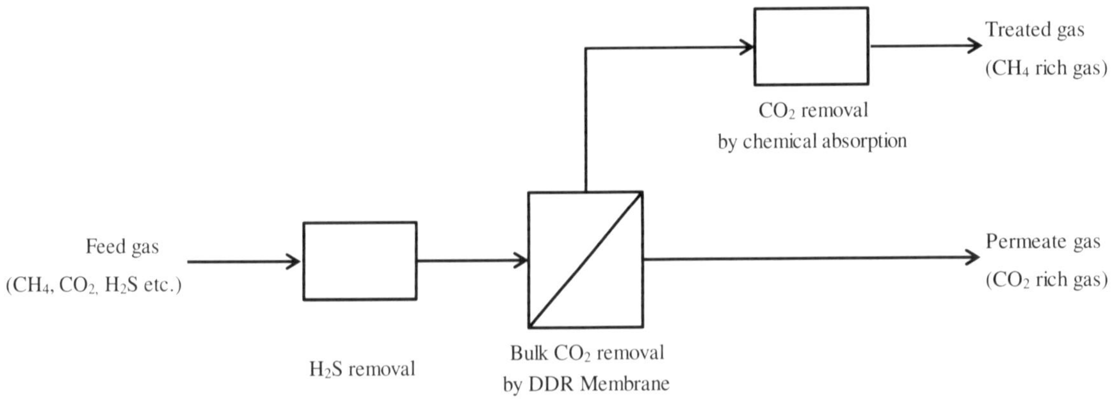

Fig. 2. Process Flow

Reference:

[1]T. Tomita, K. Nakayama, H. Sakai. (2004). Gas separation characteristics of DDR type zeolite membrane. Microporous and Mesoporous Materials. **68**. 71-75.

[2]N. Kosinov, J. Gascon, F. Kapteijn, E. J. M. Hensen. (2016). Recent developments in zeolite membranes for gas separation. Journal of Membrane Science. **499**. 65-79.

[3]A. Bos, I. G. M. Punt, M. Wessling, H. Strathmann. (1999). CO_2-induced plasticization phenomena in glassy polmers. Journal of Membrane Science. **155**. 67-78.

[4]M. E. Parker, S. Northrop, J. A. Valencia, R. E. Foglesong, W. T. Duncan. (2011). CO_2 Management at ExxonMobil's LaBarge Field, Wyoming, USA. Energy Procedia. **4**. 5455-5470.

A New Method of Fabricating Charge Mosaic Membranes by Inkjet Printing

Peng Gao, Aaron Hunter, William Phillip
Chemical and Biochemical Engineering, University of Notre Dame, South Bend, Indiana

In order to help meet the growing demand for fresh water, intense efforts have been undertaken to broaden our fundamental knowledge of existing desalination technologies as well as to develop new desalination technologies. Piezodialysis is a promising membrane-based desalination technology that removes salt from water.[1] Charge mosaic membranes, which contain unique positively-charged and negatively-charged domains, facilitate the transport of ions and are suitable candidates for piezodialysis applications.[2,3] Here, we report a novel method of preparing charge mosaic membranes by inkjet printing polymer blends on template membranes. Mosaic patterns consisting of both positively-charged and negatively-charged domains were printed on PCTE membranes using a commercial inkjet printer. Poly(diallyldimethylammonium chloride) and poly(sodium 4-styrene sulfonate) were used to generate the positively-charged and negatively-charged domains, respectively. In this talk, we demonstrate that the surface charge of the mosaic membranes can be altered by controlling the deposition of the oppositely-charged materials.[4] The mosaic membranes show promising ion enrichment results in piezodialysis experiments. Inkjet printing offers a straightforward and scalable route toward the design and fabrication of charge mosaic membranes.

(1) Bolto, B.; Hoang, M.; Tran, T. Review of Piezodialysis - Salt Removal with Charge Mosaic Membranes. *Desalination* **2010**, *254*, 1–5.

(2) Ishizu, K.; Amemiya, M. Transport of Electrolytes through Charge Mosaic Composite Membranes. *J. Memb. Sci.* **1992**, *65*, 129–140.

(3) Rajesh, S.; Yan, Y.; Chang, H.; Gao, H.; Phillip, W. A. Mixed Mosaic Membranes Prepared by Layer-by-Layer Assembly for Ionic Separations. *ACS Nano* **2014**, *8*, 12338–12345.

(4) Gao, P.; Hunter, A.; Benavides, S.; Summe, M. J.; Gao, F.; Phillip, W. A. Template Synthesis of Nanostructured Polymeric Membranes by Inkjet Printing. *ACS Appl. Mater. Interfaces* **2016**, *8*, 3386–3395.

Thin-Film Composite Membranes with Regular, Sub-micron Surface Patterns: Fabrication and Characterization

Sajjad H. Maruf, Alan R. Greenberg, and Yifu Ding*

Membrane Science, Engineering and Technology Center and
Department of Mechanical Engineering, University of Colorado, Boulder, CO 80309-0427, USA

Abstract

Thin-film composite (TFC) membranes are commonly used in reverse osmosis (RO) and nanofiltration (NF) processes for applications including desalination, wastewater treatment, and organic fractionation. Control of both chemistry and topography of the thin barrier layer atop the TFC membranes is key to improving their performance and thus energy efficiency`. Most commercial TFC membranes display "ridge-and-valley" or "nodular" structures, which are randomly distributed and attributed to the interfacial polymerization (IP) process utilized in their manufacture. In this presentation, we describe recent experimental studies on fabricating TFC membranes with submicron, periodic surface patterns by a unique approach in which the barrier layer is formed via IP on a previously surface-patterned ultrafiltration (UF) membrane support. This work was designed to quantify the influence of substrate topography and interfacial polymerization (IP) conditions on the characteristics of the patterned TFC membranes. For all conditions evaluated, TFC membranes with regular surface patterns were achieved by maintaining amine soaking time and IP reaction time within 120 s. Importantly, the surface topography of the patterned TFC membranes was determined to be independent of IP reaction time. Characterization of the morphological details suggests non-conformal growth of the barrier layer on the patterned UF substrates. Results indicate that the extent of such non-conformal growth can be reduced by decreasing the amine concentration as well as by choosing an amine monomer that produces a thinner semi-aromatic barrier layer. The overall findings of this study provide a strong basis for achieving desired surface features for specific TFC membrane applications.

* Corresponding authors: Department of Mechanical Engineering, 427 UCB, University of Colorado, 1111 Eng. Drive, Boulder, CO 80309-0427; Tel.: +1 303 492 2036; E-mail addresses: Yifu.Ding@Colorado.Edu; Fax: +1 303-492-3498.

ABSTRACT: Carbon Nanotube Multilayer Polyelectrolyte Charge Mosaic Membranes

Ghazal Vaseghi and G. Glenn Lipscomb

Chemical and Environmental Engineering

University of Toledo

Toledo, OH 43606-3390

The shortage of clean water in many parts of the world brings out the worldwide growing attention to desalination using advanced water treatment methods. Reverse osmosis (RO) and nanofiltration (NF) have been extensively developed as polymer membrane-based desalination during the last 50 years. Moreover, in the same period, ion exchange membranes (IEM) have evolved from laboratory stage to industrial products with remarkable technical and commercial impacts. The evolvement of ion exchange membranes makes the process cleaner and more energy-efficient. Today, ion exchange membranes are successfully applied for desalination of seawater and brackish water.

Anion and cation exchange membranes are ion exchange membranes classified based on the type of ionic groups attached to the membrane matrix. Cation exchange membranes contain negatively charged groups (i.g. $SO3^-$, COO^-, etc.) fixed to the membrane backbone, thereby allow the passage of cations and reject anions. Anion exchange membranes, on the other hand, contain positively charged groups (i.g. $NH3^+$, $NR3^+$, etc.) fixed to the membrane backbone, and allow the passage of anions while rejecting cations. Differentiating from one individual charged membrane with positive/negative fixed ions, a charge mosaic membrane (CMM) consists of parallel arrays of anion and cation exchange elements, each element providing a continuous pathway from one bathing solution to the other. Establishing a gradient of electrolyte concentration across the membrane, anions and cations can flow in parallel through their respective pathways without violating macroscopic electroneutrality. This results in a circulation of current between the mono-ion-exchange elements; therefore, the charge mosaic membrane shows negative osmosis and salt permeability much greater than its permeability to non-electrolytes.

Since the idea of charge mosaic membranes was first presented by Sollner in 1932, many researchers attempted to fabricate CMMs. While the feasibility of the previous work has been demonstrated, the performance was poor. Some of these membranes were made by mixing ground ion exchange resins and forming nano-ion exchange particles and others were built from creating interpenetrating networks of polyelectrolytes. Thus far, these membranes do not possess the required degree of contact between cation and anion exchange regions.

This work is another effort toward development of charge mosaic membranes with new formation techniques that offer the potential to increase the density of contact regions, and are expected to enhance the performance. Convective flow of salt with water through adjacent cation and anion exchange layers provides the potential to permeate salt ions from a low concentration solution to a higher concentration, thereby selectively removing salts and producing a less saline product. The membranes described here consist of double-walled carbon nanotubes (DWCNT) coated with alternating layers of cationic and anionic polyelectrolytes. The process is similar to that used in production of planar polyelectrolyte membranes. This provides concentric cylindrical structures of alternating cation and anion exchange polymers and greatly increases the degree of contact regions by aligning and embedding the fabricated DWCNTs within the pores of an isopore track-etched polycarbonate membrane. The challenges of producing these structures are discussed. Fibers with varying number of layers are examined using TEM, while the support membranes are visually tested using SEM images before and after embedding of fibers. Membrane performance results are reported illustrating the potential of the new membranes for desalination of seawater.

Facilitated Transport Membranes in the Solid State Containing Surface-Activated Ag Nanoparticles for Olefin/Paraffin Separation

Young Rae Kim, Gi Hyeon Moon and Yong Soo Kang*

Dapartment of Energy Engineering, Hanyang University, Seoul 04763, South Korea

**kangys@hanyang.ac.kr*

Trade-off relationship between permeability and selectivity has been commonly observed from polymeric membranes for gas separation. Facilitated transport membranes in the solid state have been attractive because they can improve both the permeability and the selectivity simultaneously to overcome the trade-off behavior. The improved separation performance of the facilitated transport membrane is due to the carrier-mediated transport, in addition to the normal Fickian transport, where the carrier reacts selectively and reversibly with a specific solute. Therefore, the carrier activity for facilitated transport plays a key role in determining separation performance.

We have reported the solid-state facilitated transport membranes for separation of olefin/paraffin mixtures, particularly propylene/propane mixture. They contain surface-activated Ag nanoparticles (NPs), as an olefin carrier, dispersed in a polymeric matrix. When Ag NPs have been positively charged on their surfaces by polarizer such as tetrathiafulvalene (TTF) and 7,7',8,8'-tetracyanoquinodimethane (TCNQ), they interact with propylene exclusively but not with propane, consequently resulting in the olefin carrier activity for facilitated transport. Interestingly, the binding energy of Ag atom in Ag NPs, representing the surface positive charge density, was linearly correlated with the propylene solubility and also with the propylene/propane selectivity. In addition, the separation performance has been maintained unchanged up to 500 hours. In conclusion, it is demonstrated that facilitated transport membranes containing surface-activated Ag NPs dispersed in poly(vinyl pyrrolidone) show both the high separation performance for propylene/propane mixture and the long-term operational stability.

Comparison of Performance between Laboratory Scale and Industrial Scale Membranes

Abstract

Cellulose acetate (CA) membranes are among the most researched polymers for separation and barrier applications because of their many useful characteristics including relative low cost, good toughness, high biocompatibility, good desalting, high potential flux. There are different methods used to fabricate porous polymeric membranes, and these can lead to irreproducible results, especially during scale up. The performance of the membrane is greatly influenced by not only the material but also the processing methods and conditions, because they affect the morphology of the membrane. Generally, a laboratory scale approach, such as knife over blade or solution casting, is used to cast a thin sheet of the polymeric membrane solution. Then, the final form of the membrane is prepared by sintering, stretching, or phase inversion. In many studies, the primary film casting process has been limited to doctor blade extrusion, however, this method is not always the most suitable for the industrial scale, because the solution exposed to the environment during the processing phase. Alternatively, slot die extrusion is a well-developed pre-metered approach that allows for continues casting of liquid film in a controlled environment, which also minimizes changes in liquid properties, such as rheology. The slot die coating process is a commonly used process for the manufacturing of plastics and polymer films, as well as coatings and optical films for liquid crystal displays in industry. For the scale up fabrication of hollow fiber CA membranes, a hollow fiber spin-line was used. The feasibility of processing cellulose acetate using a laboratory-scale doctor's blade versus a scale up slot die extrusion vs hollow fiber membrane was examined. The effects of processing methods, conditions, and substrate on the morphology and on the flux of CA membranes were studied. CA membranes were tested for their performance under the same operational conditions of pressure, feed volume, and feed concentration.

Manipulating Nanopore Selectivity in Lyotropic Liquid Crystal Polymer Nanofiltration Membranes for the Recovery of Targeted Organic Solutes

With the growing number of industries, there is an increased number of complex aqueous effluent streams in which there are both products and toxins. Membranes offer an energy efficient method to separate these various components, for example, products from toxins. One industrial example is the simultaneous concentration of glucose and removal of furfural from hydrolysis effluents before feeding the stream into a fermentation reactor for the production of 2nd-generation biofuels. However, current membrane fabrication methods limit the control on pore selectivity, especially at the nanopore scale (reverse osmosis and nanofiltration). A novel membrane material that offers an opportunity to manipulate pore character in a controlled fashion is the polymerized, nanoporous, bicontinuous cubic lyotropic liquid crystal (LLC) material recently developed by our research group, which is coated as the active top layer of a thin-film composite (TFC) membrane. This LLC polymer material has three distinct characteristics that enable the resulting TFC membrane to achieve high performance: (1) distinct pores on the size scale of 1 nm, (2) a more uniform pore size distribution due to its liquid crystalline nature, and (3) positively charged pore walls. Anions associate via Coulombic interactions to the positively charged pore walls and can be exchanged upon soaking the fabricated TFC membrane in a salt solution. By placing anions with different physical and chemical characteristics at the nanopore wall, it is possible to manipulate the pore character and effective size and thereby change the selectivity of the membrane. Anion-exchange using organosulfonates of varying bulk causes an increase the rejection of organic solutes in this membrane that is consistent with a reduction of the effective pore size from 1.4 ± 0.3 nm to 1.00 ± 0.08 nm. Current research is focused on exploring the degree to which the pore character may be manipulated as well as the possibility of limiting the most selective anion to the pore entrance such that hindrance to water transport is minimized throughout the rest of the pore, thereby increasing the flux. By correlating characterization of the material to observed filtration performance, we are able to contribute to the development of targeted solute selectivity via manipulation of the nanopores in these LLC materials.

Polymeric Micro-Spherical Particles Grafted on Water Treatment Membranes: Mechanism of Formation and Applications.

The fabrication of stable, non-leachable micro-spherical polymers on solid surfaces, particularly on water treatment membranes, has the advantage of high surface area and high permeability, which is beneficial for anti-fouling coatings. A one-step generation of micro-scale spherical polymeric particles on membrane surfaces was achieved using UV-initiated graft-polymerization of polyethylene glycol methacrylate (PEGMA). The yield of coating was determined gravimetrically (Fig. 1A), surface chemistry by ATR-FTIR and surface morphology by SEM (Fig. 1).

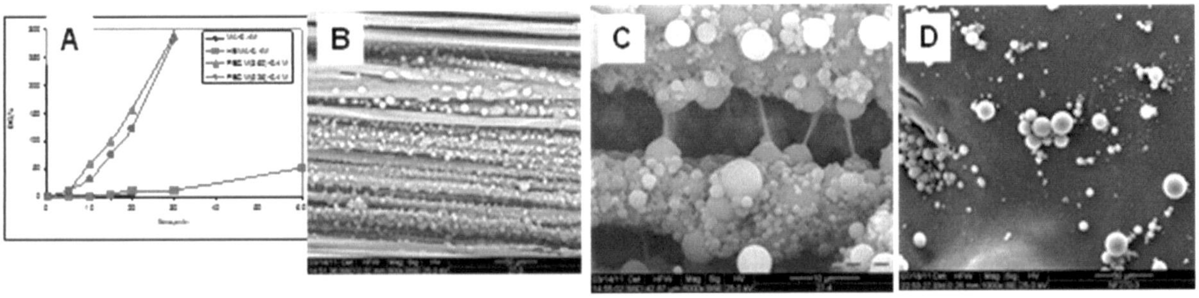

Figure 1: *Degree of graft-polymerization of polyethyleneglycol methacrylate (PEGMA) on polyethylene terephthalate fibers (A) and SEM images (B-C) after graft polymerization. (D) Graft-polymerization on polyamide membrane. Scale bar in B, D is 50 µm, and 10 µm in C.*

Mechanistic studies of the formation of spherical-particle coating was performed with membrane templates with diverse properties, and in varied solvent polarities; it was found that spherical polymeric particles were formed preferentially on hydrophobic templates and in polar solvents. IR and Raman spectroscopy suggested that the spherical shape of the coating was related to the amphiphilic nature of the grafted poly(PEGMA). The effects of grafting UF membranes with poly(PEGMA) were studied in a membrane bioreactor treating domestic wastewater, and results showed no reduction of fouling, however improvement observed in cleaning ability of the membranes.

The method presented here for coating membranes by stable polymer brushes with nano- and micro-spherical structures has also potential implications for other uses such as materials with high sorption capacity, nanoreactors, and resins for heterogeneous chemical catalysis.

UV and EB cross-linked polysulfone based membranes for versatile applications

Veysi Altun , Ivo F.J. Vankelecom*

Centre for Surface Chemistry and Catalysis, Faculty of Bioscience Engineering, KU Leuven, Kasteelpark Arenberg 23, Box 2461, B-3001 Leuven, Belgium

* ivo.vankelecom@biw.kuleuven.be

In separation technology, polymeric membranes have found wide application. Cross-linking procedures are needed to enhance the stability of common used polymers such as polyimides and poly(aryl sulfones) which are not sufficiently resistant to heat and harsh solvents, such as aprotic solvents. Chemical cross-linking of membranes has so far mainly been performed to apply them in solvent resistance nanofiltration (SRNF). However, this method is not suitable for polymers without reactive groups, e.g. poly(aryl sulfones) or poly(ether ether ketones). Hence, there is a need to design an efficient method for post-synthesis cross-linking of such polymers. It is possible to use such modified membranes, e.g. poly(aryl sulfone) membranes in SRNF[1].

Environmentally more friendly two irradiation method as UV and EB cross-linking were applied in this work. Non-solvent phase inversion method has been used to synthesize the membranes which is followed by UV or EB curing[2]. Semi-interpenetrating polymer network (SIPN) of a linear polysulfone and a cross-linked polyacrylate formed the matrix of the membranes. The choice of the photoinitiator and the cross-linker were the most crucial in all process. Therefore, the effect of the type and concentration of the photoinitiator and cross-linker functionality on the curing efficiency and membrane SRNF performance, as well as chemical stability was in-depth investigated. Also different irradiation doses were applied to tune curing parameters. Membranes with various morphologies from macrovoid- to sponge-like structure, as well as with good SRNF performance and superior chemical stability have been produced by tuning the synthesis conditions. The influence of phase inversion parameters, such as polymer concentration, co-solvent type and concentration, as well as evporation time, on SRNF membrane performance was studied additionally, for the most optimal system.

References

[1] K. Vanherck, P. Vandezande, S.O. Aldea, I.F.J. Vankelecom, Cross-linked polyimide membranes for solvent resistant nanofiltration in aprotic solvents, J. Membrane Science 320 (2008) 468-476.

[2] I. Strużyńska-Piron, J. Loccufier, L. Vanmaele, I.F.J. Vankelecom, Synthesis of solvent stable polymeric membranes via UV depth-curing; Chem. Commun. 49 (2013) 11494-11496.

Development of a Composite Membrane with a Liquid-barrier Chemical-resistant Polymer Matrix to Recover Water from Urine-containing Wastewater

Afsaneh Khosravi, Andrew Dopilka, Katelyn Kline, Mary Laura Lind*

School for Engineering of Matter, Transport and Energy,

Arizona State University, Tempe, AZ 85287, USA

Abstract:

Current osmotic processes offer the potential to increase water recovery from wastewater; however there are two problems with currently available reverse osmosis (RO) membranes. First, the polyamide-based RO membranes rapidly degrade when exposed to acid pretreated urine and urine brines. Additionally, present RO membranes have relatively low rejections (~60%) of neutral solutes such as urea[1] while they have very high rejections (>99%) of ionic solutes.

Our research focuses on the development of mixed matrix membranes that solve both problems. We developed an innovative mixed matrix membrane design that integrates water-selective zeolite nanoparticles into a liquid-barrier chemically resistant polymer film.[2] The main distinguishing characteristics of our new membrane design compared with current mixed matrix membranes include: (1) the use of an impermeable polymer matrix enabling us to use a broader range of chemical resistant polymers; (2) the use of zeolite particles with specific pore size ensureing us the high rejection of the neutral molecules since water is transported through the zeolite rather than the polymer; (3) the use of latex dispersions as the polymer matrix source. Latexes are environmentally friendly water based-solutions, sharing the qualities of low volatile organic compound (VOC), low cost, and non- toxicity. In our design, the zeolite particles are anchored to a porous substrate. Then, a liquid-barrier chemically resistant polymer matrix fills the space between the zeolite nanoparticles and binds the zeolite into a versatile nanocomposite thin film.

In this talk, we will discuss our progress on development and synthesis of this novel membrane. We have successfully anchored zeolite nanoparticles to porous substrates. We have studied the effect of the polarity of solvents, leaving group of silane agents, the temperature and time of the silanization reaction on anchoring zeolite nanoparticles to porous substrates. Furthermore, we have developed a novel fundamental understanding of the latex film formation on porous substrates. We investigated the effect that the properties of the latex solution and the properties of the porous substrate have on the formation of defect free films of latexes on porous substrates.[3] Additionally, we will report on the flux and sodium chloride rejection of brackish water, as well as the material characterization, of our new mixed matrix membrane.

References

.1 Yoon, Y. & Lueptow, R. M. Reverse osmosis membrane rejection for ersatz space mission wastewaters. *Water Res* **39**, 3298-3308, doi:10.1016/j.watres.2005.05.016 (2005).

2 Lind, M. L., Fink, S. & Sasaki, K., Lin, J. Nanocomposite membranes, US Patent Pending. US20160030893 A1 (2014).

3 Khosravi, A., King, J. A., Jamieson, H. L. & Lind, M. L. Latex Barrier Thin Film Formation on Porous Substrates. *Langmuir* **30**, 13994-14003, doi:10.1021/la502812d (2014).

How to Enhance ZIF-8 Growth on Porous Polymers?

Eva Barankova, Xiaoyu Tan, Pradeep Neelakanda, Mahendra Kumar, Klaus-Viktor Peinemann

Advanced Membranes and Porous Materials Center, King Abdullah University of Science and Technology (KAUST), Thuwal, Kingdom of Saudi Arabia

ZIF-8 became the most studied MOF for membrane's applications mainly because of its easy synthesis, good chemical and thermal stabilities, recent commercial availability and pore size of 3.4 Å. Thin dense ZIF-8 layers have been successfully synthesized on inorganic supports and offered superior selective properties. However polymeric supports have advantages, because they are much cheaper, they are light, they can be easily produced on large scale and module fabrication is easier.

The aim of our work is to develop simple methods for growing ZIF-8 thin layers on polymer supports. The challenge is to obtain well-intergrown thin compact and defect-free ZIFs layer on a flexible porous polymer. We present new methods for preparation ZIF-8 membranes on polymers. We introduced presence of zinc oxide nanoparticles in the support as a secondary metal source for ZIF-8 growth. Initially the ZnO particles were incorporated in the polymer matrix and later on the surface of the polymer by magnetron sputtering (1,2). In both cases the ZnO helped to create more nucleation opportunities for ZIF-8 and improved the ZIF-8 growth comparing to synthesis without using ZnO. We were able to obtain very thin (900 nm) ZIF-8 layer with significant separation performance. To the best of our knowledge, this is the thinnest ZIF-8 layer synthesized on polymeric substrate by secondary growth method so far.

Another promising method to facilitate the growth of MOF's on polymeric supports is the chemical functionalization of the membrane surface and pore walls with functional groups, which can complex metal ions and which can covalently bond the MOF crystals. We functionalized the surface of porous polymeric membranes with tetrazole and imidazole groups, which took part in the reaction to form ZIF-8 nanocrystals. Very thin defect-free ZIF layers could be obtained by the standard secondary growth method and by a counter diffusion method. Obtained membranes were tested for permeation of hydrogen and hydrocarbons. Promising selectivities higher 50 could be obtained. The ZIF-8 layers were analyzed by scanning electron microscope and X-ray diffraction.

References:

1. P. Neelakanda, E. Barankova, K.V. Peinemann
Polymer supported ZIF-8 membranes by conversion of sputtered zinc oxide layers
Microporous and Mesoporous Materials 220 (2016) 215-219
2. E. Barankova, N. Pradeep, K.V. Peinemann
Zeolite-imidazolate framework (ZIF-8) membrane synthesis on a mixed-matrix substrate
Chemical Communications 49 (2013) 9419-9421

Mixed-matrix membranes containing engineered nanoporous materials for biogas purification

Tae-Hyun Bae[a,b]*, Heqing Gong[a], Tien Hoa Nguyen[a], and Siew Siang Lee[b]

[a]*School of Chemical and Biomedical Engineering, Nanyang Technological University, Singapore*

[b]*Singapore Membrane Technology Centre, Nanyang Technological University, Singapore*

E-mail: thbae@ntu.edu.sg

Biogas produced in anaerobic digestion process has been considered as a renewable energy source. For the recovery and the purification of biogas, various processes for CO_2 separation including amine-scrubbing and adsorption have been employed. Although membrane-based gas separation is an ideal platform for the biogas recovery/purification, the applications have been limited by the unsatisfactory performance of commercial polymeric membranes under the biogas separation conditions such as low CO_2 partial pressure. Thus, improving the performance of current polymeric membranes, which are inexpensive and mechanically stable, may facilitate the application of membrane technology in large-scale biogas upgrading processes. To this end, various porous materials that can selectively transport CO_2 can be incorporated into polymer matrices to design CO_2-selective mixed-matrix membranes.

Among various porous materials, metal-organic frameworks which comprise metal center and organic linkers to form three-dimensional porous structure have attracted vast research interests in the recent years as potential fillers for mixed-matrix membrane fabrication owing to their large pore volumes and tunable functionalities. In this work, a submicron-sized $Zn(pyrz)_2(SiF)_6$ metal-organic framework crystal possessing both an optimum pore size for selective CO_2 permeation and a strong affinity for CO_2 molecule was synthesized and incorporated into various polymer membranes to design mixed-matrix membranes for biogas upgrading. CO_2/CH_4 mixture gas permeation tests revealed that the separation properties of mixed-matrix membranes, especially selectivities, were significantly improved compared to those of pure polymeric membrane owing to the selective CO_2 uptake and transport in the metal-organic framework crystals.

Recently, our research group have developed hierarchical mesoporous Ca-A zeolites with amine-appended mesoporous domain for application in post-combustion CO_2 capture. Owing to the contributions of both the active sites in Ca-A zeolites and the amine groups in mesoporous surfaces, the material exhibited an excellent CO_2 uptake property even at low CO_2 partial pressure. Moving forward, amine-appended hierarchical Ca-A was employed as a filler for mixed-matrix membranes to improve CO_2 selectivity. Gas permeation testing revealed that the amine-appended hierarchical Ca-A is highly efficient in enhancing CO_2/CH_4 selectivity of polymeric membranes.

References

1. H. Gong, T. H. Nguyen, R. Wang, T. H. Bae, Separations of binary mixtures of CO_2/CH_4 and CO_2/N_2 with mixed matrix membranes containing $Zn(pyrz)_2(SiF)_6$ metal-organic framework, *J. Membr. Sci.* **2015**, *495,* 169-175

2. T. H. Nguyen, S. Kim, M. Yoon, T. H. Bae, Hierarchical LTA zeolites with amine-functionalized mesoporous domain for carbon dioxide capture, *ChemSusChem,* **2016**, *DOI:10.1002/cssc.201600004*

3. T. H. Nguyen, H. Gong, S. S. Lee, T. H. Bae, Amine-appended hierarchical Ca-A zeolite for enhancing CO_2/CH_4 selectivity of mixed-matrix membranes, *submitted*

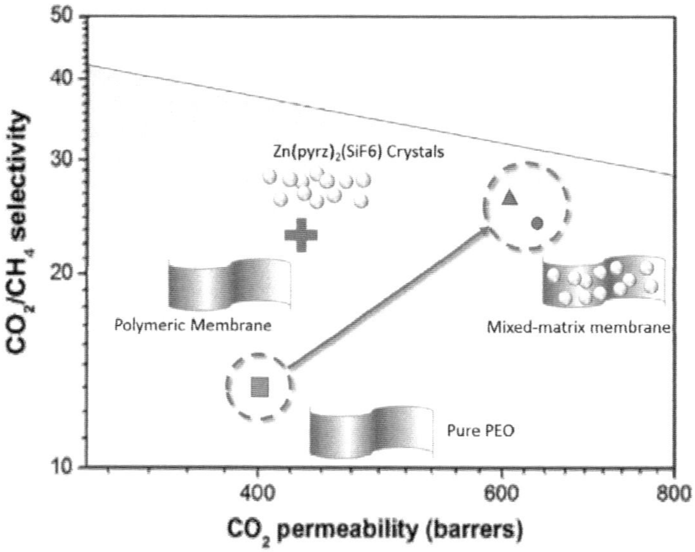

Figure 1 Enhanced CO_2/CH_4 separation property of mixed-matrix membranes containing submicron-sized $Zn(pyrz)_2(SiF_6)$ crystals.

Metal-rich polymer membranes by CIPS: from metal ions to crystals.

Luis Francisco Villalobos,[†] Suzana Nunes,[§] and Klaus-Viktor Peinemann[†]

[†]*Advanced Membrane and Porous Materials Center, KAUST, Thuwal, Saudi Arabia*
[§]*Biological and Environmental Science and Engineering Division, KAUST, Thuwal, Saudi Arabia*

The majority of state-of-the-art polymeric membranes for industrial or medical applications are fabricated by phase inversion. Complexation induced phase separation (CIPS)—a surprising variation of this well-known process—allows direct fabrication of hybrid membranes in existing facilities.[1] In the CIPS process, a first step forms the thin metal-rich selective layer of the membrane, and a succeeding step the porous support. Precipitation of the selective layer takes place in the same solvent used to dissolve the polymer and is induced by a small concentration of metal ions. These ions form metal-coordination-based crosslinks leading to the formation of a solid skin floating on top of the liquid polymer film. A subsequent precipitation in a non-solvent bath leads to the formation of the porous support structure. The thickness and morphology of the dense layer as well as the porosity of the support can be controlled over a wide range by manipulating simple process parameters.

The nature of the CIPS process facilitates a precise loading of a high concentration of metal ions that are located in only the top layer of the membrane. Moreover, these metal ions can be converted—during the membrane fabrication process—to nanoparticles or crystals. During the lecture the CIPS process will be discussed, and examples of membranes containing metal ions, metal nanoparticles and metal salt crystals presented. This simple method opens up fascinating possibilities for the fabrication of metal-rich polymeric membranes with a new set of properties. Two promising applications of such membranes will be presented: (i) catalytic membranes containing palladium nanoparticles (PdNPs),[2] and (ii) antibiofouling tight-UF membranes containing silver chloride (AgCl) crystals.

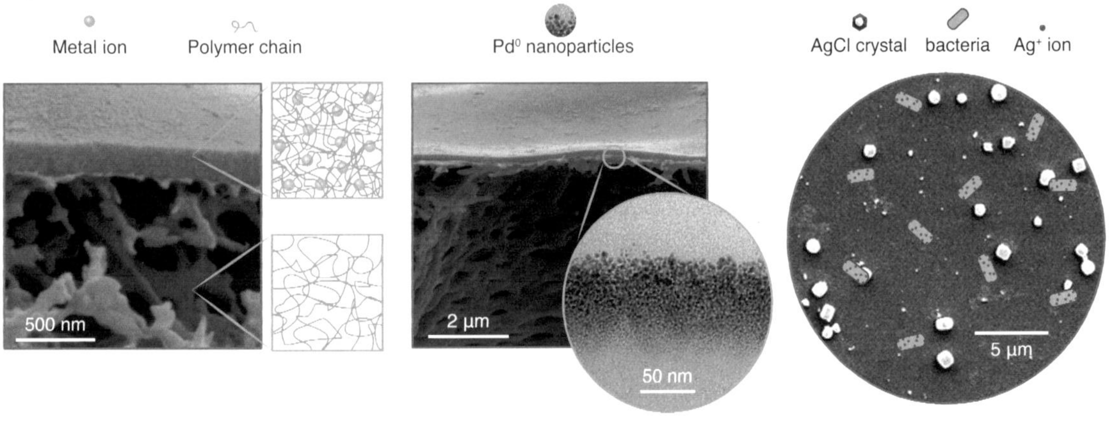

(1) Villalobos, L. F.; Karunakaran, M.; Peinemann, K.-V. Complexation-Induced Phase Separation: Preparation of Composite Membranes with a Nanometer-Thin Dense Skin Loaded with Metal Ions. *Nano Lett.* **2015,** *15*, (5), 3166-3171.
(2) Villalobos, L. F.; Xie, Y.; Nunes S.; Peinemann, K.-V. Polymer and Membrane Design for Low Temperature Catalytic Reactions. *Macromol. Rapid Comm.* doi: 10.1002/marc.201500735

The enhanced performance of iron-oxide and silica-coated-iron-oxide nanocomposite membranes during the filtration of oily emulsion

I. A. Ike[a*], J. D. Orbell[a], and M. Duke[a]

[a]*Institute for Sustainability and Innovation, College of Engineering and Science, Victoria University, Melbourne, Victoria 8001, Australia.*

* Corresponding author's e-mail address: ikechukwu.ike@live.vu.edu.au, phone number: (+61)-45-0294-694.

Abstract

The filtration of oily emulsion for water recovery is challenging due to the low surface energy of oil that often result in rapid fouling of common polymeric membranes such as poly (vinylidene fluoride) (PVDF). A number of reports have shown that the addition of inorganic fillers can improve the membrane fouling resistance leading to higher permeate flux. This work reports the performance of PVDF composite membranes in the filtration of stable oil emulsion with iron-oxide (IO) and silica-coated-iron-oxide (SIO) nanoparticles as fillers over a large filler concentration of 1 – 150 wt.% with respect to the mass of PVDF. For all composite membranes tested, TOC rejection was approx. 92% (except for 150 wt.% IO with a slightly higher value of 93%). The composite membranes rejections were significantly greater than the rejection of the pure PVDF membrane which was only 58%. With regards to membrane permeate flux, the 1 wt.% IO membrane showed the best performance improvement with approx. 530% increase in flux over the pure PVDF and was followed by the 1 wt.% SIO membrane with approx. 440% flux increase. Higher filler concentrations resulted in the steady decline in permeate flux for the IO membranes in which the 150 wt.% IO membrane recorded the lowest flux improvement (approx. 340%) for all composite membranes studied. However, for the SIO membranes, the minimum flux improvement was recorded for the 50 wt.% membrane (approx. 380% flux increase) while at 150 wt.%, the improvement in flux was approx. 420% which was close to the improvement recorded for 1 wt.% (SIO). Based on the results of a mild cleaning cycle with 0.1 M NaOH adopted in this work to test the reversibility of the fouling of the composite membranes, the 1 wt.% IO membrane showed the greatest reduction in flux (61%) due to irreversible fouling while the 150 wt.% SIO registered the least flux reduction (46%). These results suggest that dissimilar filler materials behave differently at diverse concentrations and this can be an important parameter for the design of composite membranes for challenging applications.

Sub–10 nm polyamide nanofilms with ultrafast solvent transport for molecular separation

Santanu Karan*, Zhiwei Jiang, Andrew G. Livingston*
E-mail: skaran@imperial.ac.uk; a.livingston@imperial.ac.uk

Department of Chemical Engineering, Imperial College London, South Kensington Campus, London SW7 2AZ, UK.

Membranes with unprecedented solvent permeance and high retention of dissolved solutes are needed to reduce the energy consumed by separations in organic liquids. We used controlled interfacial polymerization to form free-standing polyamide nanofilms less than 10 nanometers in thickness, and incorporated them as separating layers in composite membranes. Manipulation of nanofilm morphology by control of interfacial reaction conditions enabled the creation of smooth or crumpled textures; the nanofilms were sufficiently rigid that the crumpled textures could withstand pressurized filtration, resulting in increased permeable area. Composite membranes comprising crumpled nanofilms on alumina supports provided high retention of solutes, with acetonitrile permeances up to 112 liters per square meter per hour per bar. This is more than two orders of magnitude higher than permeances of commercially available membranes with equivalent solute retention.

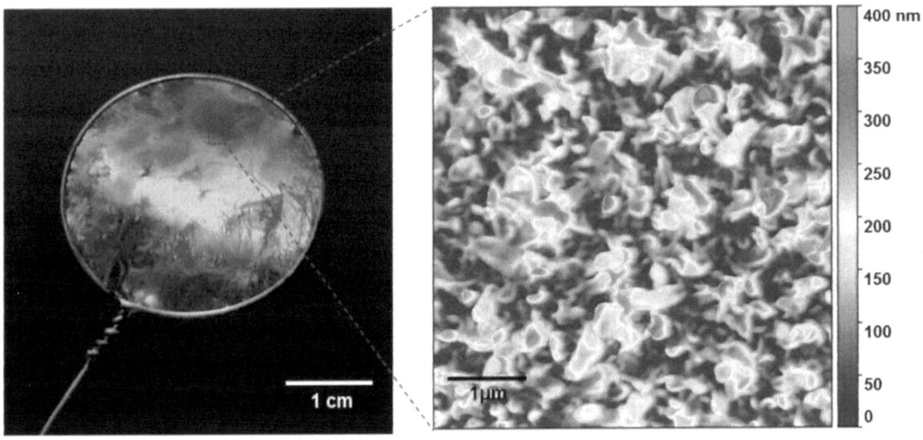

Figure 1. Freestanding sub-10 nm polyamide nanofilm on a lasso and the AFM image showing crumpled features.

References

1. Santanu Karan, Sadaki Samitsu, Xinsheng Peng, Keiji Kurashima and Izumi Ichinose, ***Science***, 335, 444 – 447, 2012.
2. Patricia Gorgojo, Santanu Karan, H. C. Wong, M. F. Jimenez-Solomon, J. T. Cabral and Andrew G. Livingston, ***Advanced Functional Materials***, 24, 4729 – 4737, 2014.
3. Santanu Karan, Zhiwei Jiang and Andrew G. Livingston, ***Science***, 348, 1347 – 1351, 2015.

Iron Nanoparticle Functionalized Membrane for PCB Degradation from Water

Hongyi Wan (Speaker)[a], *Minghui Gui*[a], *Lindell Ormsbee*[b], *and D. Bhattacharyya*[a]
[a] *Department of Chemical and Materials Engineering, University of Kentucky, Lexington, KY 40506*
[b] *Kentucky Water Resources Research Institute, University of Kentucky*

Abstract

The development of functionalized membranes with catalytic metal nanoparticles provides an effective method of environmental remediation and wastewater treatment. Polyvinylidene fluoride (PVDF) is an excellent polymer for both chemical and mechanic resistance as well as thermal stability. By immobilizing zero-valent iron (ZVI) or bimetallic nanoparticles (NPs), functionalized PVDF membranes can be used in chlorinated organic treatment. Polyacrylic acid (PAA) was polymerized inside PVDF membrane pores to maximize iron adsorption. Adsorbed iron was reduced to form NPs in-situ in order to prevent metal ion loss, NPs aggregation and iron precipitation. Furthermore, tunable membrane pore size could be achieved by changing the environmental pH because of ionization of PAA.

This study includes three major aspects: 1. Reactivity toward model toxic chlorinated organic compounds--polychlorinated biphenyl (PCBs) in both batch and convective flow study; 2. NPs characterization of its average sizes, composition, extent of oxidation and distribution inside membrane pores; 3. Simulating pH responsive behavior and PCB reaction within the membrane.

PCBs are not biodegradable and tend to cycle between air, water and soil, which might accumulation in living organisms. PCBs dechlorination can be achieved by Fe/Pd reductive pathway. The hydrogen, generated from reaction between ZVI and water, gets activated by Pd for hydrodechlorination. In batch study, 95% of PCB 126 ([Co]=15 μM) was consumed and 67% was converted to biphenyl in 5 hours. In convective flow study, 96% of PCB 126 was consumed at 26s residence time. However, the final product biphenyl still showed toxicity. In order to degrade biphenyl, the further oxidative pathway was combined to break down the aromatic ring and eventually form organic acid. Both iron oxide with H_2O_2 method (Fenton Reaction) and persulfate method significantly reduced toxicity of the products of PCB degradation.

Fe/Pd NPs size distribution was analyzed by SEM and TEM (TEM samples were prepared by Focus Ion Beam) and the average size of NPs was 19.4±3.2nm. Iron core and discontinuous palladium shell was tested by STEM-EELS line scan and elemental mapping and extent of oxidation was studied by XRD. Furthermore, membrane permeability change from 32.2LMH/bar to 1.1LMH/bar when environmental pH change from 2 to 6.5. The pH-responsive behavior was simulated and fitted with experimental data, which is a part of PCBs reactivity model.

This research is supported by the NIEHS-SRP grant P42ES007380, and by the NSF KY EPSCOR program. Full scale PVDF membranes were provided by Nanostone/Sepro (USA).

Novel supra-molecular organic framework (SOF) based mixed matrix membranes (MMMs) for CO_2/CH_4 separation.

Zoban V. Singh[1], Matthew G. Cowan[1,2], LiLi Tan[2], Douglas L. Gin[1,2], Richard Noble[1] and Wei Zhang[2]

(1) Department of Chemical and Biological Engineering, University of Colorado, Boulder, CO,
(2) Department of Chemistry, University of Colorado, Boulder, CO

Abstract: Supra-molecular organic frameworks (SOFs) represent a new class of microporous solids that display excellent CO_2-light-gas selectivity. Analogous to inorganic zeolites, SOFs conduct high-selectivity CO_2-light gas separation via a similar sorption based mechanism. However, forming large-scale membranes from pure microporous solids, such as zeolites, is challenging due to defect-spaces arising between the tight packing of zeolite crystals. We have previously presented a series of mixed-matrix membranes containing zeolites, which combine the excellent CO_2/CH_4 separation properties of the zeolite with the processability of a polymer. In contrast to zeolites, which are insoluble, SOFs may be readily dissolved into many conventional solvents, making solvent-based casting fabrication techniques for MMM formulation a convenience. In this talk, we present a new Matrimid-based MMM platform containing a SOF. Upon complete crystallization and activation of the SOF at elevated temperatures, CO_2 permeability and CO_2/CH_4 selectivity values skyrocket. Due to thermal cross-links that form in Matrimid at higher temperatures—a phenomenon pervasively reported in the literature—and due to the formation of larger crystalline surface areas in the SOF at elevated temperatures, sorption and diffusivity based permselectivity synergize. For these reasons, the separation performance of Matrimid-based MMMs created with SOFs display CO_2/CH_4 selectivity of > 150. We believe, these are amongst the most selective MMMs ever created for CO_2/CH_4 separations. In addition to the excellent separation properties of the SOF, the ability to dissolve the dispersed phase into a solvent that can co-dissolve the polymer (i.e., Matrimid), presents a significant advantage towards ease of processability and scale-up.

Fouling Control Using Temperature Responsive Membranes Activated using Superparamagnetic Iron Oxide Nanoparticles

Abstract

Membrane fouling occurs when there is reversible or irreversible accumulation of macrosolutes present in the water on the membrane surface. Membrane cleaning and eventual replacement due to fouling can add to the operating costs of membrane systems. Reversible fouling can be minimized by crossflow operation and/or backflushing. On the other hand, irreversible fouling cannot be minimized during operation, often requires chemical cleaning, and may result in permanent flux decline. Among irreversible foulants, natural organic matter (NOM) is considered to be a major contributor. NOM is composed of a wide range of hydrophilic and hydrophobic components; hence, any stagnant hydrophobic or hydrophilic membrane has the potential to become fouled. Therefore, a dynamic membrane able to alternate between being more or less hydrophilic would be expected to decrease fouling. The purpose of this study was to cast stimuli responsive membranes to control fouling made of cellulose acetate (CA) and N-isopropylacrylamide (NIPAAm). NIPAAm is a stimuli-responsive polymer, which offers the potential to reversibly collapse or expand the membrane as a function of changes in temperature. However, temperature activation of the membranes can be energy extensive since it requires heating entire housing and feed streams. Furthermore, heat transfer resistances within the housing can hinder the temperature response time of the membranes. To address these, heating was localized within the membrane matrix by embedding superparamagnetic iron oxide (SPIO) nanoparticles within the temperature-responsive NIPAAm polymer film. Nanoparticles were chemically attached to polyNIPAAm, and the resultant product was added to the dope solution. Membranes were fabricated and tested for their response to the RF heating. An alternating current (AC) electromagnetic field was used to activate the temperature responsive membrane via electromagnetic heating caused by nanoparticles. Membranes with nanoparticles were studied with and without RF heating, and the results suggested that during RF heating, CA-NIPAAm membranes with nanoparticles became less hydrophilic as compared to without RF heating. CA-NIPAAm membranes with and without SPIO nanoparticles were subjected to RF heating to investigate the effect of the presence of nanoparticles on water infiltration. Higher temperatures were recorded near the membranes with nanoparticles as compared to the membranes without nanoparticles suggesting that an oscillating magnetic field has the potential to be used for temperature activation.

Tuning the Carboxylic Groups of the Thin-Film Composite Membranes and Their Effects on the Performance

Shuren Chou

R&D Center, Bejing Origin Water, Huairou, Beijing, China, 101407

With the rapid expansion of the population and elevation of human demands in China, we are facing the pressure and challenges from the water related problems including water environment, pollution freshwater scarcity and drinking water safety. The membrane based technologies, membrane bioreactor (MBR), nanofiltration (NF) and reverse osmosis (RO), have been developed rapidly and polymeric membranes are mostly employed to solve these challenging issues. The semi-permeable functional layer of most of NF and RO membranes is synthesized on the porous support layer by interfacially polymerizing the monomers on its surface, and the residual functional groups acquire surface charge when contacting with an aqueous solution (Tiraferri & Elimelech, 2012). The residual functional groups are closely related to the physiochemical properties of the selective layer (Chuyang Y Tang, Kwon, & Leckie, 2009a, 2009b), including the degree of polymer crosslinking, the membrane hydrophobicity (Jin & Su, 2009), the surface charge density (Childress & Elimelech, 2000), the membrane performance (Coronell, Mi, Mariñas, & Cahill, 2012), and the membrane fouling propensity (Chuyang Y. Tang, Kwon, & Leckie, 2007).

In this work, we carefully control the reaction during the interfacial polymerization, and the consequent residual carboxylic groups due to the incomplete polymer condensation vary in terms of the amount per unit area. The effects of the surface charge of the NF and RO membranes based on piperazine (PIP) and m-phenylene diamine (MPD), respectively, have been systematically investigated. For the fully-aromatic membranes, it is interesting to found that the salt rejection varies little (93%~95%) while increasing the surface charge density of the membranes when we measure it with sodium chloride as the feed, but the rejection is decreasing (84%~93%) while the surface charge density increases if we measure it with calcium chloride as the feed. However, a reverse order of the salt rejection is observed for the semi-aromatic membranes synthesized by PIP monomer. Several analytical approaches, e.g., zeta potential measurement, surface charge qualification and x-ray photoelectron spectra have been employed to investigate the mechanism causing the difference between these two membrane recipes.

References

Childress, A. E., & Elimelech, M. (2000). Relating nanofiltration membrane performance to membrane charge (electrokinetic) characteristics. *Environmental Science & Technology, 34*(17), 3710-3716.

Coronell, O., Mi, B., Mariñas, B. J., & Cahill, D. G. (2012). Modeling the effect of charge density in the active layers of reverse osmosis and nanofiltration membranes on the rejection of arsenic (III) and potassium iodide. *Environmental Science & Technology, 47*(1), 420-428.

Jin, Y., & Su, Z. (2009). Effects of polymerization conditions on hydrophilic groups in aromatic polyamide thin films. *Journal of Membrane Science, 330*(1), 175-179.

Tang, C. Y., Kwon, Y.-N., & Leckie, J. O. (2007). Fouling of reverse osmosis and nanofiltration membranes by humic acid—Effects of solution composition and hydrodynamic conditions. *Journal of Membrane Science, 290*(1-2), 86-94. doi: 10.1016/j.memsci.2006.12.017

Tang, C. Y., Kwon, Y.-N., & Leckie, J. O. (2009a). Effect of membrane chemistry and coating layer on physiochemical properties of thin film composite polyamide RO and NF membranes: I. FTIR and XPS characterization of polyamide and coating layer chemistry. *Desalination, 242*(1), 149-167.

Tang, C. Y., Kwon, Y.-N., & Leckie, J. O. (2009b). Effect of membrane chemistry and coating layer on physiochemical properties of thin film composite polyamide RO and NF membranes: II. Membrane physiochemical properties and their dependence on polyamide and coating layers. *Desalination, 242*(1), 168-182.

Tiraferri, A., & Elimelech, M. (2012). Direct quantification of negatively charged functional groups on membrane surfaces. *Journal of Membrane Science, 389*, 499-508.

Extended field trials of sweep modules for carbon capture with flue gas from a power plant

Lloyd S. White[a,*], Karl D. Amo[a], Tony Wu[b], Timothy C. Merkel[a]

[a]*Membrane Technology & Research, Inc., 39630 Eureka Drive, Newark, CA 94560, USA*
[b]*National Carbon Capture Center, Southern Company, P.O. Box 1069, Wilsonville, AL 35186, USA*

Abstract

Previous trials in this program during 2012-2014 have shown that carbon capture and sequestration technology can be aided by application of industrial-scale membrane systems. A pilot plant was designed and built to continually test full-scale spiral-wound modules at a feed rate of one ton-per day carbon dioxide taken from a flue gas feed stream generated by a power plant fueled with pulverized coal. Field trials at the National Carbon Capture Center in Wilsonville, Alabama, USA continued during 2015 in a test campaign to show stable operations of the newest designs of Step One high flux cross-flow modules and Step Two high selectivity sweep modules with flue gas. These are commercial-scale spiral-wound modules specifically designed for Polaris™ membranes.

Variation of sweep rates in Step Two show increased pressure drop at higher sweep rates and increased CO_2 removal rates, but these results need to be reconciled against lower CO_2 levels returned to the boiler as the Sweep Out stream. The sweep modules in Step Two showed carbon capture rates at a substantial fraction of comparable cross-flow modules under full vacuum, and also operate without the requirement of an applied vacuum. The use of higher selectivity membrane in Step Two improved both the CO_2 and O_2 concentrations of the Sweep Out stream.

The test program concluded after six months of stable performance for both sets of modules, which included significant shutdown periods, but also successful restarts of the on-line pilot plant.

Analysis of Flux Reduction Behaviors of PRO Hollow Fiber Membranes: Experiments, Mechanisms, and Implications

Jun Ying Xiong [a], Zhen Lei Cheng [a,1], Chun Feng Wan [a,1],

Si Cong Chen [a], Tai-Shung Chung [a,b,*]

[a] Department of Chemical and Biomolecular Engineering, National University of Singapore, 4 Engineering Drive 4, Singapore 117585, Singapore

[b] Water Desalination & Reuse (WDR) Center, King Abdullah University of Science and Technology, Thuwal 23955-6900, Saudi Arabia

* Corresponding author: Tel.: +65 6516 6645, Email: chencts@nus.edu.sg

[1] These authors contributed equally to this study.

Abstract

Pressure retarded osmosis (PRO) is a promising technology to harvest renewable osmotic energy using a semipermeable membrane. However, a significant flux reduction has been always observed that severely shrinks the harvestable power to a level only marginally higher or even lower than the economically feasible value. This work focuses on the elucidation of various underlying mechanisms responsible for the flux reduction. First, both inner-selective and outer-selective thin film composite (TFC) hollow fiber membranes are employed to examine how the fundamental *internal* factors (such as the surface salinity of the selective layer at the feed side ($C_{F,m}$) and its components) interact with one another under the fixed bulk salinity gradient, resulting in various behaviours of *external* performance indexes such as water flux, reverse salt flux, and power density. Then, the research is extended to investigate the effects of the growing bulk feed salinity due to the accumulated reverse salt flux along PRO modules. Finally, the insights obtained from the prior two stationary conditions are combined with the advanced nucleation theory to elucidate the dynamic scaling process by visualizing how the multiple fundamental factors (such as local supersaturation, nucleation rate and nuclei size) interplay with one another in various membrane regimes and how they evolve during the whole scaling process. To our best knowledge, it is the first time that the advanced nucleation theory is applied to study the PRO scaling kinetics in order to provide subtle and clear pictures of the events occurring inside the membrane. This study may provide useful insights to design more suitable TFC hollow fiber membranes and to operate them with enhanced water flux so that the PRO process may become more promising in the near future.

Keywords:

Pressure retarded osmosis (PRO); Flux reduction behaviour; Outer-selective hollow fiber membrane; Inner-selective hollow fiber membrane, Gypsum scaling

Highly Proton-conductive Polyelectrolyte Membranes with Greatly Reduced Water Swelling

Joseph Aboki, Shuangjiang Luo, Ruilan Guo*

*Department of Chemical and Biomolecular Engineering, University of Notre Dame, Notre Dame, IN 46556-5637, * rguo@nd.edu*

Sulfonated poly(arylene ether sulfone) multiblock copolymers with alternating hydrophilic and hydrophobic blocks have been shown to hold great potential as alternatives to benchmark Nafion® for polyelectrolyte membrane fuel cells (PEMFCs), due to their much better proton conductivity at low relative humidity (RH) levels and thermal stability. Previous research have also shown that long hydrophilic sequences (>10,000 g/mol) are typically needed to effect nanophase separation and form continuous proton conducting nanochannels enabling high proton conductivity at low RH, which, however, invariably comes at the expense of excessive water uptake and swelling resulting in deterioration of the dimensional stability and mechanical robustness of PEM membranes.

To address this problem, we adopted an innovative molecular strategy to suppress excessive water swelling by introducing triptycene structural units into the hydrophobic sequences of multiblock copolymers, which induce strong supramolecular interactions with the hydrophilic sequences via the so-called chain threading and interlocking mechanism and effectively suppress water swelling of the PEM membranes. Specifically, a series of triptycene-containing poly(arylene ether sulfone) multiblock copolymers were synthesized via a coupling reaction between phenoxide-terminated biphenol-based hydrophilic oligomers (BPS100) and fluorine terminated triptycene-containing hydrophobic oligomers (TRP0). The molecular weight (ranging from 5,000 to 20,000 g/mol) and end group functionality of oligomers were precisely controlled by adjusting the stoichiometric feeding ratio of monomers. Additionally, the ion exchange capacity (IEC) values of the multiblock copolymers were controlled by coupling oligomers with even and uneven block lengths. The resulting multiblock copolymers yielded tough and ductile membranes via solution casting from DMAc. Fundamental membrane properties were characterized including water uptake, swelling ratio, proton conductivity and thermal stabilities. In general, the multiblock copolymer membranes showed increasing water uptakes and proton conductivities with increasing block length. Anisotropic swelling behavior was observed for all the copolymer membranes suggesting nanophase separated lamellae-like membrane morphology. Most importantly, for copolymers with long hydrophilic sequences (> 10,000 g/mol), despite of their expectedly high water uptake, they all showed greatly reduced swelling ratios as compared to Nafion and non-triptycene-containing multiblock copolymers. This observation indicates that the supramolecular chain threading and interlocking interactions induced by triptycene units in the hydrophobic sequences are instrumental in suppressing membrane swelling and improving dimensional stability. For example, the multiblock copolymer with hydrophilic block length of 15,000 g/mol had a water uptake of 107%, an excellent proton conductivity of 0.155 S/cm in R.T. liquid water (almost 3 times that of a Nafion® 212) and a volume swelling ratio of just 27% (more than 65% reduction when compared to Nafion® 212 and similar multiblock copolymer systems). In addition, a super high liquid water proton conductivity (R.T.) of 0.224 S/cm was achieved by a multiblock copolymer with uneven block length (10,000 g/mol of BPSH100 and 7,000 g/mol of TRP0). These multiblock copolymer membranes hold tremendous potential for application in proton exchange membrane fuel cells.

Polymer blend electrolyte membrane containing hydroxyl group for long term durable proton exchange membrane

Bo Hyun Kim [a,b], *So Young Lee* [a], *Jong Hyun Jang* [a], *Dirk Henkensmeier* [a], *Sung Jong Yoo* [a], *Ki Wan Bong* [b], *Hyoung-Juhn Kim* [a]*

[a] Fuel Cell Research Center, Korea Institute of Science and Technology (KIST), 39-1 Hawolgok-dong, Seongbuk-gu, Seoul 136-791, Republic of Korea

[b] Department of Chemical and Biological Engineering, Korea University, 145 Anam-ro, Seongbuk-gu, Seoul, 02841, Republic of Korea

Email: hjkim25@kist.re.kr

Polymer electrolyte membrane fuel cells (PEMFCs) continue to attract a great deal of attention owing to their strength as eco-friendly clean energy source for transportation and stationary application. However, for commercialization of these fuel cells polymer electrolyte membrane which is one of the key factors limiting the performance of PEMFCs require high proton conductivity depends on the relative humidity, and controls the cost a and durability. During last two decades, there have been numerous investigations for hydrocarbon-based polymer electrolyte membrane to replace Nafion type perfluorosulfonated polymers for proton exchange membrane fuel cells. Although some sulfonated aromatic PEMs could attained high proton conductivity and good chemical stability, the overall performance at low relative humidity (RH) and high temperature is still not satisfactory for actual applications.

In previous study, we presented a new strategy to introduce hydroxyl group to sulfonated poly(arylene ether sulfone) system. This hydroxyl group is supposed to have ability for hydrogen bonding to other hydroxyl group, sulfonated group and water. Therefore, it could improve proton conductivity as well as mechanical property. However, this polymer suffer from the complicated preparation process problems and low membrane processability in high degree of sulfonation.

In this study, we investigate novel and simple blend membrane system with sulfonated poly(ether sulfone), poly(ether sulfone) containing hydroxyl group and sulfonated poly(ether sulfone) with hydroxyl group as a compatabilizer. These blend electrolyte membranes show high proton conductivity at low humidity condition as well as dimensional stability. And it indicate unusual morphological behaviors and long-term durability properties compared to those of pristine polymer membrane. These characterization results and more detail properties of blend membranes will be discussed in the presentation.

Title: Application of trimethylamine–carbon dioxide in osmotic heat engine for power generation
Authors: Lingling Xia, Jason T. Arena, Jian Ren, Jeffrey R. McCutcheon

Abstract: Salinity gradient energy were for years touted as a new renewable energy source. Naturally occurring salinity gradients, such as where river water meets seawater, were considered untapped sources energy that could be harnessed with technologies like pressure retarded osmosis (PRO). Unfortunately, operational challenges plague industrialization of PRO. The relatively low osmotic pressure of seawater limits power generation potential. More problematic are the organic and inorganic content of the waters which pose a severe fouling threat to the membranes. Pretreatment to remove these contaminants would be cost prohibitive and could result a substantial loss in energy production [1]. The concept of the osmotic heat engine (OHE), first proposed by Loeb in 1975 [2], addresses these two problems of PRO. The OHE is a “closed loop” PRO process with a thermally regenerable draw solute. By choosing a thermolytic solute with a high solubility, osmotic pressures far in excess of seawater are possible. Furthermore, since the process is close loop (no mass inputs or outputs), the solutions are pristine and contain no foulants. OHEs therefore can produce far more power and with fewer losses than conventional PRO.

The development of the OHE was made possible by the recently developed NH_3-CO_2 thermolytic draw solute. This solute, which in water consists of ammonium salts, can be easily regenerated by conventional steam stripping and was the focus of numerous studies on forward osmosis (FO) [3,4]. The regenerable nature of the solute led to the development of the OHE concept [5]. However, one of the major challenges with the NH_3-CO_2 solute is that ammonia, which is present in the draw solute, can readily cross the membrane because of its “water-like” characteristics. This high flux of the ammonia does not mean that it is lost (the system is close loop and all of the solutes end up recycled in the process, but it does impair the water flux and power density due to lost osmotic pressure and internal concentration polarization.

A newly investigated thermolytic draw solute is trimethylamine–carbon dioxide (TMA–CO_2). Like NH_3-CO_2, TMA–CO_2 has a high solubility in water and can generate high osmotic pressure. It also features comparable volatility to ammonia and low enthalpy of vaporization, indicating comparable or less energy would be consumed for draw solute regeneration. The benefit of TMA–CO_2 lies in its larger molecular size. This makes it less likely to pass through the membrane. In this work, we explore the use of TMA–CO_2 in PRO and compare its performance to other model draw solutes (NaCl) and NH_3-CO_2. Tests were conducted with up to 5 M solutions of TMA–CO_2 using a lab-scale PRO test system with a proprietary Porifera membrane. Estimated power densities of up to 18.6 W m^{-2} were achievable at relatively low pressures (10 bar). Higher hydraulic pressures would yield higher power density if a membrane housing could be made to support a membrane without forming defects. For comparison, NaCl and ammonia-carbon dioxide (NH_3–CO_2) have been evaluated for PRO performance using Porifera membrane. The water flux of TMA-CO_2 was about 20 % lower than that of NaCl and comparable with that of NH_3–CO_2. Reverse solute flux of TMA–CO_2 ranged from 0.5 to 3.5 mol m^{-2} h^{-1} in all experiments. This value was much lower than NH_3-CO_2 (4.0 to 51 mol m^{-2} h^{-1}) and approximately half that of NaCl. Low solute flux was attributed to molecular size differences between solutes. In all, TMA-CO_2 represents an OHE draw solute option with low draw solute flux without substantial sacrifice of power density. This solute, as well as its many derivatives, may therefore represent the “sweet spot” for thermolytic OHE systems.

[1] N.Y. Yip, M. Elimelech. Thermodynamic and energy efficiency analysis of power generation from natural salinity gradients by pressure retarded osmosis. Environ.Sci.Technol., 46 (2012) 5230.

[2] S. Loeb. Method and apparatus for generating power utilizing pressure-retarded-osmosis. United States patent # 3, 906 (1975) 250.

[3] J.R. McCutcheon, R.L. McGinnis, and M. Elimelech. Desalination by ammonia–carbon dioxide forward osmosis: Influence of draw and feed solution concentrations on process performance. J.Membr.Sci., 278 (2006) 114.

[4] R.L. McGinnis, N.T. Hancock, M.S. Nowosielski-Slepowron, and G.D. McGurgan. Pilot demonstration of the NH3/CO2 forward osmosis desalination process on high salinity brines. Desalination, 312 (2013) 67.

[5] R.L. McGinnis, M. Elimelech, and J.R. McCutcheon, Osmotic Heat Engine.

Photocatalytic membrane reactors for high throughput virus removal and inactivation

B. Guo, B. J. Starr, I. Xagoraraki, V.V. Tarabara*,
Department of Civil and Environmental Engineering, Michigan State University, East Lansing, MI 48824
* Corresponding author: tarabara@msu.edu; +1 (517) 432-1755

Abstract

Photocatalytic membrane reactors (PMRs) combine membrane separation and photocatalysis in one hybrid process. PMR is a highly versatile technology due to the range of engineering designs it affords and many possible applications. Photocatalytic treatment can be highly effective with respect to viruses (e.g. [1–4]). Most of the work on the application of PMRs to water treatment, however, has focused on chemical pollutants. To our knowledge, there have been only five reports on the application of photocatalytic membranes to water disinfection [5–9]. All five studies were on *E. coli* control and three of the five [7–9] also used silver, a known bactericide, as a component of photocatalytic membranes.

In the proposed hybrid process the permeate surface a ceramic membrane (d_{pore} = 0.8 μm) is coated with photocatalytic nanoparticles (TiO_2, anatase) and exposed to UV irradiation (Figure 1) to inactivate viruses during microfiltration [10]. The membrane is operated in the in-out flow geometry to eliminate turbidity and improve efficiency of UV disinfection (Table 1). The membrane-supported photocatalyst ensures generation of reactive oxygen species that act on the pollutants that break through the membrane. Two coating methodologies – dip coating and layer-by-layer self-assembly were employed to render the highly porous permeate side of the membrane photocatalytic. The layer-by-layer self-assembly method produced sub-monolayers of photocatalyst and minimized coating-induced reduction of membrane permeability (Figure 2). Methylene Blue, bacteriophage P22, and human adenovirus 40 were used to evaluate the efficiency of the membrane reactor. Batch UV disinfection and crossflow filtration tests showed that the hybrid photocatalytic UV-microfiltration process is considerably more effective in inactivating viruses (e.g. for P22, $LRV_{P22} = 5.0 \pm 0.7$) than the constituent processes - UV disinfection and microfiltration - applied in series (aggregate $LRV_{P22} = 2.0 \pm 0.5$) (Figure 3).

To our knowledge, this is the first application of photocatalytic membrane reactors to virus removal or inactivation. Potential applications of the proposed treatment process include disinfection of turbid, high fouling, and high flow rate streams.

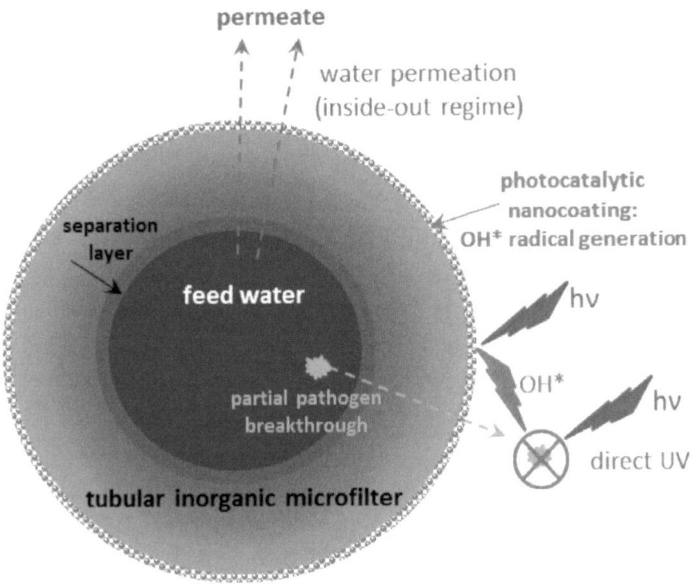

Figure 1: Conceptual illustration of the hybrid membrane filtration-UV disinfection process [10].

Table 1. Synergistic benefits of combing membrane filtration and photocatalytic UV disinfection into a hybrid process.

Technology and its benefits		Challenges	How the challenge is addressed in the proposed PMR
Photocatalytic UV disinfection	Photocatalytic UV is effective against a wide range of microbial pathogens. UV disinfection is catalytically enhanced	Some pathogens are resistant to UV or can repair UV-induced damage	Catalytic oxidation at the membrane surface by ROS complements the physical effect of the direct UV
		Catalyst needs to be recovered	Membrane-supported catalyst is immobilized and does not need to be recovered
		Efficiency is limited when turbidity is present	Turbidity is removed by the microfilter "upstream" from the UV reactor
Membrane filtration	Membranes provide absolute barrier to pathogens	Lower pore size for removal of smaller pathogens results in lower permeate fluxes	Redundancy introduced with catalytic UV disinfection enables trade-offs in pore sizes and product water fluxes

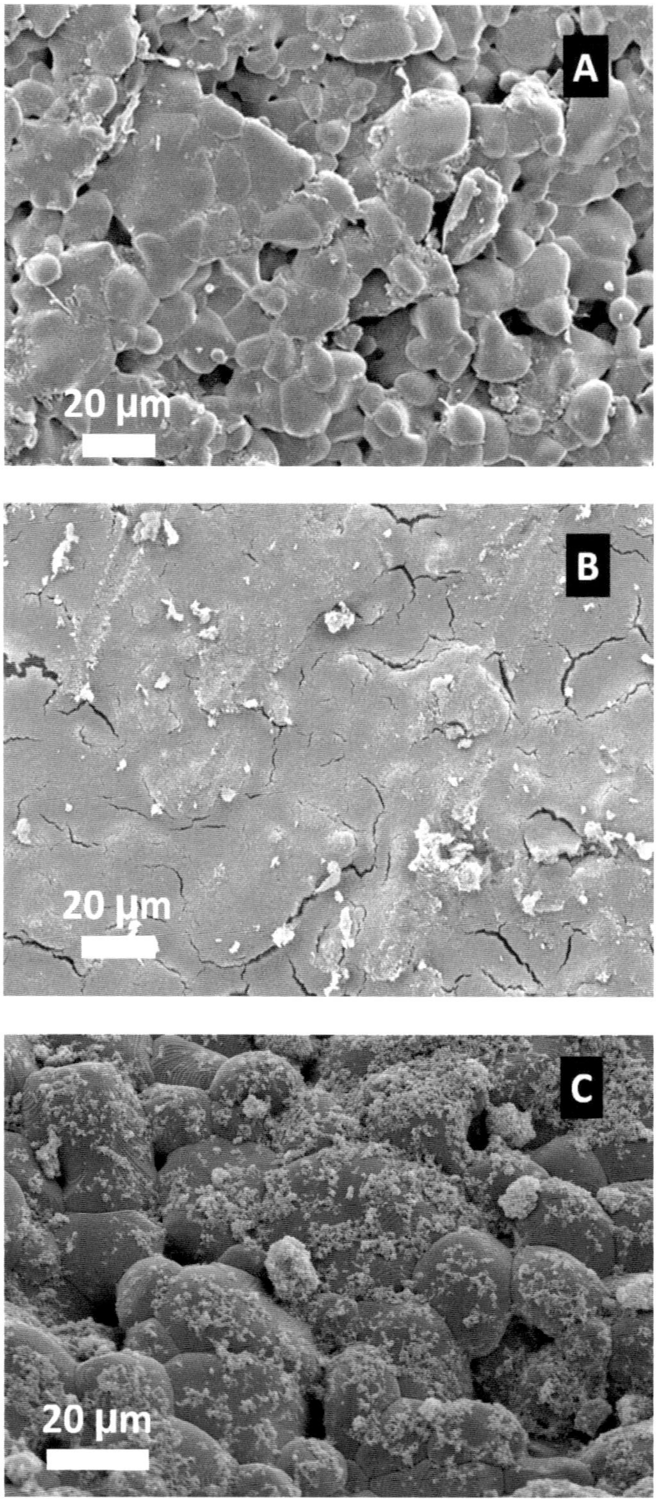

Figure 2: SEM images of outer surface of a ceramic 0.8 m nominal pore size membrane: A) uncoated, B) dip-coated by 10 coats of TiO_2, C) coated by1 layer of TiO_2 using the L-b-L method with PDADMAC and PAA as binder polyelectrolytes [10, 11].

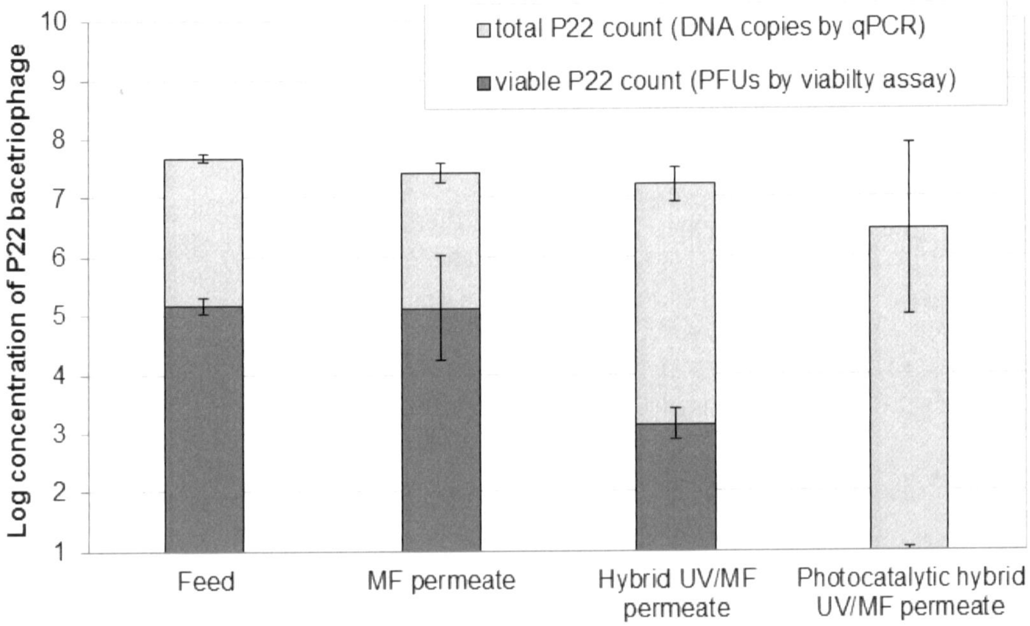

Figure 3. Log concentration of viable and total bacteriophage P22 in the feed solution and in the permeate 30 min into the filtration process. Error bars correspond to standard deviations [10].

References

1. J.C. Sjorgen, R.A. Sierka, Inactivation of phage MS2 by iron-aided titanium dioxide photocatalysis, Appl. Environ. Microbiol. 60 (1994) 344–347.
2. Q. Li, M.A. Page, B.J. Marinas, J.K. Shang, Treatment of coliphage MS2 with palladium-modified nitrogen-doped titanium oxide photocatalyst illuminated by visible light, Environ. Sci. Technol. 42 (2008) 6148–6153.
3. J. Lee, K. Zoh, G. Ko, Inactivation and UV disinfection of murine norovirus with TiO_2 under various environmental conditions, Appl. Environ. Microbiol. 74 (2008) 2111–2117.
4. D. Gerrity, H. Ryu, J. Crittenden, M. Abbaszadegan, Photocatalytic inactivation of viruses using titanium dioxide nanoparticles and low-pressure UV light, J. Environ. Sci. Health A 43 (2008) 1261–1270.
5. H. Choi, M.G. Antoniou, A.A. de la Cruz, E. Stathatos, D.D. Dionysiou, Photocatalytic TiO2 films and membranes for the development of efficient wastewater treatment and reuse systems, Desalination 202 (2007) 199–206.
6. H. Zhang, H. Zhao, P. Liu, S. Zhang, G. Li, Direct growth of hierarchically structured titanate nanotube filtration membrane for removal of waterborne pathogens, J. Membr. Sci. 343 (2009) 212–218.

7. N. Ma, X. Fan, X. Quan, Y. Zhang, Ag–TiO2/HAP/Al2O3 bioceramic composite membrane: fabrication, characterization and bactericidal activity, J. Membr. Sci. 336 (2009) 109–117.
8. L. Liu, Z.Y. Liu, H.W. Bai, D.D. Sun, Concurrent filtration and solar photocatalytic disinfection/degradation using high-performance Ag/TiO2 nanofiber membrane, Water Res. 46 (2012) 1101–1112.
9. R. Goei, T.-T. Lim, Ag-decorated TiO2 photocatalytic membrane with hierarchical architecture: photocatalytic and antibacterial activities, Water Res. 59 (2014) 207–218.
10. Guo, B.; Pasco, E. V.; Xagoraraki, I.; Tarabara, V. V. Virus removal and inactivation in a hybrid microfiltration-UV process with a photocatalytic membrane. Separ. Purif. Technol. 2015, 149, 245–254.
11. Starr, B. J. Photocatalytic membranes for virus inactivation. MS Thesis, Michigan State University, East Lansing, MI. 2015

The Latest Development of Thin-Sheet Zeolite Membranes for Building Applications

Wei Liu

Buildings consume nearly 50% of primary energies in the US, which is largely attributed to space heating and cooling. Two application opportunities of advanced H_2O-selective membranes will be introduced for significant enhancement of building energy efficiency. The first application is membrane enthalpy recovery ventilators (ERV) that precondition fresh air close to building temperature and humidity through heat and mass exchange with the exhaust air. The membrane ERV can be used for ventilated buildings to obtain 20-40% energy savings in year-around operation. The second application is evaporative cooling coupled with membrane dehumidification for air conditioning. A membrane separation device is used to remove water vapor from incoming humid air and condense it into water by rejecting latent heat into environment. A fraction of the harvested water can be re-injected back into the dried air stream and the air is cooled down as the water vaporizes. This is a complete green air cooling technology and can double the energy efficiency of convention air conditioning technologies.

Membranes of high H_2O permeance, high H_2O/air selectivity, long-term durability, and low cost are necessary to realize the above applications. Five-year development of thin (50μm)-sheet zeolite membranes will be discussed from initial product designs, preparation of test coupons, making of a number of membrane sheets of consistent quality, to fabrication of membrane modules. A robust membrane preparation process is developed to obtain almost 100% yields in preparation of a number of 13cmx 13cm membrane sheets. The membrane shows exceptionally high water permeance and H_2O/air selectivity, and its performances are not affected by the size of membranes prepared. The cross-flow module prototypes comprising 50 of 13cm x 13cm membrane sheets will be described. With proper designs, the module shows separation performances even better than the small test coupons. Finally, process economics analysis of the membrane product for industrial-scale production will be presented.

Organo-soluble Polybenzoxazoles (PBOs) as Gas Separation Membranes: New Synthesis and Gas Transport Properties

Ashish Kushwaha (akushwah@nd.edu), Shuangjiang Luo (sluo2@nd.edu), Ruilan Guo* (rguo@nd.edu)

Department of Chemical and Biomolecular Engineering, University of Notre Dame, Notre Dame, IN 46556-5637

Membrane materials featuring ultrafast and highly selective transport properties combined with good chemical, mechanical, and thermal stability in the process environment are highly desired for energy-efficient membrane gas separations. Polybenzoxazole (PBO) and PBO-based polymers possess many of the above mentioned attributes and are considered to be excellent candidates for gas separation membranes. A well-known and successful example of PBO-based polymers as gas separation membranes is TR (thermally rearranged) polymers, which possess unprecedented combination of high permeability and selectivity well beyond the upper bound limit. However, cost and difficulty in preparing PBOs and fabrication of PBO-based membranes due to their poor organo-solubility and processability have been reasons why, for many years, PBO has been known only among researchers and scientists.

In this paper, we report a novel new synthetic approach that produces organo-soluble PBOs via a one-pot, silylation-assisted in-situ solution conversion of ortho-functional polyamide precursors. Using this method, a series of organo-soluble PBOs were prepared from a wide pool of monomers, which demonstrated the feasibility and applicability of this new synthetic method. Molecular dynamics studies suggested that silylation played the key role in assisting the cyclolyzation reaction of polyamide precursors by forming a meta-stable intermediate, which significantly lowered the activation energy of ring closure reaction leading to organo-soluble PBOs. All PBOs synthesized using this novel method yielded tough and ductile films via conventional solution casting, which is not possible for the PBOs prepared using conventional single-stage method with phosphoric acid as solvent or two-stage method of solid state conversion of poly(hydroxyamide) precursors. Fundamental membrane properties were comprehensively characterized in terms of thermal and mechanical stabilities, fractional free volume and gas transport properties via pure gas permeation tests with H_2, N_2, Ar, CO_2, and CH_4 in a wide range of feed pressures (2-17 atm, 35 °C). In general, all PBO membranes showed high fractional free volume, which rendered high gas permeabilities in these membranes. Meanwhile, the highly rigid benzoxazole backbone structure enables excellent size sieving properties for these membranes. For example, the PBO membranes prepared from fluorinated monomers (6FC and 6FDA) showed excellent CO_2/CH_4 and H_2/CH_4 separation performances that were beyond the Robeson's upper bounds; the PBO membranes prepared from triptycene-based monomer showed great potential in hydrogen-related separations due to the unique bimodal size distribution of free volume in these membranes. Taking together with their excellent organo-solubility and processability, these newly synthesized PBOs represent a significant leap towards advancing the practical application of PBO-based membranes for gas separations.

Pilot-scale Development of a PEEK Hollow Fiber Membrane Contactor Process for Post-combustion CO_2 Capture

Shiguang Li,* Travis Pyrzynski, Naomi B. Klinghoffer, Timothy Tamale, James L. Aderhold, and S. James Zhou, Howard S. Meyer
Gas Technology Institute, 1700 S. Mount Prospect Road, Des Plaines, IL 60018, USA
Yong Ding and Ben Bikson
Air Liquide Advanced Separations, 35A Cabot Road, Woburn, MA 01801, USA

Abstract

A novel hollow fiber membrane technology for carbon dioxide capture from flue gases has been developed. The process combines advantageous features of both absorption and membrane processes to cost-effectively separate and capture CO_2 from emission sources. In this process, CO_2-containing flue gas passes through one side of the hollow fiber membrane contactor (HFMC), while a CO_2 selective solvent flows on the other side. CO_2 permeates through the hollow fiber membrane pores and is chemically absorbed into the solvent. The CO_2 rich solvent is regenerated and sent back to the membrane absorber.

The key component of the HFMC process is the super-hydrophobic porous hollow fiber, which is made from poly(ether ether ketone) (PEEK). PEEK is an engineered polymer with thermo-mechanical properties and chemical resistance superior to most commercial plastics including typical polymers used in membrane fabrication. The unique characteristics of PEEK allow the HFMC device to be utilized successfully in challenging and aggressive chemical environments. Compared to conventional absorption/desorption technologies, the critical advantage of the HFMC process is the high contact surface area provided by the hollow fibers that enable an increased volumetric mass-transfer rate. In the PEEK HFMC process, the specific surface area has been increased by an order of magnitude over structurally packed or trayed columns, resulting in compact systems with small footprints. The bench-scale field testing conducted at a coal-fired power plant showed greater than 90% CO_2 removal with 97% CO_2 product purity with activated methyldiethanolamine (aMDEA) solvent. The mass transfer coefficient in the absorption step was 1.2 $(sec)^{-1}$, which is over an order of magnitude greater than that of conventional column contactors.

The objectives of our pilot-scale development program (supported by the US Department of Energy) were to build a 0.5 MW_e equivalent pilot-scale CO_2 capture HFMC system, conduct tests on flue gas at the National Carbon Capture Center (NCCC) and demonstrate a continuous, steady-state operation for a minimum of two months. Under this program, the PEEK HFMC technology was scaled from the bench scale to a pilot scale utilizing larger size HFMC modules. The modules measured 8-inches in diameter by 5 feet long with a membrane surface area of about 40 m^2 per module. These modules have shown exceptionally high CO_2 capture rates due to the high membrane intrinsic CO_2 permeance as well as the structured PEEK hollow fiber module configuration. In our study, CO_2 gas permeance of 2,150 GPU [1 GPU = 10^{-6} $cm^3(STP)/(cm^2 \cdot s \cdot cmHg)$] was attained by optimizing the membrane preparation procedures. By using aMDEA solvent and a simulated flue gas, the HFMC showed greater than 90% CO_2 removal with 98% CO_2 product purity. The mass transfer coefficient was 2.0 $(sec)^{-1}$.

Preliminary design of the 0.5 MW_e equivalent CO_2 capture system was completed. A HAZOP review meeting held at the NCCC facility (Wilsonville, Alabama) indicated that no "show-stopper" or "major" modifications were identified. We are currently performing detailed engineering of the 0.5 MW_e system. It is anticipated that the construction will be completed by September 2016, the system will be installed in NCCC in December 2016 for testing, and the project will be completed by June 2018. After the current project, we plan to complete a 25 MW_e equivalent CO_2 capture system demonstration (500 Tonne/day) by 2021. Currently, we are also developing a transformational hybrid process combining a gas separation membrane configuration unit and PEEK HFMC unit that can further reduce the CO_2 capture cost.

---

* Principal presenter: Shiguang.Li@gastechnology.org; 847-544-3478 (phone)

High ionic liquid content membranes for effective biogas (CO_2/CH_4) separation

Marek Lanč[1], Ondřej Vopička[1], Kryštof Pilnáček[1], Vlastimil Fíla[1], Pavel Izák[2], Zuzana Sedláková[2], Richard D. Noble[3], Matthew G. Cowan[3], Douglas L. Gin[3] and Karel Friess[1],*

[1]*University of Chemistry and Technology, Technická 5, Prague, 166 28, Czech Republic*
[2]*Institute of Chemical Process Fundaments, Czech Academy of Sciences, Rozvojová 135, Prague, 165 02, Czech Republic*
[3]*University of Colorado, Chemical & Biological Engineering Dept., UCB 424, Boulder, CO 80309, United States of America*

* presenting author, *email*: karel.friess@vscht.cz

Among different alternative energy sources to fossil fuels, biogas has probably the greatest potential. Its upgrading can provide a renewable low-cost source of energy (methane). However, raw biogas contains up to 35% of carbon dioxide and circa 5% of other contaminants (mainly water vapor) that must be removed. Therefore, the development of new separation membranes for effective methane purification represents a real challenge also because biogas composition depends on its source (farms, landfills, sewage plants) and a year season[1].

The major issue[2] for developing of membranes with enhanced separation performance is achieving high selectivity towards the target gas purity. The polymer based mixed matrix membranes with embedded ionic liquid (IL), IL-polymer gel membranes or supported ionic liquid membranes (SILMs) with entrapped room temperature IL in can be considered as a promising approach for that[2-5].

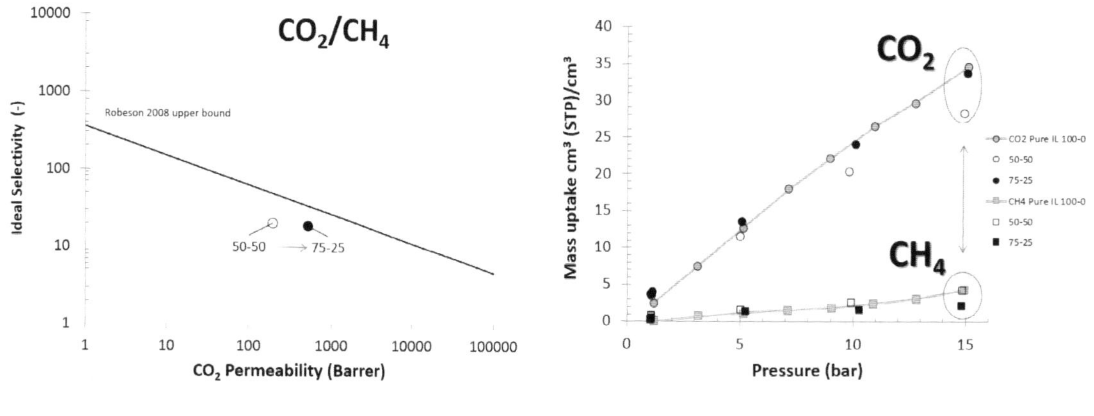

Fig. 1 The comparison of permeation and sorption properties of samples with 50 and 75 wt.% of IL. The CO_2/CH_4 Robeson plot[7] (left) and gas sorption isotherms of samples and pure IL [EmimTf$_2$N] (right).

In this work, the permeation and sorption of CO_2, CH_4 and their mixtures in membranes[5] based on polyethersulfone (PES) support containing 50 and 75 wt.% of highly CO_2 interacting ionic liquid 1-ethyl-3-methylimidazoliumbis (trifluoromethanesulfonyl)imide [EmimTf$_2$N] at 25°C is presented. The sorption experiments were performed gravimetrically[6] in pressure range of 1 to 15 bar, the permeation measurements were performed *via* time-lag fixed volume pressure increase method[4]. Dry or wet mixed gas permeation was monitored by GC. A sample with 75 wt.% ionic liquid content exhibited substantially higher CO_2 permeation and sorption and corresponding permeability and selectivity values were located close to the CO_2/CH_4 upper bound 2008[7].

Acknowledgements
The financial support of the Czech Ministry of Education, Youth and Sports (project KONTAKT II LH-14006 and a specific university research MSMT No 20/2015) are gratefully acknowledged.

References

1. M. Karaszova, K. Friess, M. Sipek, J.C. Jansen, P. Izak P: Biogas Upgrading for the 21st century, in R. Litonjua, I. Cvetkovski (Eds.), *Biogas: Production, Consumption and Applications*, Nova Science Publishers, New York 2012.
2. P. Scovazzo, *J. Membr. Sci.* **343**, 199, 2009
3. J.E. Bara, S.E. Hatakeyama, D.L. Gin, R.D. Noble, *Polym. Adv. Technol.* **19**, 1415, 2008
4. J.C. Jansen, K. Friess, G. Clarizia, J. Schauer, P. Izák, *Macromolecules* **44**, 39, 2011
5. W. M. McDanel, M. G. Cowan, N. O. Chisholm, D. L. Gin, R. D. Noble, *J. Membr. Sci.* **492,** 30, 2015
6. O. Vopička, K. Friess, V. Hynek, P. Sysel, M. Zgažar, M. Šípek, K. Pilnáček, M. Lanč, J.C. Jansen, C.R. Mason, P.M. Budd, *J. Membr. Sci.* **434**, 148, 2013
7. L. M. Robeson, *J. Membr. Sci.* **320**, 390, 2008

Dual CO_2- and N_2- selective intrinsically microporous hydroxyl-functionalized triptycene-based polyimide for natural gas upgrading

Fahd Alghunaimi, Bader Ghanem, Nasser Alaslai, Mohsin Mukaddam, Ingo Pinnau*

King Abdullah University of Science and Technology (KAUST), Advanced Membranes and Porous Materials Center (AMPMC), Division of Physical Sciences and Engineering, Chemical and Biological Engineering Program, Thuwal 23955-6900, Saudi Arabia

*ingo.pinnau@kaust.edu.sa

The synthesis and gas permeation properties of two novel PIM-polyimides (TDA1-DMN) and (TDA1-APAF) prepared from 9,10-dimethyl-2,3,6,7-triptycene tetracarboxylic dianhydride (TDA1) with commercially available 3,3 -dimethylnaphthidine (DMN) and 2,2-bis(3-amino-4-hydroxyphenyl)-hexafluoropropane (APAF) diamine monomers, respectively, are reported. The triptycene building block with bridgehead methyl side groups offers TDA1-DMN with high BET surface area of 760 $m^2\ g^{-1}$ and very high gas permeabilities (e.g. CO_2 = 3700 Barrer) but with only moderate selectivity ($\alpha CO_2/CH_4$ = 17). The *ortho*-hydroxyl-functionalized polyimide TDA1-APAF exhibited lower BET surface area of 260 $m^2\ g^{-1}$ and was designed to enhance selectivity, while maintaining high gas permeability. A freshly prepared sample exhibited excellent gas permeation properties: (i) CO_2 permeability of 40 Barrer coupled with a CO_2/CH_4 selectivity of 55 and (ii) H_2 permeability of 94 Barrer with a H_2/CH_4 selectivity of 129. Physical aging over 250 days resulted in significantly enhanced CO_2/CH_4 and H_2/CH_4 selectivities of 75 and 183, respectively with only ~ 25% loss in CO_2 and H_2 permeability. Aged TDA1-APAF polyimide exhibited 5-fold higher pure-gas CO_2 permeability (30 Barrer) and 2-fold higher CO_2/CH_4 permselectivity over conventional dense cellulose triacetate membranes at 2 bar. In addition, TDA1-APAF polyimide had a N_2/CH_4 selectivity of 2.3, thereby making it possible to simultaneously treat CO_2- and nitrogen-contaminated natural gas. Gas mixture permeation experiments with a 1:1 CO_2/CH_4 feed mixture demonstrated higher mixed- than pure-gas selectivity and plasticization resistance up to 30 bar. These results suggest that intrinsically microporous hydroxyl-functionalized triptycene-based polyimides are promising candidate membrane materials for *simultaneous* removal of CO_2 and N_2 for natural gas sweetening and hydrogen purification in petrochemical refinery applications.

NEW MEMBRANE BASED HYBRID UNIT OPERATION FOR GAS SEPARATION

F.Doghieri[1], F.Miccio[2], M.Minelli[1], G.C.Sarti[1], E.Landi[2], V.Medri[2]

(1) DICAM-University of Bologna – via Terracini 28, Bologna, Italy
(2) ISTEC- National Research Council (CNR) –via Granarolo 64, Faenza (RA), Italy

A lab-scale test has been completed for the implementation of a new membrane based separation process for gaseous streams, described in an application for patent recently submitted by the University of Bologna and CNR [MISE 102015000037715]. The new process can be run on the same modules currently used for conventional membrane based gas separations and it counts on the periodic swing of downstream pressure while upstream conditions are kept constant. The entire process is managed without switching streams and allowing for constant flow rate and composition of both feed and retentate, while permeate is collected from the module at assigned frequency, resulting in an overall on-average steady state process. Time based on-off control of permeate collection ultimately produces alternate sorption/desorption steps in the membrane, associated to increase/decrease cycles of downstream pressure in the process, similarly to what occurs in pressure swing adsorption (PSA) processes. Indeed, consistently with the hybrid nature of the new process, the specific dynamic control of downstream pressure allows to take advantage of both resistive (permeation) and capacitive (sorption) properties of the membrane, as well as associated characteristics for selectivity. By changing the duration of pressure swing period in the new process, from very low to very high values, the resistive or capacitive properties of the membrane can be distinctively stimulated to explore different balances between key component recovery and separation factor in the process. In order to prove the above concept, with reference to the problem of post-combustion carbon capture a two-layer membrane has been prepared in which a dense polydimethylsiloxane (PDMS) coating was applied to the external surface of a tubular porous ceramic membrane according to 1:50 ratio between thicknesses of polymeric and ceramic layer. Polymer and ceramic layers confer desired resistive and capacitive properties, respectively, to the membrane, which is ultimately endowed with relatively high permeance to carbon dioxide, as well as sorption capacity. In fact, characteristic times for sorption and permeation phenomena in the membrane are considerably different and interesting results for effective membrane performances in CO_2/N_2 separation may be obtained through the dynamic control of the process when period Tp fot the imposed pressure swing cycle is increased from negligible to significant fraction of characteristic time Ts for the sorption process in the membrane. Effective permeance Π_E and selectivity α_E measured in the lab-scale test for the organic-inorganic membrane prepared in this work, for the case dilute atmospheric CO_2/N_2 mixture at room temperature was used as feed stream, are reported in table below as function of pressure swing frequency. Experimental results are expressed as ratio between effective performances in dynamic regime and corresponding value for steady state processes (permeance Π_0 and selectivity α_0) and table relates the latter to the ratio between pressure swing period and characteristic time for sorption. The results clearly show the potential of the new process and its additional flexibility with respect to traditional membrane based gas separation, as rather different performances for permeate to selectivity ratio can be established by properly setting the pressure swing period, for the case of assigned loading and fixed characteristics of the membrane module.

Results for effective membrane permeance and separation factor in dynamically controlled gas separation process

Tp/Ts	0.021	0.042	0.083	0.17	0.33	0.5	1.2
Π_E/Π_0	0.96	0.91	0.86	0.68	0.47	0.36	0.15
α_E/α_0	1.1	1.6	2.9	4.5	6.4	6.9	7.3

Steric Hindrance Effect Demonstrated in Polyvinylamine Facilitated Transport Membranes for CO_2 Separation

Zi Tong, The Ohio State University, tong.87@osu.edu
W.S. Winston Ho, The Ohio State University, ho.192@osu.edu

This abstract consists of a brief description of CO_2 Facilitated Transport Membranes, Sterically Hindered Amine, and New Membranes based on Sterically Hindered Polyvinylamines in the following paragraphs for CO_2 separation.

CO_2 Facilitated Transport Membranes: CO_2 capture from flue gas (main component: N_2) is essential for eliminating greenhouse gas emission, and CO_2 separation from synthesis gas is effective for hydrogen purification. Due to the inherent advantages of membrane separation, e.g., high energy efficiency, operational simplicity, small footprint, and the elimination of thermodynamic equilibrium capacity limitations, membrane technology is receiving more and more attention for CO_2 separation. For most conventional polymer membranes, CO_2/N_2 separation follows the solution-diffusion mechanism. The key challenge of solution-diffusion membranes is the trade-off between permeability and selectivity, i.e., the Robeson's Upper Bound. However, for facilitated transport membranes, the non-reactive gas component, e.g., N_2, diffuses through the membrane via the solution-diffusion mechanism, whereas CO_2 follows the facilitated transport mechanism via its reversible chemical reaction with carriers (often amines). Therefore, CO_2 transport is facilitated, and both high CO_2 permeability and high CO_2/N_2 selectivity can be achieved.

Sterically Hindered Amine: Amine carriers are the key for the CO_2 separation performance of a facilitated transport membrane. Amine carriers exhibiting a high CO_2 loading capacity and reaction rate are always desirable. A sterically unhindered amine forms a stable carbamate ion with CO_2, as shown in Eq. (1), and the stoichiometric CO_2 loading is one mole of CO_2 per two moles of the unhindered amine. However, for a sterically hindered amine, its carbamate ion is not stable due to the incorporation of a bulky substituent group to the nitrogen atom and thus is converted to the bicarbonate ion as the sink for CO_2 with the resultant reaction shown in Eq. (2). Therefore, the stoichiometric CO_2 loading is doubled, with one mole of CO_2 per mole of the sterically hindered amine.

$$CO_2 + 2\,R\text{-}NH_2 \rightleftharpoons R\text{-}NH\text{-}COO^- + R\text{-}NH_3^+ \quad (1)$$
$$CO_2 + R\text{-}NH_2 + H_2O \rightleftharpoons HCO_3^- + R\text{-}NH_3^+ \quad (2)$$

A sterically hindered amine is defined as either a primary amine in which the amino group is attached to a tertiary carbon or a secondary amine in which the amino group is attached to at least one secondary or tertiary carbon. The effects of sterically hindered amines have been reported in a lot of literature on aqueous amine solutions for CO_2 absorption. Those works focus on the small molecules of sterically hindered amines in the aqueous phase. For solid membranes, amine carriers anchored to the polymer matrix are more desired, since it can eliminate the carrier loss. Zhao *et al.* in our group were the first to demonstrate the steric hindrance effect in solid polyamine-containing membranes. In that work, the polyallylamine was modified into sterically hindered amine, and the steric hindrance effect on separation performance was investigated for syngas purification applications.

New Membranes based on Sterically Hindered Polyvinylamines: The goal of this work on sterically hindered polyamines is to extend the research of steric hindrance effect demonstrated in polyallylamine-based membranes to other polyamines for the synthesis of new, improved membranes for carbon capture applications. In this work, polyvinylamine was chosen as the amine carrier and modified into sterically hindered polyamines since it possesses the highest content of amino groups among all the common polyamines for CO_2 facilitated transport. The membranes synthesized in this work have been investigated mainly for CO_2 capture from flue gas in coal-fired power plants.

The amine modification reactions were conducted as follows: to synthesize poly(N-methyl-N-vinylamine), one equivalent amount of iodomethane as well as potassium hydroxide was added into the aqueous solution of polyvinylamine (PVAm) and reacted with the PVAm at 45°C for 48 h. After the reaction, one methyl group should be attached onto the nitrogen atom of the amino group in each repeating unit of polyvinylamine to form poly(N-methyl-N-vinylamine), with one equivalent amount of potassium iodide salt and water generated as byproducts. The salt was removed from the polyamine solution by dialysis. Similarly, poly(N-isopropyl-N-vinylamine) and poly(N-tert-butyl-N-vinylamine) were synthesized by reacting PVAm with 2-bromopropane and 2-bromo-2-methylpropane, respectively, under the same reaction conditions. The steric hindrance degree can be quantitatively described by Taft's steric hindrance parameter, $\sum E_s$, which is determined by the group contribution method. The $\sum E_s$ values of polyvinylamine, poly(N-methyl-N-vinylamine), poly(N-isopropyl-N-vinylamine) and poly(N-tert-butyl-N-vinylamine) were calculated to be 0.93, 1.00, 1.86 and 2.67, respectively. The obtained sterically hindered polyamine solutions were concentrated by N_2 purging to have an appropriate viscosity for coating onto nanoporous polyethersulfone (PES) substrates and dried in the fumed hood at ambient conditions.

The membranes were all tested at 57°C, the typical flue gas temperature. The dry feed gas consisted of 20% CO_2 and 80% N_2, and the sweep gas was purified argon. Both feed gas and sweep gas were humidified and saturated by water vapor at the testing temperature (the saturated water vapor content in each of the feed and sweep sides was 17 mol% at 57°C). The pressure of the feed side was controlled at 1.5 psig while the sweep side pressure was maintained at 1.4 psig. Under the testing conditions, polyvinylamine membranes demonstrated an average CO_2 permeability of 213.9 Barrers and CO_2/N_2 selectivity of 48.5. Although the content of amino groups reduced by incorporation of alkyl groups, CO_2 permeability was improved by 24% and CO_2/N_2 selectivity was increased by 14% by modifying polyvinylamine into poly(N-methyl-N-vinylamine). These results have demonstrated the effect of steric hindrance on CO_2 separation performance with the incorporation of the methyl group for enhancing CO_2 loading capacity. Poly(N-isopropyl-N-vinylamine) and poly(N-tert-butyl-N-vinylamine) membranes showed 15% and 11% improvements on CO_2 permeability, respectively, with no significant changes in CO_2/N_2 selectivity. The improvement on selectivity was not as significant as CO_2 permeability, which could be attributed to the reduced polarity caused by the incorporation of bulky alkyl groups, thus increasing the solubility of N_2 in the membranes. Comparing the CO_2 separation performances of the above polyamine membranes and their corresponding steric hindrance degrees, we found that the CO_2 permeability increased first significantly with increasing the steric hindrance degree and then less pronouncedly with further increasing the steric hindrance degree. This could be explained by the reduced amino group content in the membranes and the lower reaction rate due to the incorporation of bulky groups onto amino groups. It should be noted that the new poly(N-methyl-N-vinylamine) membrane can give higher CO_2 permeability than the best sterically hindered polyamine membrane, i.e., poly(N-isopropyl-N-allylamine), reported in the literature.

Transferring Membrane Polymer Properties into Gas Separation Process Performance.

Abstract

In membrane separation processes value for the customer is often directly linked to separation performance of the membrane module. Naturally, the key component of the membrane module is the membrane material itself. For polymeric membranes this will usually be either in form of flat sheets or hollow fibers. While module concepts and membrane types can be different, obtaining good separation performance will rely on two factors. One factor is the ability to control intrinsic properties of the polymer and the second factor is the ability to control the membrane formation process (fig. 1). With the control of both factors it is possible to come up with a superior membrane product. This is where competencies of a specialty chemicals company like Evonik can contribute to unique membrane solutions for the market which will be addressed in the first part of the presentation.

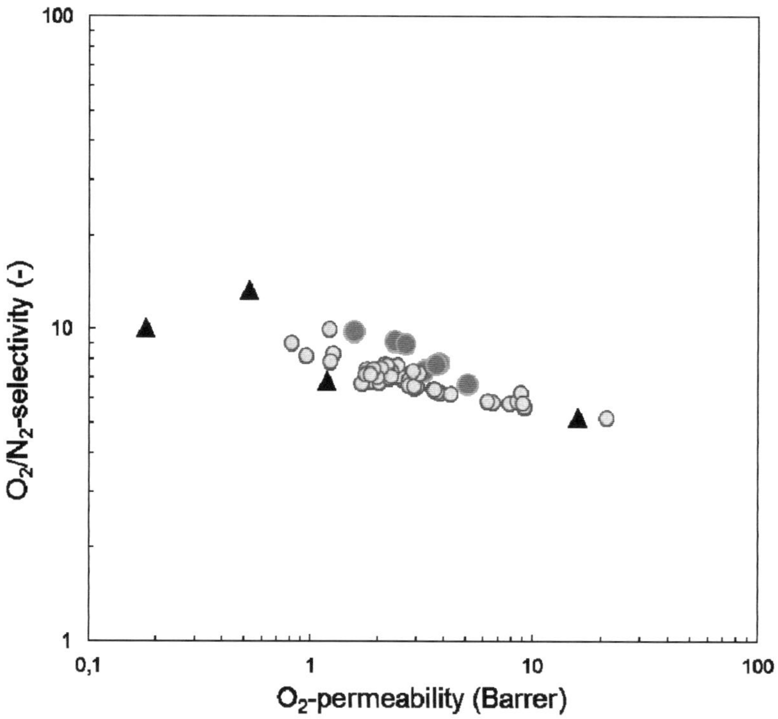

Figure 1: Tailor intrinsic properties according to application needs

The second part of the presentation will show the influence of high performing membranes on different industrial applications of gas separation membranes. Currently, the most important applications of gas separation membranes are in the generation of nitrogen, hydrogen recovery, air drying, and removal of sour gases from natural gas. These applications continue to show significant annual growth rates. Newer applications are the upgrading or recovery of helium as well as the upgrading of biogas which will be described in more detail.

Helium

Helium is an inert gas, with many specialized uses. It is often used in applications because of its special properties: it is an inert gas with high thermal conductivity and is lighter than air; for example in lifting and ballooning, electronics, MRI's and medical applications, welding and leak detection.
World demand for helium is increasing, thereby increasing the price for high purity helium as well. This makes the upgrading economically more and more interesting. Helium is found in natural gas in low concentrations, typically up to a few percent, and it is obtained as a by-product of natural gas processing. After removal of impurities, a cryogenic distillation method is applied to produce crude helium (50-70% purity), and ultimately pure helium (>90%) or high purity helium (>99.99%). The conventional upgrading process is highly energy consuming, and therefore of less economical interest. Using highly selective membranes, it is possible to replace the energy-consuming multi-step fractional distillation. Depending on the helium grade required, this can be done in a hybrid process with, for example, PSA. The high selectivity of SEPURAN® Noble membranes enables a high Helium recovery, which is very important because of the value per m3 of helium.

Biogas

The main sources for biogas are landfills and anaerobic digestion of different feed stocks including waste water and manure. Crude biogas contains about 55 to 65% methane, and 30 to 45% carbon dioxide. In case of landfills smaller amounts of nitrogen and oxygen may also be present, and trace gases such as hydrogen sulfide. These impurities must be removed from the methane to generate renewable natural gas (RNG). The RNG can be compressed to RCNG (Renewable Compressed Natural Gas) or can be fed into the natural gas grid.

Historically, biogas upgrading to RNG and injection into the gas grid or usage as a fuel has been accomplished using separation techniques other than membrane technology. The simplicity of a membrane brings many advantages compared to other separation techniques. The product gas, i.e. RNG, is produced on the high-pressure side of the membrane and the CO2 waste gas permeates to the low pressure side. The advantages of membrane technology are many: membrane based separation processes are easily scalable to any needed capacity, and can be started and stopped within minutes. Up times are maximal and not limited by the membrane. There is no need for any auxiliaries such as amines or water or regeneration. Down time causing issues like bio fouling in water based separation technologies cannot occur.

SEPURAN® Green membranes with high selectivity allow a wide range of beneficial process solutions. Single compressor but multistage systems with high methane recoveries of 99.5%+ and minimized recycle stream became feasible. At the same time extremely simple single stage systems with no recycle at all are reaching recoveries of 93%+.

In this presentation we will show important factors for a successful membrane product and its implementation to applications like helium upgrading and recovery as well as biogas upgrading.

Bio-mimetic Hollow Fiber Membrane Contactor Process for Low-Cost High-Purity Oxygen Production

Travis Pyrzynski, Naomi B. Klinghoffer, Howard S. Meyer, and Shiguang Li,*
Gas Technology Institute, 1700 S. Mount Prospect Road, Des Plaines, IL 60018

Mahdi Fathizadeh, and Miao Yu
Department of Chemical Engineering, University of South Carolina, Columbia, SC 29208

Abstract

Oxygen production from air has wide applications in oxygen-intensive industries, such as energy production (e.g., IGCC and oxy-combustion systems), fuels, chemicals, and other industries. Cryogenic distillation, pressure swing adsorption, polymeric and ionic transport membrane technologies have been used for air separation. Among these technologies, cryogenic distillation is the most mature technology for large scale and high purity (>99%) O_2 production. However, cryogenic distillation-based air separation is costly and energy-intensive to operate, accounting for up to 15% of the total gasification plant capital cost, and consuming over 35% of in-plant power use.

Supported by the US Department of Energy (No. DE-FE0024080), we are developing a novel technology to produce greater than 95% oxygen from air via a single-stage gas-liquid membrane contactor with oxygen carrier solutions as solvent. In the process, air is sent to a membrane absorber and passes through small-diameter membrane tubes, roughly 15 mil inside diameter, while a lean O_2 carrier solution flows counter currently on the shell side of the membrane. Unlike the other production alternatives, the air stream needs to be compressed to only a few psi and does not require heating as for the ion transport membranes or cooling for cryogenic separations. The O_2 permeates through the membrane pores and is absorbed in the O_2 carrier solution. The O_2-rich carrier solution can be regenerated in a second membrane module (desorber) operated in a reverse process. In that case, the O_2-rich carrier solution is fed to the shell side of the hollow fibers and vacuum is used to draw O_2 into the pore side of hollow fibers. In this bio-mimetic process, a membrane module with hollow fibers and O_2 carrier solution mimics "blood vessels" and "blood".

In our study, low-cost poly(ethyleneimine)-cobalt (PEI-Co) complex solutions with high O_2 absorption capacity (as high as 750 mL O_2 (STP)/L solution), low viscosity (< 3 times of water), and fast O_2 absorption rate have been developed. The O_2 equilibrium absorption capacity increases with the PEI/Co ratio and reaches a plateau at ratio of 10. Two different approaches were employed in the desorption process: 1) vacuum, and 2) using argon as sweep gas. By using the second approach, O_2 concentration as high as 99% was observed in desorbed gas stream on an argon free basis. The production of concentrated O_2 has been validated experimentally in the bio-mimetic process. The estimated cost including capital, operating, maintenance, and energy use for the bio-mimetic O_2 separation technology is $19.96/ton O_2, which is only about 56% of the benchmark cryogenic distillation ($35.80/ton O_2). This technology has potential to produce oxygen with purity as high as 99.9% for applications in IGCC, oxy-combustion, and other advanced power generation technologies.

---

* Presenter: Mr. Travis Pyrzynski, Travis.Pyrzynski@gastechnology.org
Corresponding author: Dr. Shiguang Li, Shiguang.Li@gastechnology.org; 847-544-3478 (phone)

Development of multibore hollow fiber membranes for water deoxygenation

Lu Mao, Fan Wang, Rui Chin Ong, Shuning Lin, Mohd Abid Dhamiri Bin Atan, Jincai Su*

School of Life Sciences & Chemical Technology, Ngee Ann Polytechnic, 535 Clementi Road, Singapore 599489
**Corresponding author, Email: sji2@np.edu.sg, Fax: (65) 6467 9109*

Key words: Deoxygenation; Multibore hollow fiber; Membrane structure; Robust

Abstracts

The concentration of dissolved oxygen (DO) is critical to the quality of high-purity water required by semiconductor and pharmaceutical manufacturing processes, the injection liquid in oil/gas offshore industry for oil production as well as boiler water in power plants. Membrane contactor is an emerging technology for water degassing which easily reduces the DO level to below 10 ppb. Membrane contactors have a smaller footprint, they are simple to operate and maintain, they are modular and therefore to scale up easily, and their operating cost is much lower than that of conventional technologies such as vacuum tower or forced draft degasifier. With these advantages, however, membrane contactor is not replacing conventional technologies for oxygen removal. The main concern from industry is the limited throughput of the membrane contactor available in the market. As well, the membrane configuration, i.e., small in diameter and thin in wall thickness, is another issue for the cases where foulants are available. Robust and highly permeable degassing membranes are in need.

To obtain robust and more universal membrane contactor, we have developed multibore hollow fiber membranes. Specifically, 3-bore and 7-bore hydrophobic hollow fiber membranes have been prepared using poly(vinylidene fluoride). The multibore hollow fibers have been extensively characterized in terms of morphology, hydrophobicity, tensile strength, pore structure, and permeation properties. It is observed that the newly-developed hollow fibers are much more robust as compared to conventional single-bore hollow fibers. Experimental studies are carried out to explore how the membrane characteristics (e.g., membrane structure and fiber dimension) and operating conditions (e.g., temperature and flow rate) influence the efficiency of oxygen removal. Theoretical studies are also conducted to understand the factors dominating the mass transfer of the deoxygenation process. The experimental and theoretical studies are ongoing and the results will be presented in the conference.

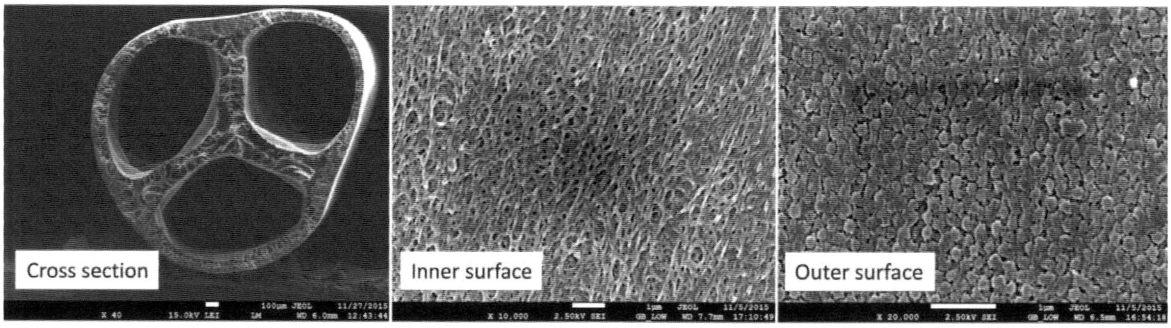

Figure 1. Morphology of 3-bore hollow fiber

Fabrication & Performance Study of Pd-Au PSS Supported Membrane Reactor Via Natural Gas Steam Reforming

Bryce Anzelmo[1], Simona Liguori[1], Ivan Mardilovich[2], Yi-Hua Ma[2], Jennifer Wilcox[1]

[1]Department of Energy Resources Engineering, Stanford University, Stanford, CA
[2]Department of Chemical Engineering, Worcester Polytechnic Institute, Worcester, MA

One possible use for hydrogen, without direct greenhouse gas emissions, is as feed for a fuel cell (FC), with the most readily available technology being a proton exchange membrane FC (PEMFC). In order to avoid poisoning the PEMFC's Pt-based catalyst due to the presence of ppm levels of CO, the hydrogen feed needs to be ultra-pure. The industrial process for hydrogen production, which is a multi-step energy intensive process followed by further separation/purification, can be a potential source [1]. However, as an alternative method a Pd-based membrane reactor (MR) can be used owing to its ability to provide the pure hydrogen without any further separation/purification step. Moreover, the MR works at milder operating conditions compared to the traditional system.

In recent years, Pd-based composite membranes, i.e., thin metallic layer supported on porous materials such as ceramics or stainless steel, have been considered due to their lower cost (thin Pd layer) and higher mechanical strength (porous support) compared to dense Pd-based ones [2]. Moreover, the use of Pd-alloys, in particular, with Ag, Au, Pt, has grown in the last years given their ability to resist sulfur poisoning with increased hydrogen permeability [3]; therefore, the aim of this study is to fabricate a Pd-Au membrane supported on porous stainless steel (PSS) and evaluate its performance in a MR with the intent to produce pure hydrogen from natural gas steam reforming.

The membrane in this study was fabricated via an electroless plating technique for depositing Pd and electro-plating for Au. The steps for fabricating this membrane are described in the following steps: 1) abrasive sanding to obtain a smooth Zr surface 2) electroless plating of Pd 3) another electroless plating of Pd 4) electroplating of Au 5) final eletroless plating of Pd [4-8]. Each fabrication step provides a benefit relating to pore size reduction, which mitigates the defects of the support. Moreover, the initial Zr layer not only mitigates pore size, but also provides an intermediate layer that limits the intermetallic diffusion. This multistep process allows for realizing a membrane characterized by a Pd-Au layer of ~ 12 μm with 4.4 wt% Au content (Fig. 1). At each step of the fabrication process, permeation tests with pure He are carried out, at room temperature and pressure difference of 1 bar . Fig. 2 shows the results of the permeation tests in which it is evident the progression and reduction of He flux during the membrane fabrication process.

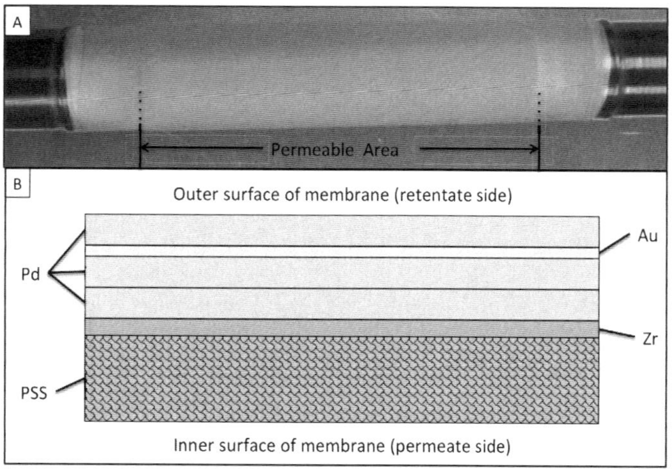

Fig. 1. Pd-Au PSS composite membrane: [A] Picture of completed membrane [B] Detailed schematic of completed metallic membrane.

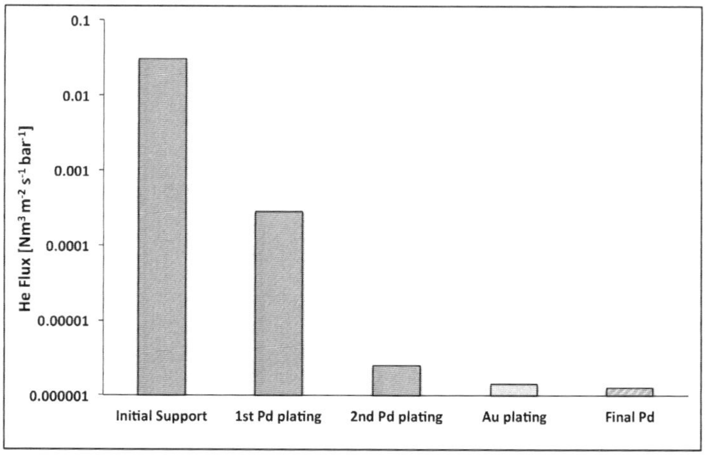

Fig. 2. Helium permeating flux during membrane progress, $\Delta P = 1 bar$, $T = 25°C$.

The initial characterization of the membrane by way of ideal selectivity took place at 400°C with H_2, He and Ar with ΔP in the range of 0.5 - 3.0 bar. After ideal selectivity characterization of the Pd-Au/PSS membrane, natural gas steam reforming reaction testing is carried out and the MR performance in terms of natural gas conversion, hydrogen production and hydrogen permeate purity is analyzed and discussed.

REFERENCES

[1] Rostrup-Nielsen, J.R., *Catalytic steam reforming*. 1984: Springer.

[2] Liguori, S., et al., *Performance of a Pd/PSS membrane reactor to produce high purity hydrogen via WGS reaction.* Catalysis Today, 2012. 193(1): p. 87-94.

[3] Chen, C.-H. and Y.H. Ma, *The effect of H2S on the performance of Pd and Pd/Au composite membrane.* Journal of Membrane Science, 2010. 362(1–2): p. 535-544.

[4] Ma, Y.H. and I.P. Mardilovich, *Composite structures with porous anodic oxide layers and methods of fabrication*, 2013, Google Patents.

[5] Ma, Y.H., I.P. Mardilovich, and E.E. Engwall, *Thin composite palladium and palladium/alloy membranes for hydrogen separation.* Annals of the New York Academy of Sciences, 2003. 984(1): p. 346-360.

[6] Ma, Y.H., I.P. Mardilovich, and E.E. Engwall, *Method for curing defects in the fabrication of a composite gas separation module*, 2007, Google Patents.

[7] Ma, Y.H., I.P. Mardilovich, and E.E. Engwall, *Composite gas separation modules having high Tamman temperature intermediate layers*, 2007, Google Patents.

[8] Ma, Y.H., I.P. Mardilovich, and E.E. Engwall, *Composite gas separation modules having intermediate porous metal layers*, 2007, Google Patents.

Twisted Hollow Fiber Membranes for Enhanced Mass Transfer.

Computational fluid dynamic simulations are used to examine incompressible flow through twisted elliptical and helically-coiled cylindrical fibers for typical flow Reynolds numbers employed in membrane filtration processes (Re < 200). The resulting flow fields are compared in terms of the distributions of vorticity and wall shear rate, the pressure drop, and the azimuthally-averaged normal velocity in the vicinity of the wall. The secondary flow in twisted elliptical fibers generates significant recirculation flow leading to near-wall velocities directed normal to the surface of the membrane (below figure). The azimuthally-averaged normal velocity near the wall of a twisted elliptical fiber is found to be two to four orders of magnitude larger than typical ultrafiltration velocities. For Re < 75, the average wall shear rate and normal velocity near the wall are found to be significantly higher in twisted elliptical fibers compared to their helical counterparts. These findings highlight the potential for mass transfer and flux enhancement using twisted elliptical hollow fiber membrane systems, which could provide a low cost, and linearly scalable module with improved performance.

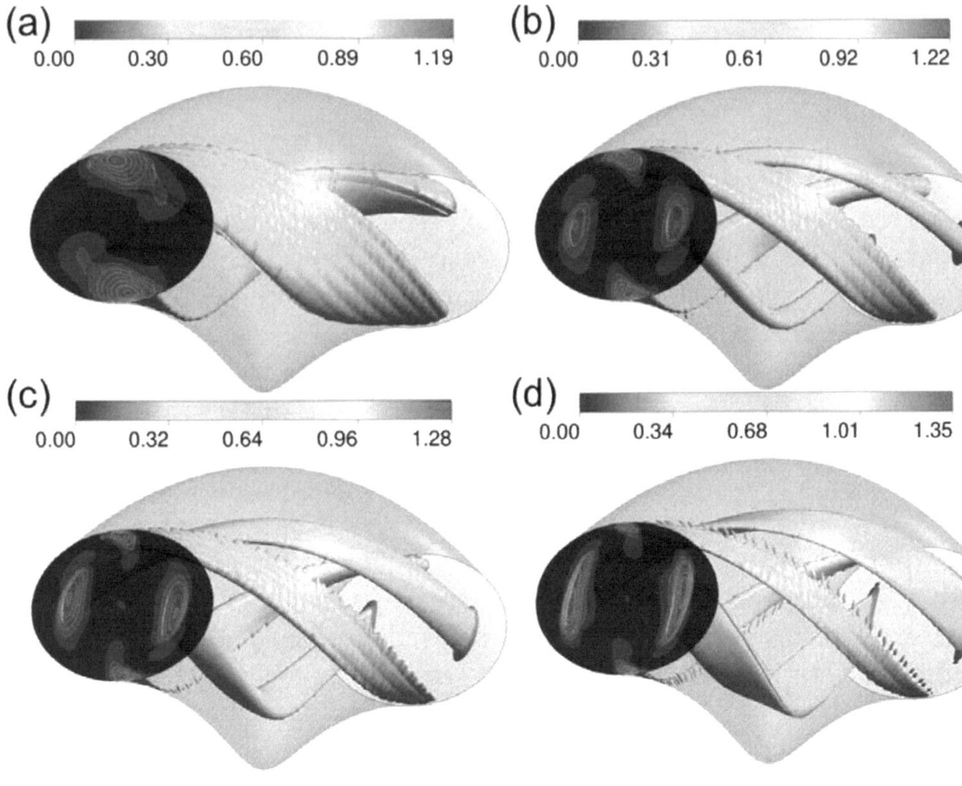

Figure 1 - Vortical structures in a twisted elliptical fiber identified by iso-surfaces corresponding to 60% of maximum swirling strength for (a) $Re = 10$, (b) $Re = 50$, (c) $Re = 100$, and (d) $Re = 200$.

How do hydrodynamic conditions in spiral wound membrane elements translate into shear forces effective for fouling control at membrane surface?

Sepideh Jankhah[1], Pierre Berube[2]

1. Sterlitech Corporation, Kent, WA, USA
2. University of British Columbia, Vancouver, Canada

Feed spacers are commonly used in commercially available spiral wound membrane elements to create turbulence at the vicinity of membrane surface to minimize formation of concentration polarization and fouling rate. However, turbulence induced by the spacers results in energy dissipation and pressure loss in the spiral wound membrane elements which should be compensated. This energy compensation translates into higher energy demand and higher operational costs. The turbulence intensity, energy dissipation rate, pressure loss and fouling rates are proved to be functions of spacer design and geometry. These are more specifically functions of spacer thickness, distance between filaments in a spacer, the angle between filaments and spacer orientation in relation to the direction of the flow. An effective spacer for a spiral wound element must provide a high mass transfer rate away from the membrane surface toward the bulk to reduce the wall concentration while maintaining a low-pressure loss along the element.

For determining an optimum spacer geometry, one should take into consideration the effects of feed flow pressure drop which affects energy dissipations and energy requirements, mass transfer rates of foulants away and toward membrane which affect concentration polarization and fouling rate, and shear forces at the membrane surface, which affect membrane fouling rate.

In the present study, shear force induced at the membrane surface in a spiral wound configurations was measured using electrochemical diffusion method (EDM). Effect of feed spacer geometry on the shear forces induced at membrane surface was investigated. It was demonstrated at the same feed flow rate thicker feed spacers result in lower magnitude of shear forces at the membrane surface with higher variation of shear force magnitude. With an increase in feed flow rates the efficiency of feed spacers in terms of fouling control decreased; demonstrating that there is an optimum feed flow rate for each spacer above which increasing the feed flow rates does not result in higher fouling control efficiency. Knowing the magnitude and variation of shear forces induced at membrane surface under different hydrodynamic conditions will minimize the time and effort for running pilot testing to find the efficiency of feed spacer in terms of energy requirements for fouling control.

Rapid Uranium Isotopic Analysis using Ultrafiltration and Alpha Spectroscopy

Christine E. Duval[A], Timothy A. DeVol[B], Scott M. Husson[A]
[A]*Department of Chemical and Biomolecular Engineering, Clemson University, SC, USA*
[B]*Department of Environmental Engineering and Earth Sciences, Clemson University, SC USA*

Clandestine activities involving the separation, concentration or manipulation of special nuclear material for the express purpose of developing a weapon of mass destruction is likely to result in the contamination of environmental water sources. The capability to conduct isotopic analyses for waterborne special nuclear material (SNM) would be a powerful nuclear forensics tool. Despite widespread interest, there currently is no fieldable measurement system available for low-level quantification or isotopic determination of waterborne uranium.

This contribution describes our efforts to develop a high throughput analytical technique for waterborne isotopic analysis with the following procedure: (1) deposition of ligand-containing, water-soluble polymer film on an ultrafiltration membrane; (2) separation of the uranium from near-neutral pH water by filtration through the selective film; and (3) isotopic analysis of the uranium-containing polymer film with high resolution alpha spectroscopy. The uranium-selective, water soluble polymer was prepared by the copolymerization of 2-hydroxylethyl methacrylate and 2-ethylene glycol methacrylate phosphate. The polymer was characterized by Fourier-transform infrared spectroscopy and nuclear magnetic resonance spectroscopy to support the incorporation of the ligand into the polymer backbone. Film morphology and thickness were observed by atomic force microscopy and laser scanning profilometry after deposition of the polymer onto the ultrafiltration membrane.

The effect of membrane type, membrane pore size and polymer film thickness on peak resolution in the alpha spectrum was investigated for pure uranium containing solutions at pH = 6. To mimic more realistic conditions, the selectivity of the membrane was tested using natural uranium in simulated groundwater. Uranium-coated membranes prepared from distilled water and simulated groundwater both showed resolutions of 60 keV in the alpha spectrum and detection efficiencies of 12% for uranium-233. The high resolution is attributed in part to the concentration of uranium on the surface of the thin polymer coating. Using a thin coating also was important for maintaining high membrane permeability. The permeability of the polymer coated, polyethersulfone ultrafiltration membranes (MWCO 50 kDa) was determined to be 780 LMH/bar. The results of this research are laying the foundation for the development of a novel rapid, field portable method for isotopic identification of SNM in environmental media.

A Method to Evaluate Membrane Protein Compatibility with Biomimetic Membranes

Tingwei Ren[a], Mustafa Erbakan[a,b], Mariusz Grzelakowski[c], Peter Butler[d], Manish Kumar[a]

a. Department of Chemical Engineering, The Pennsylvania State University, University Park, PA 16802; b. Current address: Department of Biosystem Engineering, Bozok University, Yozgat 66000, Turkey; c. Applied Biomimetic, Cincinnati, OH 45237, USA; d. Department of Biomedical Engineering, The Pennsylvania State University, University Park, PA 16802

Cell membrane matrices are mainly composed of lipids and membrane proteins. Membrane proteins, present in the membrane, performing a vast majority of functions including signal reception[1], material transport and biochemical reactions[2,3], that occur at membrane-liquid interfaces. High compatibility between cell membranes and membrane proteins are necessary for living organisms to finish complicated tasks and thrive in the nature[4]. Inspired by living cell systems, researchers are using lipids and lipid-like membrane matrices as models for functional biomimetic membranes development and the number of these membrane matrices are increasing. However, only compatible biomimetic membranes specific to different membrane proteins will provide desired membrane performance[5]. Highly compatible biomimetic membranes have high membrane protein reconstitution density without compromising membrane protein single molecule properties. However, there are few accurate methods for compatibility evaluation[4].

Here, we propose and test a method to evaluate membrane protein compatibility with biomimetic membranes. We use aquaporin (AQP) reconstituted biomimetic membranes (ABMs) as examples to verify the accuracy of this evaluation method. ABM is a widely accepted model system for novel biomimetic membrane research. Previous researchers found that ABMs have higher water transport rate and better salt rejection performance than traditional reverse osmosis membranes[6,7]. However, AQPs compatibility with different biomimetic membranes hasn't been studied systematically. We combined Stopped Flow Light Scattering technology and Fluorescence Correlation Spectroscopy technology for compatibility evaluation[8]. AQPs reconstitution density and single channel water permeability were accurately measured at the single molecule level. By reconstituting AQPs into different ABMs (both block copolymers and lipids), we show that AQP single channel water permeability stays similar in different membranes. However, reconstitution density varies in different materials, and AQPs have higher reconstitution density in block copolymers. We have also implemented this evaluation method on other membrane proteins and artificial water channels[9] and these results will be summarized as well.

References:

1. Broselid, S. et al. G protein-coupled Receptor 30 (GPR30) Forms a Plasma Membrane Complex with Membrane-associated Guanylate Kinases (MAGUKs) and Protein Kinase A-anchoring Protein 5 (AKAP5) That Constitutively Inhibits cAMP Production. Journal of Biological Chemistry 289, 22117-22127,(2014).
2. Parker, M. D., Myers, E. J. & Schelling, J. R. Na+-H+ exchanger-1 (NHE1) regulation in kidney proximal tubule. Cell. Mol. Life Sci. 72, 2061-2074,(2015)
3. Krall, J. F. & Korenman, S. G. Soluble Precursor of Membrane-Associated Protein-Kinase in Uterine Smooth-Muscle Cells. Arch. Biochem. Biophys. 199, 479-487,(1980).
4. Shen, Y.-x., Saboe, P. O., Sines, I. T., Erbakan, M. & Kumar, M. Biomimetic membranes: A review. Journal of Membrane Science 454, 359-381,(2014)
5. Nichols, S. P., Koh, A., Storm, W. L., Shin, J. H. & Schoenfisch, M. H. Biocompatible Materials for Continuous Glucose Monitoring Devices. Chem. Rev. 113, 2528-2549,(2013)
6. Kumar, M., Grzelakowski, M., Zilles, J., Clark, M. & Meier, W. Highly permeable polymeric membranes based on the incorporation of the functional water channel protein Aquaporin Z. Proceedings of the National Academy of Sciences of the United States of America 104, 20719-20724,(2007)
7. Kowal, J., Zhang, X., Dinu, I. A., Palivan, C. G. & Meier, W. Planar Biomimetic Membranes Based on Amphiphilic Block Copolymers. ACS Macro Letters 3, 59-63,(2014)
8. Erbakan, M., Shen, Y.-x., Butler, P. J., Kumar, M. & Curtis, W. R. Molecular cloning, overexpression and characterization of a novel water channel protein from Rhodobacter sphaeroides. PLoS ONE 9, e86830,(2014)
9. Shen, Y. X. et al. Highly permeable artificial water channels that can self-assemble into two-dimensional arrays. Proceedings of the National Academy of Sciences of the United States of America 112, 9810-9815, (2015).

Fabrication of Ultrathin ZnO Nanocrystal-nanocellulose Composite Paper

Jingda Wu and **Lih Y. Lin**
Department of Electrical Engineering, University of Washington, Seattle, WA 98195-2500
E-mail: albuswu@uw.edu, lylin@uw.edu

Abstract: We demonstrated a method to fabricate an ultrathin (~2 μm) ZnO nanocrystal (NC)-nanocellulose composite paper through solvent exchange. The paper is extremely transparent in visible light and shows photoluminescence under ultraviolet(UV) excitation. It can be applied towards photodetection applications.

I. Introduction

Driven by consumer needs and technology breakthroughs, we are entering an era of flexible devices, which greatly eases our daily life. While we are witnessing the commercialization of devices such as flexible OLED displays and foldable solar cells, they are more based on conventional device structures and materials. Recently, nanocellulose paper[1, 2] has been recognized for its superior performance in light transmittance, material strength and eco-friendliness, and applied towards a lot of optoelectronic applications. Being nano-sized and hydrophilic, nanocellulose can also be readily mixed with other nanomaterials in solution and create composite structures for different functionality.[3] Previously we have demonstrated quantum dot-cellulose composite structures for flexible photodetection in vertical device structures[4, 5]. However, the commercial or natural cellulose structure such as tracing paper or reed membrane have large thickness and fixed porosity, which make it hard to achieve devices with high performance. We discovered that with nanocellulose, we are able to make ultrathin transparent freestanding semiconducting film with the incorporation of water-soluble ZnO nanocrystals and use it for UV detection.

For nanopaper making, people commonly start with filtering the nanocellulose out from a suspension to form a gel cake and then dry the gel to form a paper. The gel can either be dried single-sidedly with one side in contact with a flat surface and another touching a porous structure for water extraction, or dried double-sidedly with both sides in contact with porous materials. For a film that is fairly thick, tens of micrometers, either way works well since the strength of the paper is able to hold the gel structure. However, when it comes to ultrathin (~2 μm) paper making, the adhesion between the gel and filter or drying surface is much stronger than the inner tension between nanocelluloses. Hence it's almost impossible to dry the nanopaper single-sidedly. Also the strong surface tension of water will easily crank the freestanding film. Thus we also need to modify the double-sided drying method.

II. Fabrication

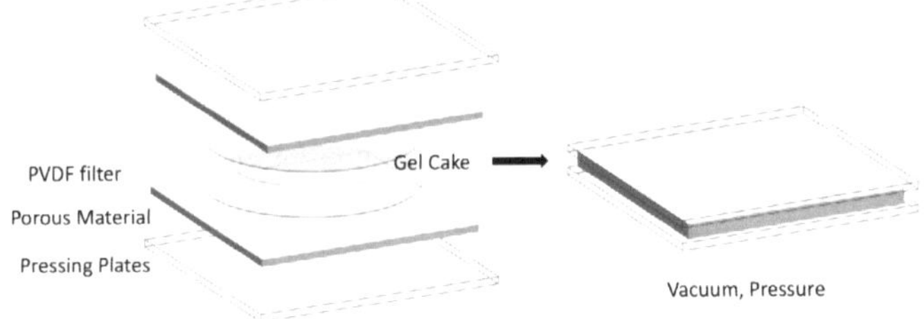

Fig. 1 Drying setup of ZnO NC-nanocellulose paper

The ZnO nanocrystals are synthesized through well-developed wet-chemistry methods. They have a size of 6-8 nm and are suspended in DI-water. Nanocelluloses are also suspended in DI-water after TEMPO-mediated oxidation and high pressure microfluidizing. These two materials are mixed proportionally depending on the need of performance, typically 1:1 in mass ratio in our case, with ultrasonication. The mixed aqueous suspension's vacuum filtered to leave a thin gel cake which is later quickly peeled off from the filter on a plain acrylic plastic. Then the gel/acrylic is soaked in isopropanol alcohol to do a solvent exchange and to suspend the gel cake which would be otherwise difficult to achieve in free space or in water. Eventually the freestanding gel is sandwiched between two PVDF filters and placed under vacuum with pressing, as is shown in Fig. 1, for drying overnight.

III.Results and Discussion

The results for the end product are shown in Fig. 2. More than 97% of visible light is transmitted through the nanopaper. We can clearly see that the paper is more transparent than cover glass slides and tracing paper. With it's ultrathin edge, it's actually hard to notice without careful searching. The paper shows uniform photoluminescence(PL) under UV excitation. The PL wavelengths are in cope with the PL of ZnO NCs and thus indicating that the ZnO NCs are still functional. The non-uniformity and bright spots in the picture are more due to the background table surface because the nanopaper is very transparent.

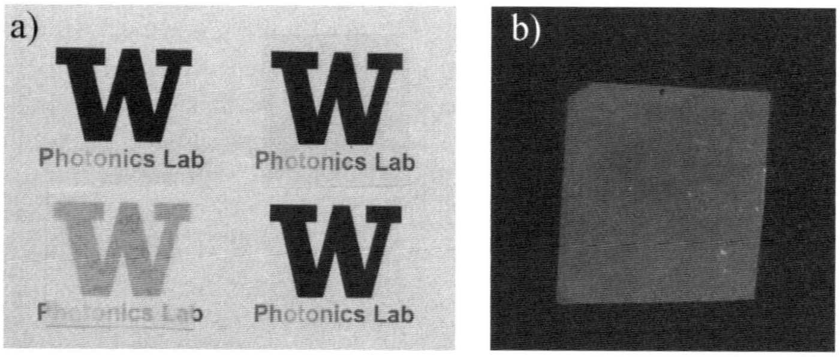

Fig.2 a) Transparency comparison of between air, cover glass slides, tracing paper and ultrathin ZnO NC embedded nanopaper (from left to right, up to down); b) Yellowish green color photoluminescence of the ultrathin nanopaper under 365nm UV excitation

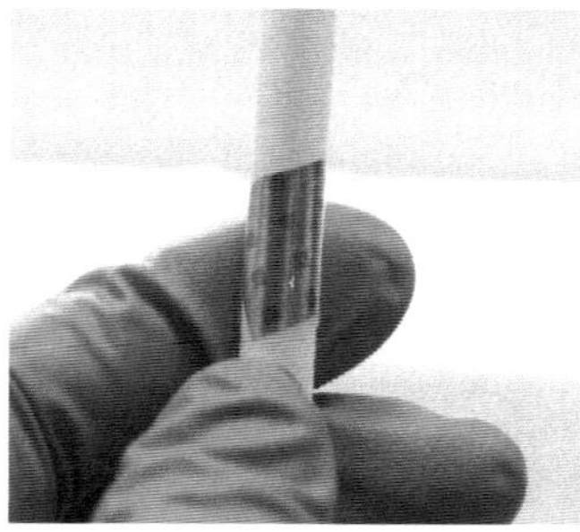

Fig. 3 Ultrathin flexible UV photodetector on nanopaper wrapped around a pen

We also managed to fabricate flexible photodetectors on the ultrathin nanopaper. The image of an array of photodetectors on nanopaper is shown in Fig. 3. The devices can be bent freely around cylindrical surfaces.

IV.Conclusion

We have developed a new way of fabricating nanocellulose papers through solvent exchange to achieve ultrathin film thickness. ZnO NCs can be readily mixed with nanocellulose to prepare ZnO NC-nanocellulose composite papers. The ultrathin paper appears to be extremely transparent in visible wavelength range and shows photoluminescence under UV excitation. The nanopaper can also be used to fabricate photodetector arrays for UV detection.

References

[1] A. Dufresne, *Nanocellulose: From Nature to High Performance Tailored Materials*, De Gruyter, Germany 2012.

[2] L. Hu, G. Zheng, J. Yao, N. Liu, B. Weil, M. Eskilsson, E. Karabulut, Z. Ruan, S. Fan, J. T. Bloking, M. D. McGehee, L. Wagberg, Y. Cui, *Energy & Environmental Science* **2013**, 6, 513.

[3] E. Morales-Narváez, H. Golmohammadi, T. Naghdi, H. Yousefi, U. Kostiv, D. Horák, N. Pourreza, A. Merkoçi, *ACS Nano* **2015**.

[4] J. Wu, L. Y. Lin, *Photonics Technology Letters, IEEE* **2014**, 26, 737.

[5] J. Wu, L. Y. Lin, *Advanced Optical Materials* **2015**, 3, 1530.

Comparison of Polymeric and Ceramic MF/UF Membranes for the Treatments of produced Water

Abstract

In recent years, the extraction of unconventional resources has driven an immediate increase in the development of oil and gas (O&G) production. Hydraulic fracturing is the key technology that has enabled the extraction of unconventional resources. The management of hydraulic fracturing flowback and produced water is a major challenge for unconventional O&G exploration and production. The expedited rate of hydraulic fracturing development coupled with increased demand for fresh water, water scarcity, limited disposal options, and concern of environmental impacts has elevated the O&G industry to minimize fresh water consumption and increase their reuse application. Ceramic membranes have been rapidly emerging in water and wastewater treatment as an alternative to polymeric membranes, but yet little information on their performance and fouling mechanism is available. The purpose of this study was to design and build a MF ceramic membrane system in order to evaluate the fouling characteristics and effluent quality of ceramic membrane and compare it with UF polymeric membrane in produced water treatment

An inside-out tubular mono channel ceramic membrane with a nominal pore size of 0.2 μm was chosen for the designed system due to its ability to operate at higher flux. The system was operated at three different cross flow velocities (CFV) of 0.2, 0.5, and 1.0. Instead of using backwashing, backpulsing was evaluated for fouling mitigation. With adjustable duration, backpulsing delivered high pressure (>100 psi) burst of reversed permeate to the membrane in an adjustable time interval. The performance of the ceramic membrane system was evaluated based on membrane fouling mitigation and treated water quality.

Finding results for the ceramic membrane indicated that increasing the CFV from 0.2 to 0.5 and to 1.0 at two different flux (40 and 60 LMH) lowered the total fouling index (TFI). The hydraulic irreversible fouling index (HIFI) appeared to decrease with increasing the CFV at 40 LMH, however, this was not observed at 60 LMH where HIFI showed an increase with increasing of the CFV which could be possibly due fewer backpulses for the same accumulative volume at the higher flux. The operation at 80 LMH for all three CFV showed a rapid increase in both TFI and HIFI and reached the critical flux for CFV 0.2 and 1.0 m/s.

The results of the comparison between ceramic and polymeric polyvinylidene fluoride (PVDF) membrane with the pore size of 0.03 μm at CFV 0.2 m/s and flux of 40 LMH indicated that the ceramic membrane had a lower TFI than the polymeric membrane. The polymeric membrane, however, appeared to have a significantly lower HIFI than the ceramic membrane. This could possibly be due to weaker interaction of the foulant with the polymeric membrane since it had a smaller pore size than ceramic membrane. Another plausible explanation is that backwashing was a more effective cleaning method compared to backpulsing.

Desalination of brackish inland surface water using nanofiltration: Fouling control and water quality improvements

Shankar Chellam and Mutiara Ayu Sari

Department of Civil Engineering, Texas A&M University, College Station, TX 77843-3136

Reverse osmosis (RO) of seawater and brackish groundwater are two options often discussed to combat drinking water shortages arising from persistent drought and population growth. In contrast to the high salinities of these waters, there are several lakes with lower total dissolved solids concentration in the range 2,000-5,000 mg/L, which can potentially be desalinated using nanofiltration (NF). Since NF membranes are operated at lower pressures and higher flux they are more economically viable than RO. However, the higher concentrations of colloids (i.e. turbidity), organic matter (i.e. dissolved organic carbon, DOC), and microorganisms in brackish surface waters and our limited experience with desalinating such waters present significant challenges to NF implementation, especially with respect to pretreatment.

The principal objectives of this research are to identify dominant foulants and systematically elucidate NF fouling mechanisms by thoroughly investigating interactions of various feed water constituents with the membrane surface to select appropriate pretreatment. We evaluate microfiltration (MF) alone and electrocoagulation-MF as pretreatment options for the NF270 membrane. Experiments were performed using water samples from the Washita Basin in West-Central Oklahoma (i.e. Foss Reservoir), which serves over 40,000 people. Results were benchmarked with synthetic feed water to isolate precipitative, organic, and colloidal fouling. Membrane morphology, organic functionalities and elemental composition were analyzed using scanning electron microscopy-energy dispersive spectroscopy (SEM-EDS), atomic force microscopy (AFM), Fourier Transform Infrared (FTIR) spectroscopy, and X-ray Photoelectron Spectroscopy (XPS).

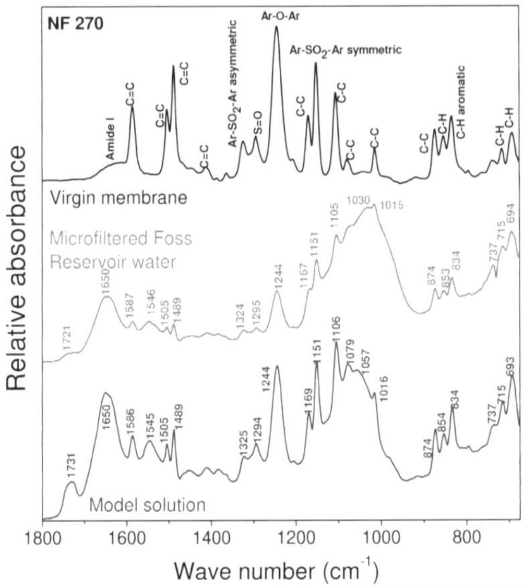

Figure 1. ATR-FTIR spectra of membranes before and after NF of model solution and pretreated natural water from Foss Reservoir

As shown in Figure 1, the virgin membrane exhibited bands typical of thin film composite polyamide membranes. The sharp and prominent peaks at 1151, 1243, and 1323 cm^{-1} corresponds respectively to Ar-SO_2-Ar, Ar-O-Ar, and asymmetric SO_2 stretching vibrations from the polysulfone support layer. The C=C aromatic in-plane ring bend stretching vibrations at 1586, 1504, and 1488 cm^{-1} are also attributed to the support layer. Following NF of microfiltered Foss Reservoir water, the virgin membrane characteristic peaks were either attenuated or reduced in intensities as a result of foulant deposition. The more intense amide I peak (νC=O + δC-N + δN-H) and the appearance of amide II peak (δN-H + νC-N) which was previously absent in the virgin membrane might indicate the presence of proteinaceous compounds.

The elemental composition of virgin and fouled membranes was obtained using survey and high-resolution XPS scans. The C1s peak in virgin membranes was deconvoluted into three components whereas the O1s was resolved into two peaks. The changes in C1s features were not very significant suggesting that only a thin foulant layer was formed. An increase in the area percentage of the peak around 286 eV indicates an increase in the number of C-(O, N, OH) bond on the nanofilter surface that might come from organics (proteins, polysaccharides, and humics) and/or carbonates. The increase in the number of carbonyl and carboxyl groups as indicated by the increase in peak at ~288.1 eV and the appearance of a small peak at ~288.9 eV might also originate from carbonates and/or organics.

As expected, natural organic matter and major divalent ions including Ca, Mg, Sr, and SO_4^{2-} were highly rejected (> 99%). The very high rejection of SO_4^{2-} is attributed to anion repulsion since the membrane is known to be negatively charged at the operating pH of 7.8. Interestingly, the dominant monovalent anion, i.e. chloride, was negatively rejected, which is attributed to its preferential transport across the membrane due to the Donnan effect necessary to maintain electroneutrality. Chloride ions compensate for the very high removal of sulfate (due to electrostatic ion-membrane repulsion) and correspondingly lower removals of cations (due to electrostatic ion-membrane attraction). More details, will be presented at the time of the conference, including fouling control results from electrocoagulation pretreatment as well as electron and force microscopy and high resolution XPS scans of membrane surfaces.

Influence of chemical coating combined with nanopatterning on alginate fouling during nanofiltration

Steven Weinman and Scott Husson

Department of Chemical & Biomolecular Engineering, Clemson University, Clemson, SC 29634

Treatment of impaired waters for beneficial use exposes membranes to feed waters containing biological and abiotic species, which leads to fouling and loss of membrane productivity over time. Since reduction in flux due to fouling is one of the largest costs associated with membrane processes in water treatment, new membrane modification methods that limit fouling would have significant economic and societal impacts. The focus of our work is to develop such a method.

Prior studies in this area largely have focused on chemical modifications to the membrane surface to limit fouling. Based on considerable literature and experience, we know that chemical modification strategies alone can be effective but not sufficient for controlling fouling. The discovery that sub-micron patterning of a membrane surface can improve its fouling resistance provides an orthogonal membrane design parameter. Our hypothesis is that combining chemical modification and patterning will yield membrane surfaces that are more effective at fouling control than either method alone. This strategy is unique in its use of both chemical coating and patterning to combat fouling.

This contribution describes our method to increase the fouling resistance of nanofiltration membranes by applying both a chemical coating and a nanoscale pattern to the membrane surfaces. A line and groove nano-pattern was applied by thermal embossing directly onto a commercial polyamide thin-film composite nanofiltration membrane. Poly(ethylene glycol) diglycidyl ether (PEGDE) was reacted onto the patterned membrane surfaces by an epoxide ring opening reaction with unreacted carboxyl groups on the polyamide selective layer. Membrane performance was evaluated by measuring flux and salt rejection using a magnesium sulfate solution and by measuring flux reduction using an alginate solution. Surface characterization showed successful patterning and chemical modification of the membrane surface. Flux and salt rejection properties were unaffected by patterning the polyamide membrane surface directly. The filtration results with alginate solutions show that combining line and groove nano-patterning with PEGDE chemical modification yields a membrane that is more resistant to fouling than either method alone.

The impact of slickwater fracturing fluid composition and shale interactions on membrane fouling of flowback water

Boya Xiong[†], *Selina Roman-White*[‡], *Travis Tasker*[†], *Benjamin Farina*[‡], *William Burgos*[†], *Andrew L. Zydney*[‡], *Manish Kumar*[‡]

[†]Department of Civil and Environmental Engineering, [‡]Department of Chemical Engineering, The Pennsylvania State University, Fenske Laboratory, University Park, Pennsylvania 16802, United States

Abstract

Although there is broad interest in the development of membrane systems for treatment of flowback water from natural gas production, there is currently little information on the critical components governing the fouling behavior of these complex fluids. Synthetic flowback water was generated by subjecting mixtures containing friction reducer (polyacrylamide – PAM), breaker (ammonium persulfate), and surfactants (etoxylated alcohol) to the high pressures and temperatures in typical fracking operations, using a Parr reactor that contained pulverized shale outcrops. Dead-end filtration experiments were performed using 0.2 μm PVDF microfiltration membranes to evaluate the impact of fluid composition on the fouling characteristics. Polyacrylamide-based friction reducer wasthe most significant membrane foulant. Size exclusion chromatography results suggested that the addition of ammonium persulfate reduced the peak molecular weight of PAM from 6 million Da to 3.5 kDa after reaction at a fracking temperature of 80℃ , resulting in a significant reduction in fouling. The interactions between fluid and shale formation reduced fouling of MF by adsorbing organics and lowering PAM concentration in flowback water. The greatest removal of organic matter by shale was obtained with friction reducer alone (70% TOC reduction); nevertheless, this solution caused the greatest fouling due to the large particle size. These results demonstrate that there are significant interactions between the components used in hydraulic fracturing operations, providing valuable insights for the development of efficient membrane-based strategies for treatment of flowback waters.

Removal of human adenovirus 40 in membrane bioreactors: Roles of membrane pore size, fouling, pressure relaxation and membrane cleaning.

Z. Yin, V.V. Tarabara*, I. Xagoraraki
Department of Civil and Environmental Engineering, Michigan State University, East Lansing, MI 48824
* Corresponding author: tarabara@msu.edu; +1 (517) 432-1755

Abstract

Virus removal in membrane bioreactors is of concern since membrane pore size can be larger than the size of certain viruses. In the present work, human adenovirus 40 (HAdV 40) removal experiments were carried out using PVDF ultra- and microfiltration membranes with nominal pore sizes of 0.04 µm and 0.45 µm simulating membranes commonly used in full scale MBRs.

In the first part of the study, HAdV40 removal from model feeds was explored to elucidate individual and combined impacts of dissolved and particulate foulants on membrane fouling and virus removal [1]. Membrane fouling by SiO_2 colloids led to 1.2 log decrease in the removal of HAdV 40 by 0.04 µm membranes but had much smaller effect (δLRV=0.2) on virus removal by the microfilters. The decreased removal was attributed to the cake-enhanced accumulation of viruses at the membrane surface. In contrast, fouling by humic acid significantly enhanced virus removal by 0.04 µm membranes (δLRV=0.8) but not by 0.45 µm membranes (Figure 1). The trend can be explained by intrapore fouling and pore blocking by humic acid. When membranes were challenged by virus-seeded feeds containing both humic acid and suspended SiO_2, the reductions in virus removal for 0.04 µm and 0.45 µm membranes were 0.04 log and 0.53 log, respectively. The overall removal appeared to be a superposition of contributions due to individual feed components fouling the membrane.

In the second part of the study, we applied pressure relaxation and permeate backwash to 0.45 µm microfilters in a submerged MBR treating synthetic wastewater [2]. The experimental variables included durations of the filtration ($t_{\Delta P>0}$), pressure relaxation ($t_{\Delta P=0}$) and backwash ($t_{\Delta P<0}$) stages. Both pressure relaxation and permeate backwash led to significant reductions in virus removal (Figure 2). For the same value of $t_{\Delta P>0}/t_{\Delta P=0}$, longer filtration/relaxation cycles (i.e. larger $t_{\Delta P} + t_{\Delta P=0}$) led to higher transmembrane pressure (ΔP) but did not have a significant impact on virus removal. A shorter backwash ($t_{\Delta P<0}$ = 10 min) at a higher flow rate (Q= 40 mL/min) resulted in more substantial decreases in ΔP and virus removal than a longer backwash ($t_{\Delta P<0}$ = 20 min) at a lower flow rate (Q = 20 mL/min) even though the backwash volume ($Qt_{\Delta P<0}$) was the same. Virus removal returned to pre-cleaning levels within 16 h after backwash was applied. Moderate to strong correlations (R^2 = 0.63 to 0.94) were found between ΔP and virus removal (Figure 3).

References

1. Yin, Z.; Tarabara, V.; Xagoraraki, I. Human adenovirus removal by hollow fiber membranes: Effect of membrane fouling by suspended and dissolved matter. J. Membr. Sci. 2015, 482, 120–127.
2. Yin, Z.; Tarabara, V.; Xagoraraki, I. Effect of pressure relaxation and membrane backwash on adenovirus removal in a membrane bioreactor. Water Res. 2016, 88, 750–757.

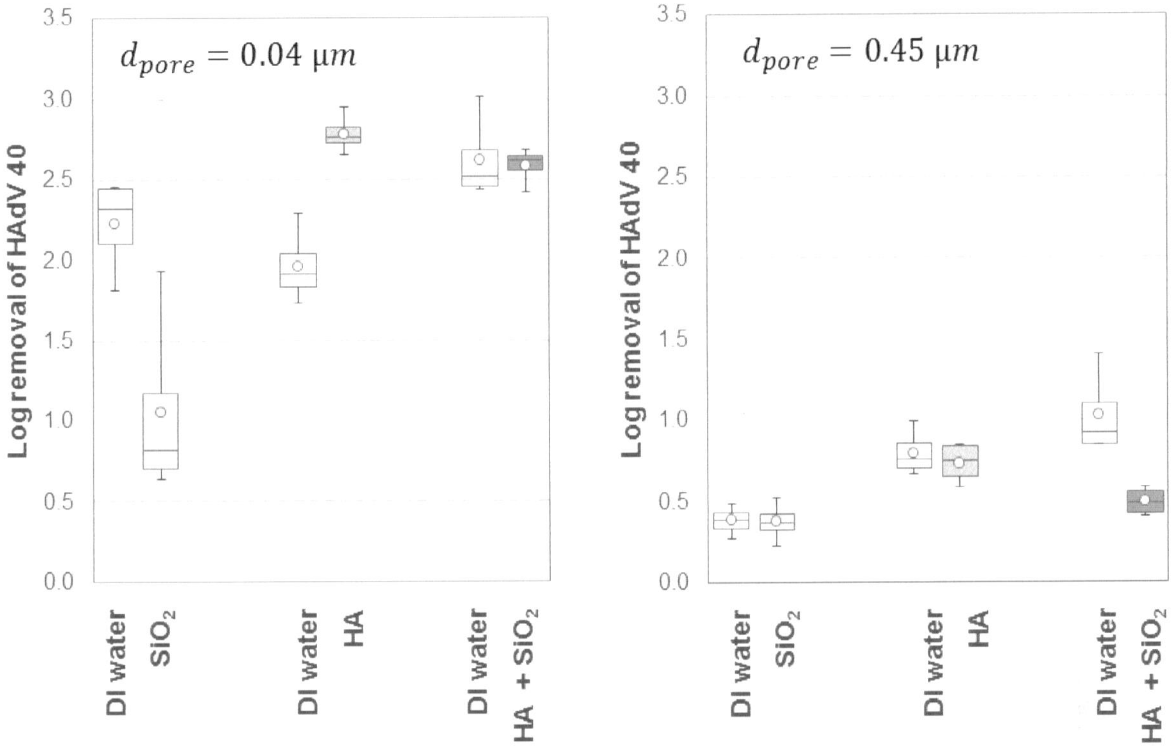

Figure 1: Comparison of HAdV 40 removal from DI water, suspension of SiO_2 microspheres, solution of humic acid, and SiO_2/HA mixture by two membranes of different nominal pore sizes: 0.04 µm, and 0.45 µm [1].

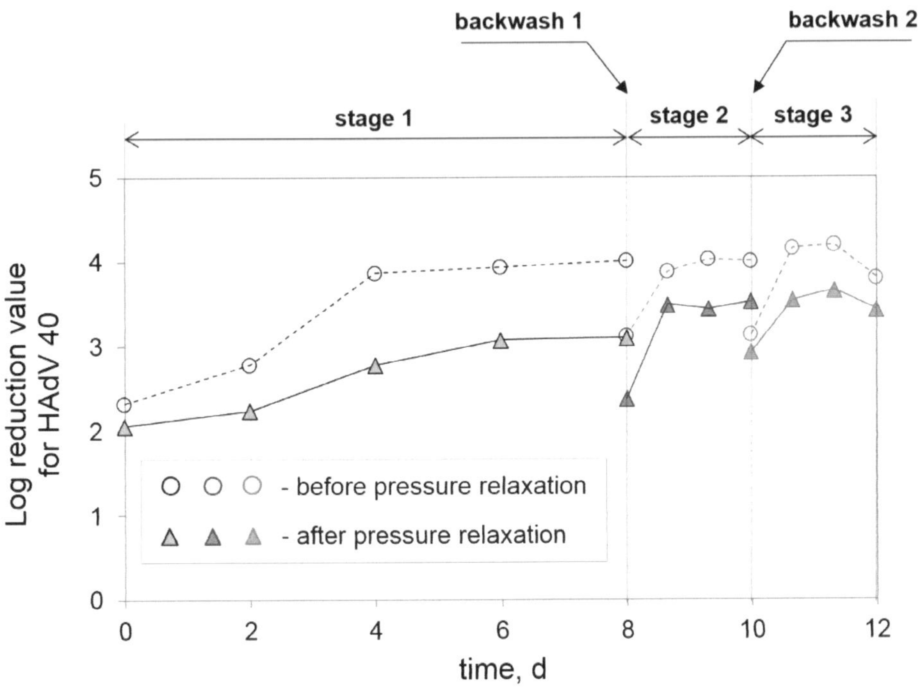

Figure 2: Effects of pressure relaxation and backwash on the removal of HAdV 40 in a submerged MBR [2]. Pressure relaxation format: $t_{\Delta P>0}/t_{\Delta P=0}$ = 25 min / 5 min.

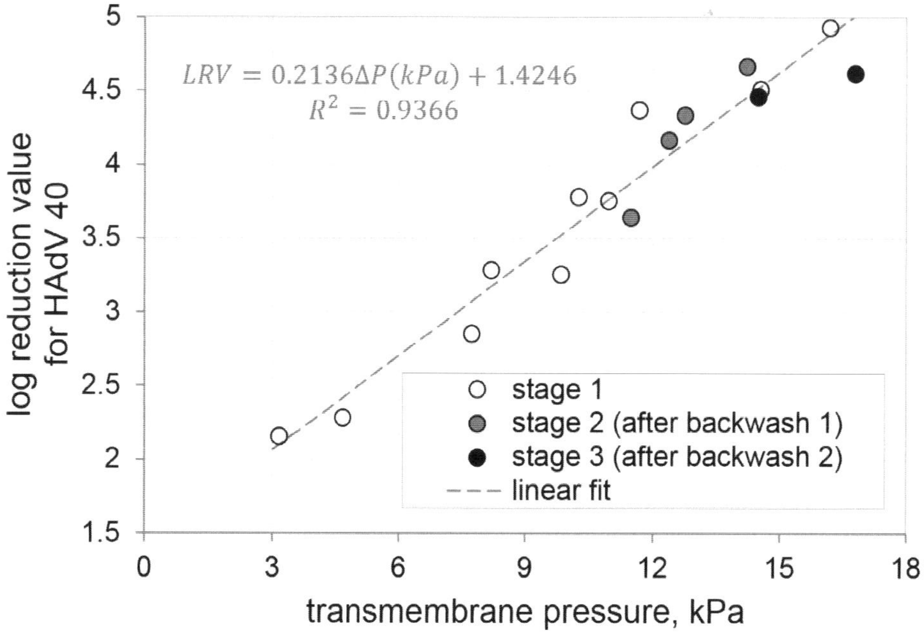

Figure 3: Correlation between HAdV 40 removal and transmembrane pressure in MBR [2]. Pressure relaxation format: $t_{\Delta P>0}/t_{\Delta P=0}$ = 50 min / 10 min.

Fouling Behavior of Charged Nanofiltration Membranes and the Influence of Graphene Quantum Dot Surface Modification

Andrew Colburn (Speaker)[a], Namal Wanninayake[b], Doo Young Kim[b], and D. Bhattacharyya[a]

[a] Department of Chemical and Materials Engineering, University of Kentucky, Lexington, KY 40506

[b] Department of Chemistry, University of Kentucky, Lexington, KY 40506

Nanofiltration membranes have been widely utilized for many applications including water softening, pharmaceutical purification, and dairy processing. In systems with organic foulants or sparingly soluble salts present, fouling of the membrane surface leads to reduced performance and shortened membrane lifespans. Preventing initial nucleation of foulant on the membrane surface is critical to preventing the formation of a fouling layer.

This work seeks observes fouling phenomenon on the surface of charged NF membranes with the particular objective of understanding the role of membrane surface charge and ion concentration in the nucleation of foulant on the membrane surface. Both BSA fouling and $CaSO_4$ scaling were studied for the Dow NF-270, NF-90 membranes, and Nanostone charged NF membranes. Chloride presence was shown to delay $CaSO_4$ scaling during water recovery compared to pure $CaSO_4$. Negative membrane charge was determined to reduce BSA fouling, but not $CaSO_4$ scaling, due to the positive charge of the Ca^{2+} rich gypsum nucleation regions.

Graphene oxide (GO) has been shown to have antifouling properties due to its hydrophilicity, and consequently there have been multiple investigations into GO modified membranes. Graphene quantum dots (GQDs), two-dimensional nanoparticles between 2-5 nm, have greater functionality than GO due to increased edge area, high water solubility, and photo-reactive properties. Through EDC/NHS coupling, GQDs were covalently bonded to the NF membrane surface. Zeta potential analysis shows higher negative surface charge properties after GQD coupling and increased rejection of Na_2SO_4 (97% compared to 76% for the standard membrane). The GQD coupled membrane exhibits double the normalized flux after 2 ½ hours of recovery of solution containing 400 ppm BSA when compared to the standard membrane.

Acknowledgements

The author acknowledges the support of the Southern Company in funding and collaboration, and additional funding from the National Science Foundation EPSCoR for funding, and from the National Institute of Environmental Health Sciences (P42ES007380). Nanostone Membrane of Oceanside, CA has provided support and collaboration in the development of membranes and spiral wound membrane modules.

REAL-TIME FOULING MONITORING AND CHARACTERIZATION FOR EFFECTIVE OPERATIONAL CONTROL OF ULTRAFILTRATION AS PRETREATMENT FOR SEWATER DESALINATION

Han Gu, Larry X. Gao, Anditya Rahardianto, Panagiotis D. Christofides, and Yoram Cohen

Abstract

Effective feed pre-treatment is vital for fouling mitigation in seawater Reverse Osmosis (RO) desalination plants. In recent years, the integration of Micro and Ultrafiltration (MF&UF) membrane systems with RO system has become increasingly popular for improving seawater RO pretreatment. However, UF fouling by particulates, colloidal, organic, and biological materials is a major challenge when treating water of fluctuating feed quality. Natural events such as algal bloom in coastal and littoral seawater can result in rapid increases of UF filtration resistance and decrease the efficiency of UF backwash. Excessive fouling can lead to reduction of UF productivity and fouling of the RO membrane. As a result, frequent UF membrane chemical cleaning and eventual costly membrane replacement may be required to restore plant performance. In this work, an integrated real-time monitoring approach was developed and implemented to quantify the fouling level in the UF filtration system. Instead of relying on a single or set of feed water quality indices as is traditionally the case, the present study demonstrated the benefit of utilizing a multiplicity of soft sensors for characterizing and effectively controlling the operation of the UF feed pretreatment to the RO plant. Post-backwash UF initial filtration resistance, UF filtration period fouling rate and foulant removal efficiency via UF backwash were autonomously evaluated in real time by the onboard plant embedded computing system. Here we note that the soft sensors enabled real-time characterization of the effectiveness inline coagulation at various dose of $FeCl_3$ without the needs of extensive jar testing experiments. The feasibility of this approach was demonstrated using a 12,000 gallons/day (RO permeate production capacity) integrated UF-RO plant for costal seawater desalination at the US Naval base in Port Hueneme, California. The plant was equipped with various process sensors (flow rates, pressure, conductivity, pH, temperature, *Chlorophyll a* and DO). Extensive field studies showed that feed water quality and coagulant dose had a direct impact on both UF fouling rate (during the filtration period) and effectiveness of foulant removal by UF backwash. Higher coagulant dose resulted in higher rate of cake formation and in turn actually increased backwash effectiveness, a behavior which was particularly observed during storm events in which the raw water feed was of a high fouling propensity. However, there is a maximum coagulant dose beyond which backwash effectiveness is not improved. Also, it is important to note that excessive coagulant dosing can lead to coagulant passage through the UF and fouling of the downstream RO membranes. Results of the current work provide the fundamental basis and guidance for implementing self-adaptive UF operational control for optimal RO pre-treatment strategies via UF filtration.

Application of Ceramic Membranes in Treatment of Oil Sands Produced Water.

Water and energy production, especially in the oil and gas sector, are directly linked; therefore, development of innovative technologies and processes for treatment and recycling of produced water (PW) is of great importance. In recent years due to increasingly stringent water management regulations in the province of Alberta, Canada, a great deal of attention has been given to ceramic membranes. Although promising, ceramic membranes, in PW treatment applications, are highly susceptible to fouling and severe loss of flux during operation. This issue renders most commercially available ceramic membranes uneconomical in large scale produced water treatment applications. Ceramic membranes with good resistance to organics fouling and high operating temperatures are uniquely poised to address many of the water management challenges in the oil and gas sectors. Consequently, successful and economically viable applications rely on a better understanding of membrane fouling in order to optimize the selective layer's chemical and physical properties for a sustainable economic range of water flux over the life of the membrane.

Natural Resources Canada in collaboration with the University of Ottawa, Environment Canada and a number of oil sands and membrane companies are involved in developing a water treatment process that utilises ceramic membranes for the deoiling of oil sands PW with focus on Steam Assisted Gravity Drainage (SAGD) operations. The process is aimed at treating SAGD produced water (SAGD-PW) at high temperature (90°C) for recycling.

A selected number of microfiltration and ultrafiltration ceramic discs and commercial size tubular multi-channel membranes, with different selective layer materials, were tested for deoiling PW samples from several oil sands operations. The water samples included SAGD-PW and oily PW samples with a bitumen content of as high as 20 g/L. Highlights of the approach and results will be summarized. In particular, the PW characterization and separation data will be presented for membranes with different selective layer materials and pore sizes. Additionally, flux data and results of the modeling of fouling for selected membranes will be discussed. Finally, characterization of the ceramic membranes using microanalytical techniques will be briefly discussed in the context of the effect of the harsh process conditions on the membrane matrix and performance.

Role of Permeate Flux and Specific Membrane-Foulant-Solute Affinity Interactions (°¼ΔG°½_slm) in Transport of Trace Organic Solutes through Fouled Nanofiltration (NF) Membranes.

T.O. Mahlangu[a,b,*], K.V.K.M. Schoutteten[b], A. D'Haese[b], J.M. Thwala[c], B.B. Mamba[a], A.R.D. Verliefde[b]

[a] University of South Africa, Nanotechnology and Water Sustainability Research Unit, College of Engineering, Science and Technology, Florida Science Campus, Roodepoort

[b] Ghent University, Faculty of Bioscience Engineering, Particle and Interfacial Technology Group (PaInT), Coupure Links 653, B–9000 Ghent, Belgium

[c] University of Swaziland, Faculty of Science, Department of Chemistry, Kwaluseni Campus Private Bag 4, Kwaluseni, M201, Swaziland

* Corresponding author; E–mail: mahlanguoranso@yahoo.com, Tel: +32 9 264 60 01, Fax: +32 9 264 62 42

Abstract

In the literature, sieving effects, charge interactions and adsorptive/hydrophobic interactions have been named as major mechanisms influencing the rejection of trace organic compounds (TOrCs) or pharmaceuticals. However, dominant mechanisms influencing TOrC rejection by fouled membranes are not yet fully understood. Most studies have attributed the decline in rejection to cake-enhanced concentration polarisation (CECP) effects and have overlooked the influence of flux decline and attractive membrane-TOrC interactions. Therefore, the aim of this study was to investigate the role of various factors in TOrC rejection for nanofiltration (NF) membranes fouled in cross-flow filtration. To further gain insight into the role of affinity interactions on TOrC rejection, membrane-solute interaction energies (ΔG_{slm}) were computed from the contact angle measurements based on the Lifshitz-van der Waals/acid-base approach. A novel fouling approach (termed sequential fouling) is used to investigate CECP effects for different fouling layers. For a virgin membrane, TOrCs are rejected through sieving mechanisms and repulsive non-electrostatic membrane-TOrC interactions. Upon fouling (especially in the presence of calcium), there was an improvement in the rejection of some TOrCs (solute permeation decreased) due to an improvement in sieving effects. The rejection of some TOrCs decreased over time (solute permeation increased) but the rejection values were close to those of the virgin membrane at similar water fluxes showing the influence of flux. When solute permeation increased as a result of fouling, TOrC transport was mainly influenced by non-electrostatic membrane-TOrC affinity interactions and CECP effects (as evidenced by novel sequential fouling experiments).

Keywords: pharmaceuticals; membrane-solute affinity; nanofiltration

Behavior of oil droplets at the membrane surface during crossflow microfiltration of oil-water emulsions

Emily N. Tummons[1], Volodymyr V. Tarabara[1], Jia Wei Chew[2,3], Anthony G. Fane[3,4]

[1] Department of Civil and Environmental Engineering, Michigan State University, East Lansing, MI 48824, USA
[2] School of Chemical and Biomedical Engineering, Nanyang Technological University, Singapore, 637459
[3] Singapore Membrane Technology Centre, Nanyang Environment and Water Research Institute, Nanyang Technological University, Singapore, 637141
[4] School of Civil and Environmental Engineering, Nanyang Technological University, Singapore, 639798

Keywords: oil-water separation; microfiltration; crossflow filtration; direct observation through membrane (DOTM); force balance.

Many industries struggle to separate waste streams containing oil-water emulsions because few technologies are capable of separating micron sized oil droplets while remaining cost effective. Microfiltration and ultrafiltration membranes are often the best option, but overtime a fouling layer of oil develops on the membrane surface that causes the permeate flux to decline. The way in which oil fouls the membrane surface is not well understood, as the blocking laws (standard blocking, intermediate blocking, and complete blocking) do not apply to deformable particles such as oil droplets.

We utilized the Direct Observation Through the Membrane (DOTM) technique to gain insights into oil droplet behavior at the membrane surface. To our knowledge, this is the first application of a direct visualization technique used to capture real-time images of a membrane surface under conditions of fouling by emulsified oil in the presence of crossflow. DOTM experiments with hexadecane-water-SDS emulsions revealed three characteristic stages of membrane fouling by oil: 1) droplet attachment and clustering, 2) droplet deformation, and 3) droplet coalescence.

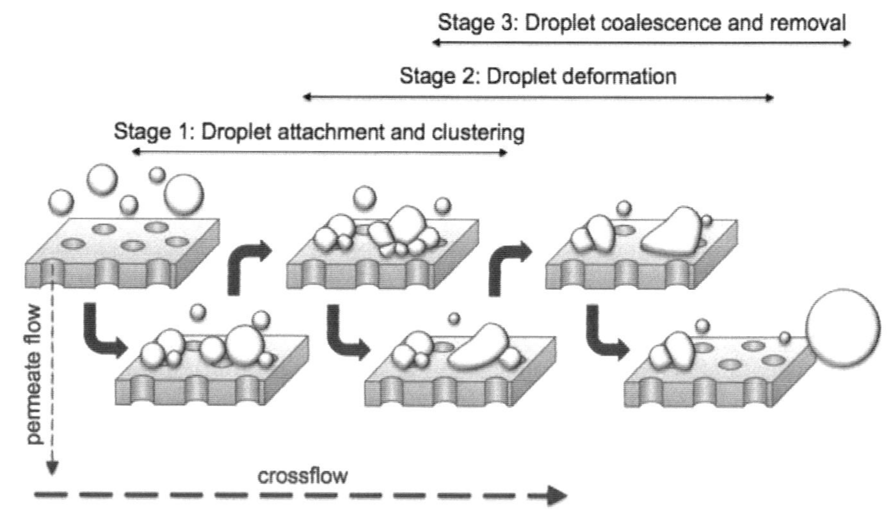

Figure 1: Three characteristic stages of membrane fouling by emulsified oil in the presence of crossflow.

The DOTM tests were conducted to screen for a range of fouling conditions; the sequence and duration of the three fouling stages are demonstrated in Figure 1. While the three stages occur sequentially for individual oil droplets and their clusters, the stages could overlap significantly when an ensemble of droplets on the entire membrane surface is considered. Figure 2 depicts a permeation event for a hexadecane oil droplet pinned at a pore entrance on the surface of a polycarbonate track-etch membrane where d_{pore} = 5 µm.

The effects of emulsion stability and membrane pore size were experimentally tested, while a force balance on a droplet helped elucidate the fate of an oil droplet pinned at a pore entrance. In qualitative agreement with visualization results, the force balance shown in Figure 3 predicted that droplets ≲ 36-40 µm would remain pinned at d_{pore} = 5 µm pores while larger droplets would be swept off the surface by the crossflow drag.

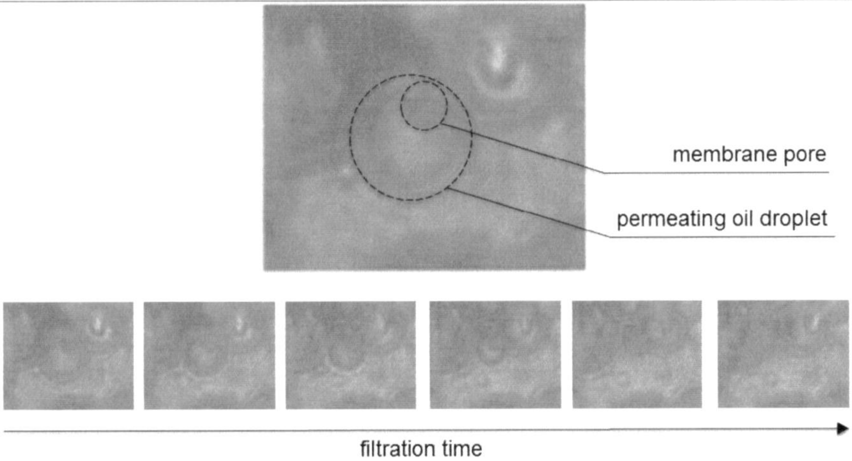

Figure 2: Different stages of oil droplet permeation through a 5 µm membrane pore in the case when $d_{drop} > d_{pore}$. The top image illustrates the initial condition of a droplet pinned at an entry to the membrane pore. The sequence of images below shows the evolution of the permeating droplet; the left most image in the sequence is the same as the large image with the pore and the droplet marked. This sequence took place in less than 2 seconds.

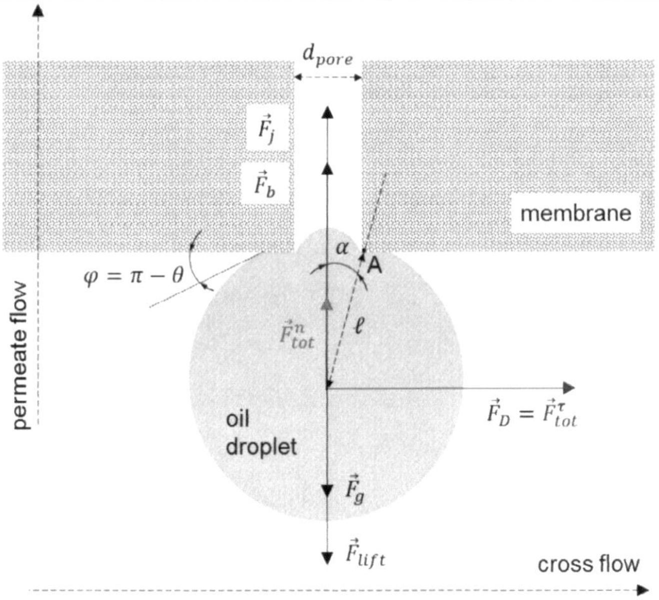

Figure 3: Hydrodynamic forces acting on an oil droplet positioned at the entry to a cylindrical pore of a membrane. The angles θ and α are the droplet's contact angle and angle of repose, respectively. For the definitions of forces see [1]. Forces are not drawn to scale. In DOTM tests, the microscope was located above the semitransparent membrane (i.e. on the permeate side) and the focal plane of the microscope was on the feed side where oil droplets were accumulating.

The results indicate that membrane fouling by emulsified oil is controlled by droplet coalescence and crossflow shear: the transport of oil to the membrane surface by the permeate flow is balanced by the shear-induced removal of the droplets that coalesce to exceed a critical size. Understanding the surface coalescence of droplets can help make optimal choices for the membrane type and operational conditions to minimize fouling and improve flux and rejection.

Reference:

[1] Tummons, E. N.; Tarabara, V. V.; Chew, J.-W.; Fane, A. G. Behavior of oil droplets at the membrane surface during crossflow microfiltration of oil-water emulsions. *J. Membr. Sci.* 2016, 500, 211-224

Sub-Angstrom Molecular Differentiation of Organic Liquids using Carbon Molecular Sieve Membranes

Dong-Yeun Koh,[a] Melinda L Jue,[a] Benjamin A McCool,[b] Harry W. Deckman,[b] Ryan P Lively[a,*]

[a]School of Chemical & Biomolecular Engineering, Georgia Institute of Technology, Atlanta, GA 30332, USA
[b]Separations & Process Chemistry, Corporate Strategic Research, ExxonMobil Research and Engineering, Annandale, NJ 08801, USA

High-performance membranes derived from carbon materials have shown excellent chemical resistance, high molecular selectivity and fast mass transport across the membrane. It is well-known that carbon molecular sieve (CMS) membranes are able to surpass the "polymer upper bound trade-off line" for many molecular pairs. CMS membranes with tailored ultramicropore and micropore dimensions show both high processability of polymeric membrane and the high selectivity of inorganic membranes. Moreover, slit-like transport pathways through the CMS materials translate into higher productivity (i.e., higher mass transport across a diffusion barrier) than that of cage-like zeolite structures due to the longer jump lengths between pore windows (**Fig. 1**). CMS membranes have been proven to be effective in a variety of gas separation processes such as olefin/paraffin separation, natural gas separation and air separations. However, the low fluxes observed in CMS hollow fibers (due to porous substructure collapse during pyrolysis) hinder scale-up of CMS membranes for industrial separation applications. The performance of carbon molecular sieve separation membranes, which exploit the effect of mass transport across a selective diffusion barrier in order to separate molecules, can be improved by reducing the thickness of the membrane. In this study, a novel post-spinning cross-linking treatment is introduced to maintain the asymmetric porous morphology in CMS hollow fiber membranes even after high temperature pyrolysis processes (**Fig. 2**). A dual-layer dry-wet spinning process was used to prepare asymmetric poly(vinylidene fluoride) (PVDF) hollow fibers with thin selective skin layers. Cross-linking of as-spun hollow fibers "locked-in" the porous substructure and the fiber morphology remained unchanged after the cross-linking process (**Fig. 2**). Moreover, this technique effectively prevented substructure collapse in asymmetric CMS hollow fiber membranes. These asymmetric CMS hollow fiber membranes have significant potential for utilization in organic solvent separations, as they exhibit essentially zero solvent-induced swelling and possess high selectivity for similarly sized molecules (**Fig. 3, 4**).

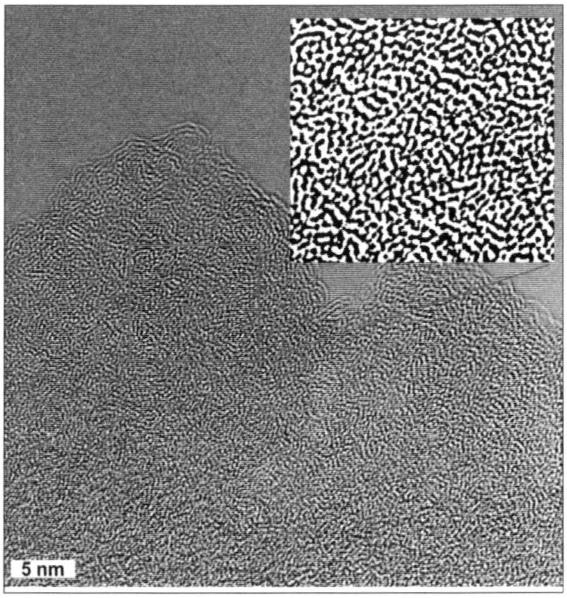

Figure 1: Nanostructure of carbon molecular sieve membranes from this work

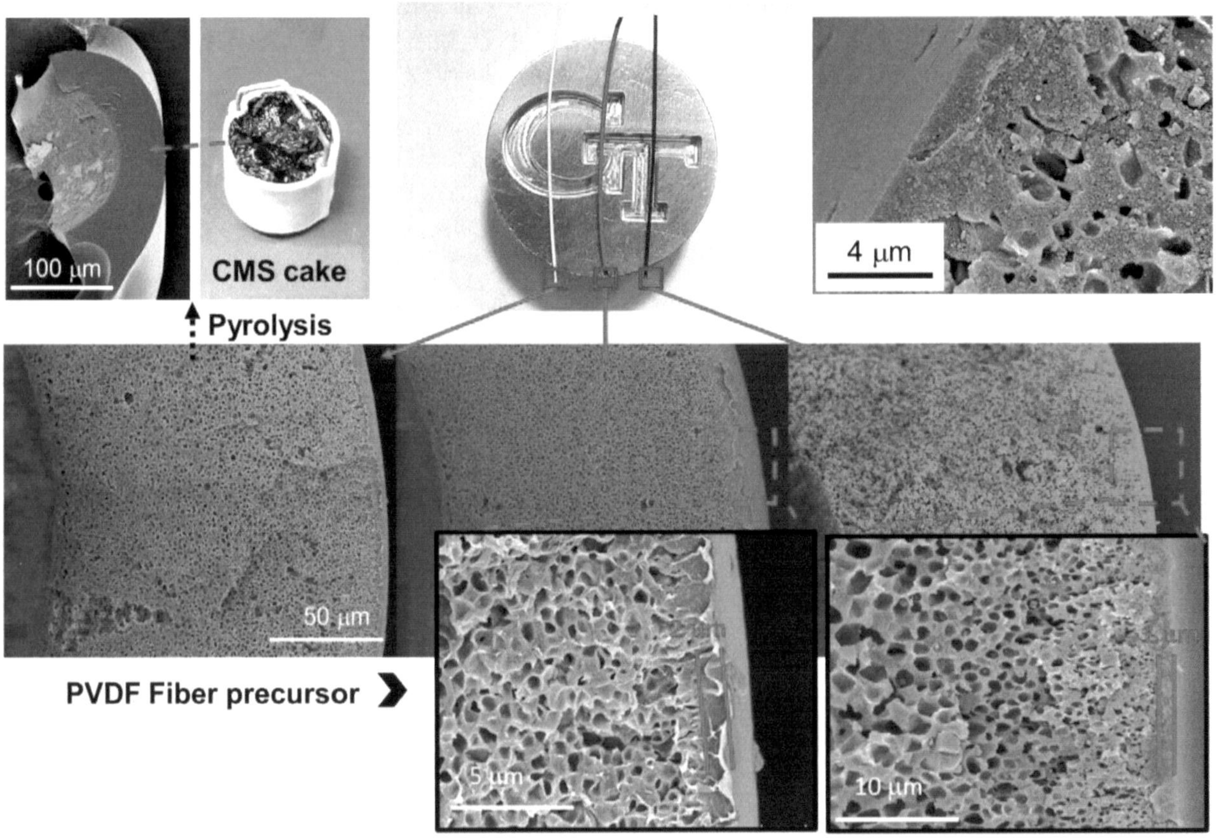

Figure 2: Neat PVDF hollow fibers (*left*) are crosslinked (*middle*), which preserves the porous asymmetry of the fibers when converted to carbon molecular sieve membranes (*right*)

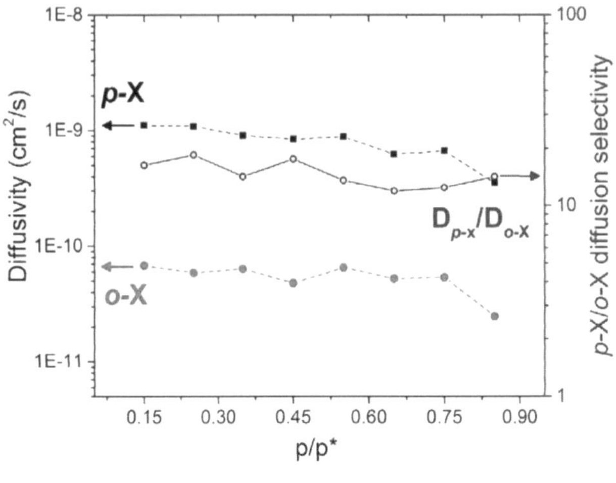

Figure 3: Diffusion selectivity of carbon molecular sieve membranes for p-xylene (p-X)/o-xylene (o-X)

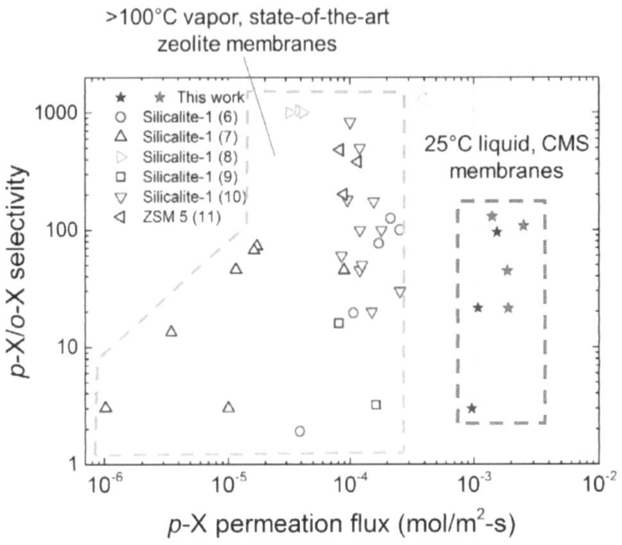

Figure 4: Mixture permeation performance of CMS membranes when challenged with p-xylene (p-X) and o-xylene (o-X)

Spent IPA Dehydration for In-house Reclamation – An Efficient, Complete Process to Solve the Environment Issues

Yu (Ivy) Huang*, Richard Baker, Xuezhen Wang, Ramin Daniels and Yan Gao

Membrane Technology and Research, Inc.
39630 Eureka Drive, Newark, CA 94560, USA

ivy.huang@mtrinc.com

Many commercial hydrophilic solvents form azeotropes or have vapor-liquid equilibrium (VLE) pinch points with water, making solvent recovery by conventional distillation difficult. Solvents that form such azeotropes or have such pinch points include isopropyl alcohol (IPA), acetone, methyl ethyl ketone (MEK), ethyl acetate, ethanol, butanol and tetrahydrofuran etc. Molecular sieve drying is possible, but requires large amount of energy to vaporize the solvent/water mixture, and is generally not economical for situations where water concentration is more than 5 wt%. A simple alternative process is pervaporative dehydration to reclaim and reuse the solvent on-site.

This talk will use IPA as an example to present a pervaporation apparatus and process for spent IPA dehydration. Currently, spent IPA generated in our company is disposed of as hazardous waste, at a cost of about $18/gal. Virgin, high-purity IPA costs about $23/gal, so the total cost of IPA used as a cleaning/drying agent is in the range of $41/gal. There is therefore a considerable economic and environmental driving force to recover and reuse the IPA solvent. A pilot system was constructed for IPA reclamation which has a treatment capacity up to 200 gal/day. Batches of spent IPA (50-100 gallons for each batch) with an average water concentration of 15 wt% are reclaimed using this pilot system via a single 4-inch module, the reclaimed dry IPA contains less than 1 wt% water. Stable module performance has been demonstrated. We will show this simple batch system can be configured to achieve the separation performance comparable to a one-stage, two-stage, two-step and more complicated pervaporation system designs.

Aromatic Polyimide and Crosslinked Thermally Rearranged Poly(benzoxazole-co-imide) Membranes for Isopropanol Dehydration via Pervaporation

Yi Ming Xu[1], Ngoc Lieu Le[1,2], Jian Zuo[1], Tai-Shung Chung[1,2]*

[1] Department of Chemical and Biomolecular Engineering,
National University of Singapore, 4 Engineering Drive 4,
Singapore 117576

[2]Water Desalination & Reuse (WDR) Center,
King Abdullah University of Science and Technology,
Thuwal 23955-6900, Saudi Arabia

In this study, novel crosslinked thermally rearranged polybenzoxazole (C-TR-PBO) membranes, which show impressive results for isopropanol dehydration, have been obtained via in-situ thermal conversion of hydroxyl-containing polyimide precursors. The polyimide precursors are synthesized by the polycondensation of three monomers; namely, 4,4'-(hexafluoroisopropylidene) diphthalic anhydride (6FDA), 3,3'-dihydroxybenzidine diamine (HAB) and 3,5-diaminobenzoic acid (DABA). Due to the incorporation of the carboxylic-group containing diamine DABA into an ortho-hydroxypolyimide precursor, the thermal induced crosslinking reaction can be achieved together with the thermal rearrangement process. Consequently, a synergistic effect of high permeability and high selectivity can be realized in one step. The resultant C-TR-PBO membrane exhibits an unambiguous enhancement in permeation flux compared to their polyimide precursors. Moreover, the newly developed C-TR-PBO membrane displays stable isopropanol dehydration performance at 60 °C throughout the continuous 200 hours. The promising preliminary results achieved in this study may offer useful insights for the selection of membrane materials for pervaporation and new methods to molecularly design next-generation pervaporation membranes.

Graphic abstract:

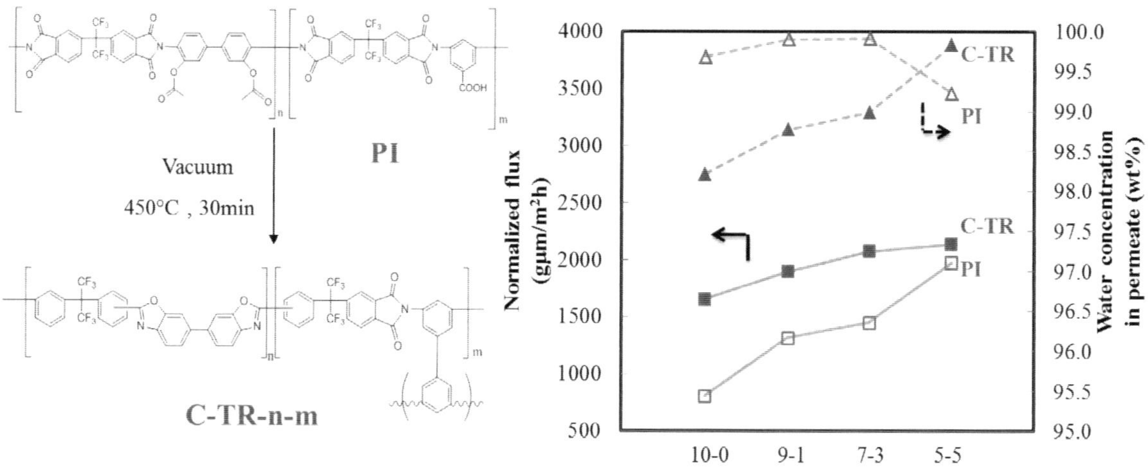

Session: Molecular Separation and Pervaporation

Session Chairs: Edgar S. Sanders (Air Liquide), JR Johnson (SABIC), Rajinder P. Singh (Los Alamos N.L.)

A New Molecular Approach to Tuning Separation Properties of Carbon Molecular Sieve Membranes

Graham B. Wenz and William J. Koros

Georgia Institute of Technology

Atlanta, GA 30332

Natural gas separations have become increasingly important as worldwide natural gas production has increased in the past decade. Common natural gas impurities, such as CO_2, must be removed in order to protect equipment and meet pipeline specifications for transmission. Carbon molecular sieve (CMS) membranes, formed by high temperature pyrolysis of polymers under controlled conditions, have emerged as a new class of materials that can surpass the polymer "upper bound". This work outlines a new post-synthetic modification technique where CMS membranes are exposed to a solution containing PPM levels of amine dopants. The original goal of this work was to modify the morphology in order to prevent "free volume" relaxation that results in physical aging of CMS materials. While long term CO_2 and CH_4 permeation experiments showed aging-resistant membranes did not result, attractive performance resulted from the *amine doped materials*. The results of these experiments suggest *amine doping* as new technique to selectively modify the large pores present within the CMS microporous structure, resulting in increased size and shape based selectivity.

Sessions: Gas separation membranes, molecular separation and pervaporation, design and processing of polymeric and inorganic membranes

Electric field-assisted membranes for molecular separation

Oleg Polyakov

Principal Engineer, Thermosolv LLC, Laramie, Wyoming, USA

Using anodic aluminum oxide (AAO) as a material platform, Thermosolv is developing new nanostructured ceramic membranes for molecular separation; the field of organic solvent nanofiltration is one example of the potential applications. AAO – a self-organized material formed by anodizing aluminum metal – has highly uniform and ordered nanoscale pores, Fig. 1. Through precise control of anodization parameters, pore diameter and period are tunable from several nm to several hundred nm, and can be varied along the pore length. The asymmetric membrane structure includes (i) relatively thick 50-100 µm support layer with large-diameter (>100 nm) pores and (ii) thin sub-µm active layer with monodisperse nanometer-scale pores, Fig. 2.

Depositing electrically conductive metallic layer on the membrane surface allows applying an electric field to the pore entrances. Small diameter of the pore openings will render the field inside the pores highly non-uniform. While not affecting noticeably the flow of solvent molecules, the field introduces dielectrophoretic force affecting the motion of the solutes, Fig. 2. For a given system of solutes/solvent, and for specific membrane pore size, the applied potential will control the molecular weight cut-off of a membrane across a wide and tunable range.

Case studies for fractioning organic dyes with the molecular weights in the range 100-3000 Da will be presented, Fig. 3.

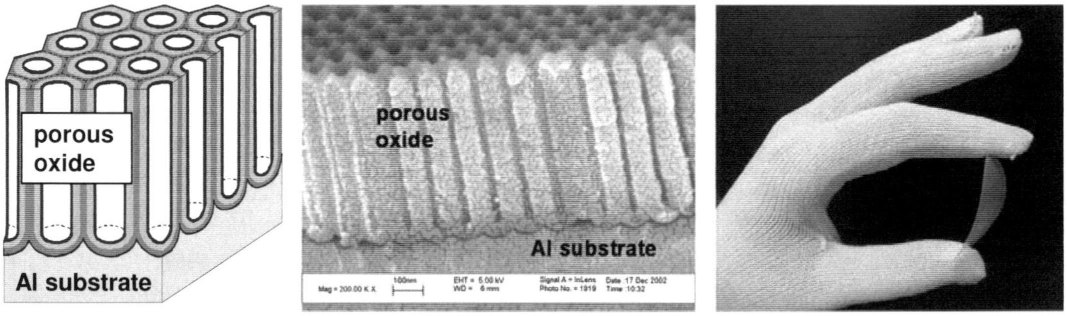

Figure 1. Artist's concept and SEM image of AAO, and free-standing AAO film.

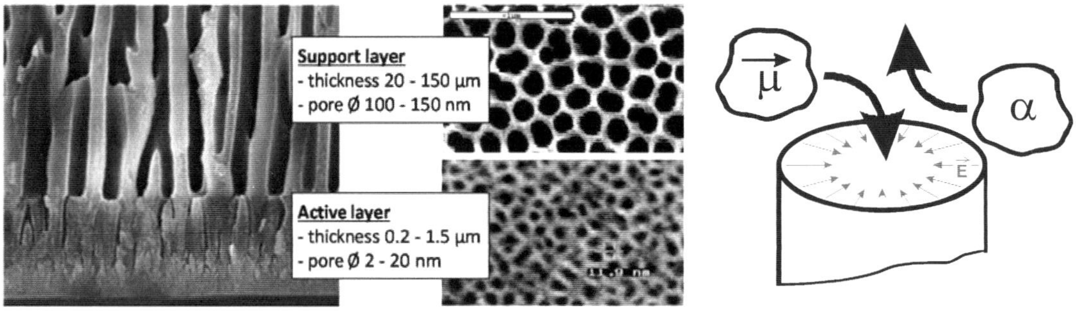

Figure 2. Asymmetric membrane structure (left) and dielectrophoretic separation (right).

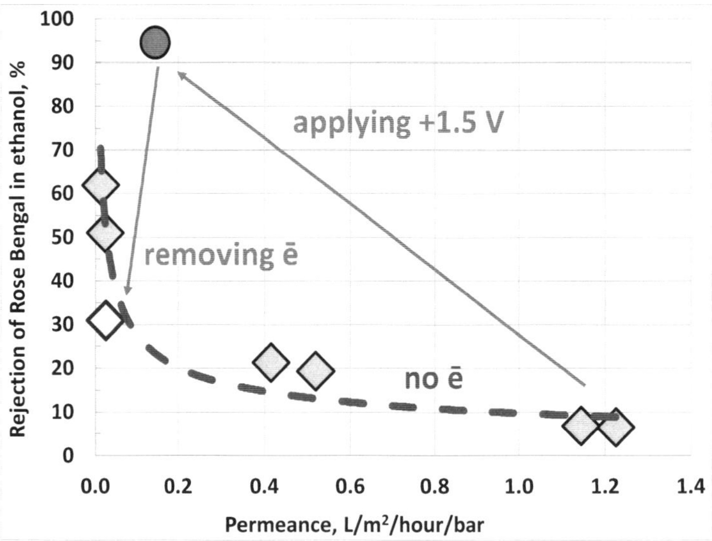

Figure 3. Applied potential changes the membrane selectivity and permeance.

Optimizing Hollow Fiber Membranes and Modules for Osmotic Processes: From Models to Membranes to Modules

Jian Ren, Jeffrey R. McCutcheon*

Department of Chemical and Biomolecular Engineering,
Center for Environmental Sciences and Engineering,
University of Connecticut,
191 Auditorium Rd. Unit 3222, Storrs, CT 06269-3222, USA
* jeff@engr.uconn.edu

Hollow fiber membranes have long been considered a valuable platform for membrane separations because of their high packing density relative to flat sheet membranes. Recently, the osmotic process community has developed hollow fiber membranes intended for forward osmosis (FO) and pressure retarded osmosis (PRO) applications in order to capitalize on these same advantages. Many of these hollow fiber membranes were made using a thin film composite (TFC) approach with the focus on the design of support layer toward a thin, highly porous, minimally tortuous supporting structure to minimize the mass transfer resistance during osmotic processes. These hollow fiber FO membranes demonstrated excellent FO performance, but also suggested a need for intensively and delicately tailored membrane support layers which can lead to compromised membrane properties.

This study evaluates an approach to make thin film composite hollow fiber membranes for forward osmosis by simply employing commercial hollow fiber ultrafiltration (UF) membranes as support material. A thin polyamide film with excellent selective properties was synthesized on the inner surface of hollow fibers via interfacial polymerization. Besides demonstrating the feasibility to be used as good TFC FO membrane supports, the commercial hollow fiber UF membranes also provide a systematic platform with consistent properties to study structure-performance relationship of FO hollow fiber membranes. A series of commercial hollow fiber membranes were used to evaluate how molecular weight cutoff (MWCO) impacted the properties of the polyamide layer and overall performance of the TFC membrane.

Aside from using commercial hollow fiber UF membranes as FO membrane supports at bench scale, we also demonstrate that the TFC hollow fiber FO membranes can be made on existing hollow fiber modules at pilot scale. A series of commercial hollow fiber modules with different fiber size were used to make TFC hollow fiber FO membrane modules. The resultant TFC hollow fiber membranes were evaluated under various operating conditions (membrane orientation, cross flow arrangement, cross flow velocity, and draw solution concentration). While we evaluate how basic performance metrics (water and solute flux) are impacted by module operating conditions, overall the modules demonstrate impressive FO performances. The facile approach for modification may promote exploration of other hollow platforms for even better performance.

With the availability of reproducible membranes and modules, we have developed a computational fluid dynamics (CFD) model with COMSOL Multiphysics to study the impacts of hollow fiber and module properties in order to optimize hollow fiber module design for FO application. Properties like module dimensions (length, width) as well as fiber dimensions (inner diameter and outer diameter) were modeled to better elucidate how such features impact performance. The FO process in a hollow fiber membrane was simulated in a 2D axis-symmetry geometry and described by fluid dynamics coupled with mass transfer across the membrane. We verified the models accuracy by constructing modules with the same dimensions and fibers and testing under pertinent conditions. The agreement between the model and experimental results provided insight into both how more accurate models can be developed and how these models can be used to design better modules without costly experimental testing.

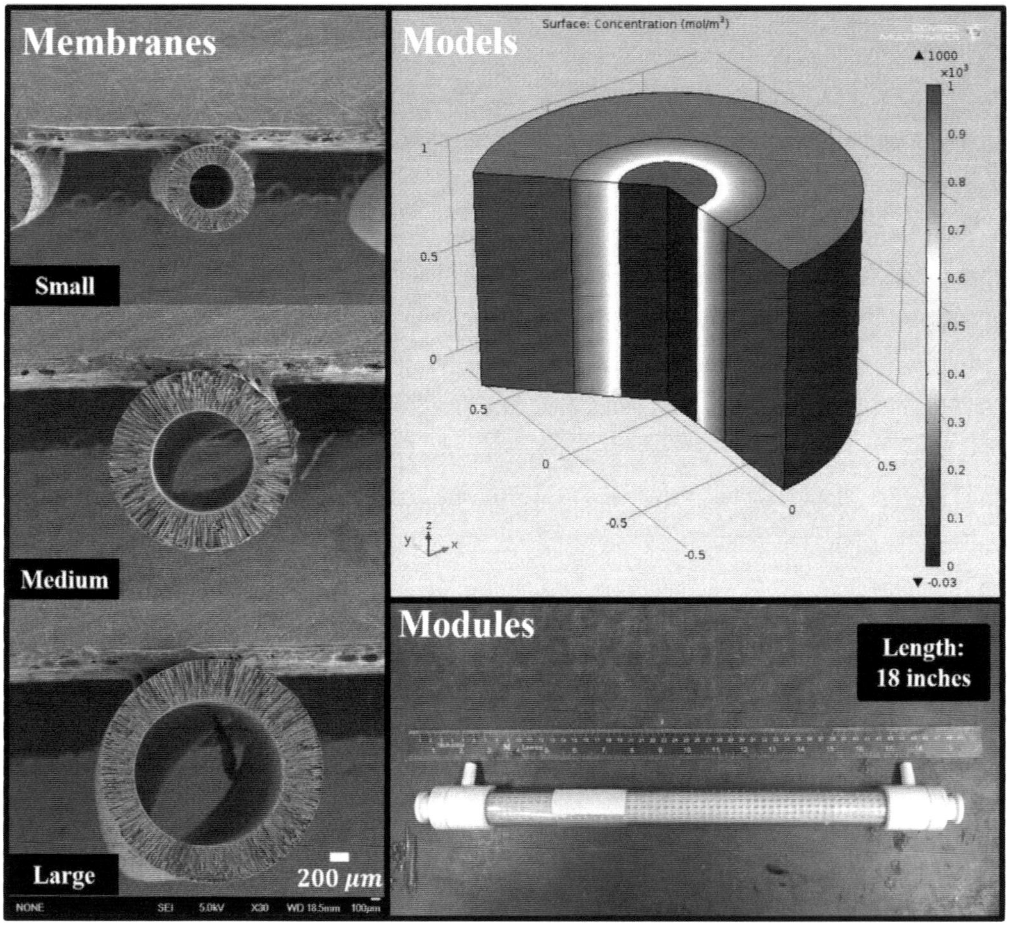

Fundamental Studies of Ion Sorption and Transport in Ion-Exchange Membranes

Jovan Kamcev, Benny D. Freeman*, Donald R. Paul

McKetta Department of Chemical Engineering, The University of Texas at Austin, USA

freeman@che.utexas.edu

Fundamental understanding of ion sorption and transport in ion-exchange polymers is critical for developing high performance membranes for water purification and energy generation applications. Despite the long history of literature on the topic, the current state of understanding is not sufficient to build the appropriate connections between polymer structure and transport properties. Experimental techniques for characterizing ion sorption and transport in charged membranes have been established, however, a simple theoretical framework to describe the experimental findings is missing. In this work, a framework for interpreting ion sorption and diffusion in ion-exchange membranes has been formulated and tested against experimental data. For the membranes used in this study, the framework could describe, and in some cases predict, concentration gradient (i.e., salt permeability) and electric field (i.e., ionic conductivity) driven ion transport in charged membranes.

Equilibrium portioning of ions between the membrane and solution phases was modeled using the classical Donnan theory. Previous studies have often overlooked the importance of including ion activity coefficients (i.e., non-ideal effects) in this model, mainly because no reliable models or experimental techniques were available for obtaining these values in the membrane phase. In this study, Manning's counter-ion condensation theory, originally developed for polyelectrolytes, was extended to ion-exchange membranes and used to calculate membrane ion activity coefficients. Solution ion activity coefficients were obtained by the Pitzer model. This framework was validated using experimental data for ion sorption of various electrolytes in commercial ion exchange membranes. Remarkably good agreement between theory and experiment was observed without using adjustable parameters.

Ion diffusion coefficients in charged membranes are affected by at least two factors: (1) the geometric impedance by the polymer chains and (2) electrostatic effects due to the fixed charge groups. The geometric factor was modeled using an approach developed by Mackie and Meares, which presumes ion diffusion occurs in the solvent in the swollen polymer matrix and the polymer chains act as impenetrable obstructions. This model has no adjustable parameters. The electrostatic influence of the fixed charge groups on the diffusing mobile ions was modeled using an approach set forth by Manning. Manning's model on diffusion is a direct extension of his counter-ion condensation theory used to predict ion activity coefficients. The framework for modeling ion sorption and diffusion in charged polymers developed in this study was validated using experimental salt permeability and ionic conductivity data for commercial charged membranes. Remarkably, this approach was able to predict salt permeability coefficients in charged membranes with no adjustable parameters and accurately describe ionic conductivity data using only one adjustable parameter. The generality and shortcomings of this framework are discussed.

Abstract

Antibody Purification from Cell Culture Supernatant Using New Multimodal Membranes

Juan Wang, Scott M. Husson

Department of Chemical and Biomolecular Engineering, Clemson University, Clemson, SC 29634 USA

This presentation will describe our latest research efforts on devising improved methods to capture and purify antibodies from cell culture supernatant using overload and elute chromatography with new multimodal membranes (MMMs). The membranes incorporate functional groups that provide orthogonal modes of interactions, which dramatically increases the range of ionic strength for operation of these membranes relative to conventional ion-exchange and hydrophobic interaction chromatography media, and makes them attractive candidates to face emerging challenges in downstream processing of biologics...

Protein binding experiments with IgG show the MMM has a high equilibrium binding capacity (180 mg IgG/ml) and excellent salt tolerance. It maintains binding capacities in excess of 100 mg IgG/ml at salt concentrations that are typical for elution buffers used in multi-stage chromatographic bioseparation processes. Protein dynamic binding capacities in excess of 50 mg IgG/ml were determined from breakthrough curve analysis.

In one study, the MMMs were used as the stationary phase following a Protein A purification step in the purification human IgG1 from Chinese hamster ovary (CHO) cell culture supernatant. The MMM stationary phase greatly enhanced load productivity compared to resin-based ion-exchange media. Further, no buffer exchange step was required between the Protein A and MMM chromatography purification steps.

In a second study, the MMMs were used as the stationary phase for the primary capture step, in place of Protein A resin chromatography. By loading the same column volumes of feed solution, the MMMs demonstrated a high selectivity for capturing human IgG from CHO cell culture supernatant. A high quality protein product was obtained from the MMM chromatography purification step.

This work shows that the new MMM chromatography operation can effectively decrease the number of purification steps and shorten the purification time and/or increase load productivity in downstream processing of biologics. It further shows that the new MMMs may be suitable for capture step purification of non-mAb proteins or when Protein A cannot be used.

Poly(ionic liquid)/ionic liquid ion-gel membranes for CO_2/light-gas separations

Matthew G. Cowan[a, b], William M. McDanel[a], Zoban V. Singh[a], Trevor K. Carlisle[a], Douglas L. Gin[a, b], Richard D. Noble[a]

[a]Dept. of Chemical & Biological Engineering, and [b] Dept. of Chemistry & Biochemistry, University of Colorado, Boulder, CO 80309, USA

The main focus of this talk is the development of a new poly(ionic liquid)/ionic liquid (PIL/IL)-based membrane materials platform (Figure 1) that has the potential to reduce the cost of post-combustion CO_2 capture from coal-fired power plants to meet the U.S. Department of Energy performance targets. This talk will illustrate how blending PILs and ILs can produce physically stable PIL/IL ion-gel composite materials that have CO_2/N_2 separation properties (CO_2 permeability up to 500 barrers; CO_2/N_2 selectivity up to 35) better than most conventional polymer membrane materials. These materials have been fabricated as 100-nm-thick defect-free active layers on a porous support. The resulting membranes display CO_2 permeance of up to 6100 GPU and a CO_2/N_2 selectivity of 22. We also present methods for chemically and structurally modifying the PIL/IL ion-gel platform as a means of optimizing CO_2/N_2 separation properties (Figure 1). These methods include the development of cross-linkable PILs that allow the inclusion of higher amounts of IL to produce membranes with a CO_2 permeability of up to 600 barrers; the inclusion of fixed-site facilitated CO_2 transport groups that produce membranes with a CO_2 permeability of up to 900 barrers and a CO_2/N_2 selectivity of up to 140 (at low CO_2 partial pressures and high relative humidity); and mixed-matrix ion-gels that contain microporous solids that produce membranes with CO_2/CH_4 selectivities of up to 90. The CO_2/N_2 separation performance ranges of these PIL/IL ion-gel materials are visually compared in a Robeson plot (Figure 2) relative to typical membrane materials and the empirical 'upper bound' line that represents the best existing membrane materials.

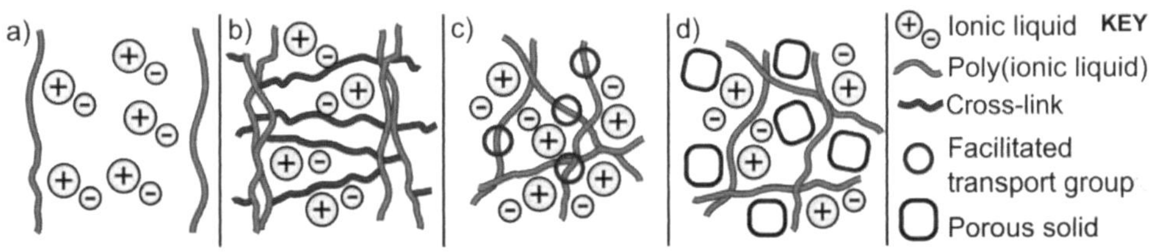

Figure 1. The range of membrane materials that have been developed from the PIL/IL ion-gel platform: a) Linear PIL/IL ion-gels (prepared as 100-nm-thick active layers); b) Cross-linked PIL/IL ion gel materials with improved *CO_2 permeability*; c) PIL/IL ion-gels that incorporate facilitated transport groups to improve *CO_2 permeability* and *CO_2/N_2 selectivity* at low CO_2 partial pressures; and d) Mixed-matrix membranes (MMMs) that incorporate a microporous solid incorporated into the PIL/IL ion-gel platform to improve *CO_2 permeability* and *CO_2/light gas selectivity*.

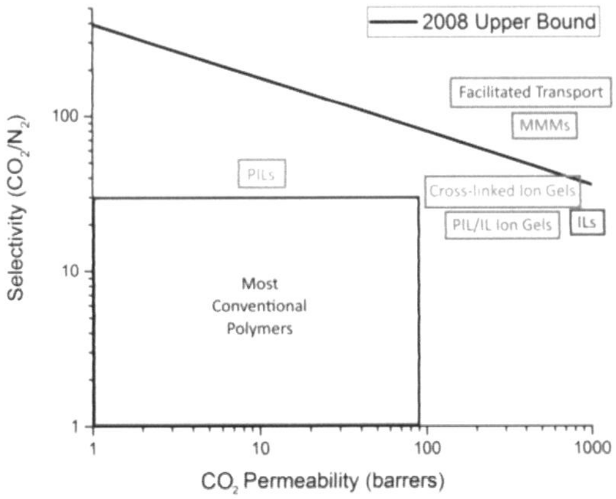

Figure 2. A Robeson plot showing the performance of the PIL/IL ion-gel materials as compared to typical membrane materials and the 2008 'Upper Bound', which represents the best CO_2/N_2 gas separation membranes.

Membrane performance characterization and process prediction in OSN: Challenges, achievements and perspective

Patrizia Marchetti [1] , Binchu Shi [1,2], Dimitar Peshev [3], Andrew G. Livingston [1,2] *

[1] *Department of Chemical Engineering, Imperial College London, Exhibition Road, London SW7 2AZ, United Kingdom*

[2] *Evonik Membrane Extraction Technology Limited, Unit 6 Greenford Park, Ockham Drive, Greenford, London, UB6 0AZ, United Kingdom*

[3] *Department of Chemical Engineering, University of Chemical Technology and Metallurgy, 8, Kl. Ohridsky Blvd, Sofia 1756, Bulgaria*

* a.livingston@imperial.ac.uk

Recent development of new solvent stable materials and continuous improvement of fabrication techniques have permitted the application of membrane filtration to different processes in organic solvent and solvent mixtures. The development of membrane processes usually involves several stages, starting from feasibility tests at membrane scale, passing through fluid dynamics and mass transfer in membrane modules and finishing with large industrial scale process design (Figure 1) [1,2].

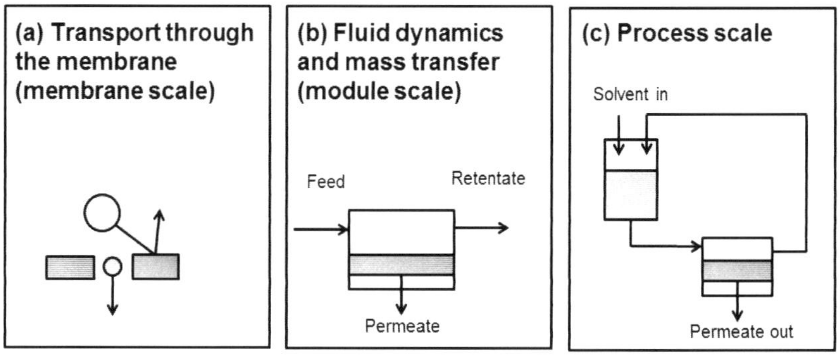

Figure 1. Modelling levels for the development of a membrane process [1].

Substantial research has focused on the description of the transport mechanism through membranes (Figure 1(a)) aiming at: *parameter estimation*, when experimental data for standard solute/solvent systems are available, and it is desired to estimate relevant parameters for as yet uncharacterized systems (Figure 2(a)); and *prediction,* when the model parameters for the solvent-solute are available, and modelling can be applied to describe performance of a different solute/solvent system (Figure 2(c)) [1,3]. Most of these studies have used flat sheet membranes. On the other hand, there is lack of studies on the fluid dynamics and mass transfer characteristics in OSN spiral-wound modules (Figure 1(b)), and attempts to implement fundamental transport models on a process level (Figure 1(c)) are few, and are based on simple, non-predictive membrane transport models.

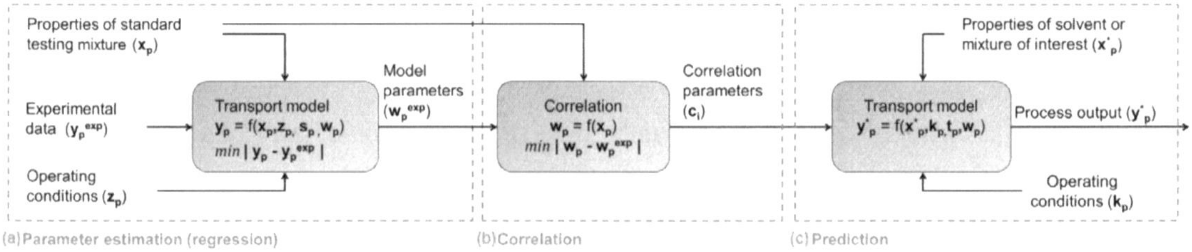

Figure 2. Modelling framework for Organic Solvent Nanofiltration. (a) Parameter estimation (regression); (b) Correlation; (c) Prediction.

This paper reports a systematic comparison of different transport models using selected experimental data for various solutes, solvents and membranes [3]. Solution-diffusion, pore-flow and transient models were used to perform regression of experimental data, and prediction, based on the regressed model parameters. It was observed that solution-diffusion models described permeation through glassy membranes better than pore-flow models, against both positive and negative rejection data (Equation 1).

$$J_i = P_i^{SD}\left[x_i^R - x_i^P \cdot \exp\left(-\frac{v_i}{R \cdot T} \cdot \Delta p\right)\right] \quad \text{Equation 1}$$

(R = retentate; P = permeate)

$$P_i^{SD} = f\left(\mu_i, \sigma_i, \varepsilon_i, MW_i, \Delta\delta_i\right) \quad \text{Equation 2}$$

Afterwards, the permeability coefficient of the classical solution diffusion model (P_i^{SD}) was correlated with the solvent properties (viscosity, surface tension, dielectric constant, molecular weight and Hansen solubility parameter) and the importance of each property was assessed (Equation 2). Finally, the classical solution-diffusion model combined with the film theory was applied to model the performance of different rubber-coated spiral-wound modules (1.8"x12", 2.5"x40" and 4.0"x40") under various pressures and retentate flowrates. Correlations for the fluid dynamics and mass transfer in the modules were derived by regression of experimental data. OSN unit operations at membrane and module levels were used in the Aspen Plus process modelling environment (using the OSN Designer [4,5]), to model common batch and steady-state membrane processes.

This paper provides also critical insight and future perspective into the state-of-the-art research, with special emphasis on how transport and process modelling in OSN are advancing and how the technology is leading to the benefits of membrane processing being applied in organic solvents.

References

[1] P. Marchetti, M.F. Jimenez Solomon, G. Szekely, A. Livingston, *Chem. Rev.* 114 (2014) 10735

[2] D. Peshev, A. Livingston, *Chem. Eng. Sci.* 104 (2013) 975

[3] P. Marchetti, A. Livingston, *J. Membr. Sci.* 476 (2015) 530

[4] B. Shi, P. Marchetti, D. Peshev, S. Zhang, A. G. Livingston, *J. Membr. Sci.* 494 (2015) 8

[5] B. Shi, D. Peshev, P. Marchetti, S. Zhang, A. G. Livingston, *Chem. Eng. Res. Des.* In press, DOI: 10.1016/j.cherd.2016.02.005

Polymer nanofilms with enhanced microporosity by interfacial polymerisation for molecular separations

Maria F. Jimenez-Solomon[1], Qilei Song[1], Kim E. Jelfs[2], Marta Munoz-Ibanez[1], and Andrew G. Livingston[1*].

*a.livingston@imperial.ac.uk

[1]*Department of Chemical Engineering, Imperial College London*

[2]*Department of Chemistry, Imperial College London*

Many of the conventional separation and purification processes in oil and gas, chemical, and pharmaceutical industries use large quantities of organic solvents and entail high energy consumption. Such processes could be totally or partially replaced by membrane technology, with an order of magnitude lower energy consumption [1]. To enable this, membranes need to provide high permeance, in order to process large volumes of solvent or gas with a viable area and in a viable timeframe.

Existing state of the art polymeric Organic Solvent Nanofiltration (OSN) membranes are either integrally skinned asymmetric (ISA) or thin film composite (TFC) membranes [2]. However, ISA membranes age as the polymer chains relax towards equilibrium packing, losing permeance. Polymers of intrinsic microporosity (PIMs) are a new class of rigid polymers with great potential in separation due to their extremely high free volume, which results from inefficient polymer packing. Linear PIMs are soluble in certain solvents, making them suitable to cast a membrane by phase inversion, or fabricate a TFC membrane via dip-coating. However, their solubility in a range of solvents restricts their applications in OSN. Several efforts to make dip-coated TFC-PIMs membranes solvent resistant have been reported, including photo-crosslinking, blending with thermally reactive polymers and chemical crosslinking.

In this work we fabricated ultra-thin highly crosslinked PIM-like nanofilms down to 20nm in thickness by interfacial polymerisation. To increase permeance we sought to design and control the polymer nanofilm structure at a molecular level by incorporating contorted monomers during the interfacial reaction. This results in enhanced microporosity and higher interconnectivity of intermolecular network voids, as rationalised by molecular simulations. Composite membranes comprising these nanofilms with enhanced microporosity fabricated in-situ on crosslinked polyimide UF membranes show outstanding separation performance in organic solvents, with up to two orders of magnitude higher solvent permeance than membranes fabricated with nanofilms made from non-contorted, planar monomers, and higher permeance than commercial membranes and most polymeric membranes reported in the literature.

[1] G. Szekely et al., *Green Chem.*, **16** (2014) 4440-447.

[2] P. Marchetti et al., *Chem. Rev.*, **114** (21) (2014) 10735–10806.

Highly permeable peptide-appended pillar[5]arene artificial water channel in block copolymer membranes

Yue-xiao Shen, Tingwei Ren, Daniel Tsai, Ian T. Sines, Manish Kumar*
Department of Chemical Engineering, Pennsylvania State University, University Park, PA, 16802, USA
*Corresponding author: 155 Fenske Laboratory, Pennsylvania State University, University Park, PA, 16802, USA, (814) 865-7519, manish.kumar@psu.edu

000We have reported that peptide-appended pillar[5]arene (PAP) artificial water channels (Fig. 1) combine the advantages of biological water channels Aquaporins (AQPs) and carbon nanotubes (CNTs).[1] It can be synthesized using simple chemistry at gram scale in the lab and provide stable structure and good solvent compatibility. The water conductance of PAP is in the range of AQPs. It can also be vertically aligned in lipid membranes, with much higher packing density than current synthetic CNTs-based membranes.

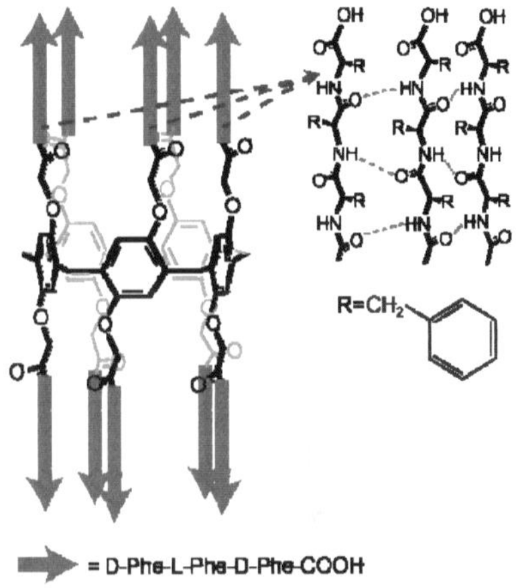

Fig. 1 Structure of the peptide-appended pillar[5]arene (PAP) channel.

Although the initial testing of the PAP channels was conducted in lipid systems, this is not an ideal format for commercial separation materials. We continue to investigate the functionality of these channels in block copolymer (BCP) membranes. BCPs, with lipid-like structures, are most commonly synthesized for biomimetic applications because of the advantages of BCPs membranes including high mechanical and chemical stability,[2,3] low water and gas permeability[4] and customizable properties (e.g., a larger range of membrane thickness[2] and end groups[5]). We are testing a series of poly(butadiene)-*b*-poly(ethylene oxide) (PB-PEO) diblock copolymers and polymethyloxazoline-polydimethylsiloxane (PMOXA-PDMS-PMOXA) triblock copolymers. These two kinds of BCPs have different hydrophobicities and thicknesses, and been shown to functionally incorporate water channel proteins aquaporin 0,[6] potassium channel[7] and α-hemolysin.[8]

We incorporated fluorescently labeled channels into polymersomes using the film rehydration method. The number of channels per vesicle was measured based on a fluorescence correlation spectroscopy (FCS) technique.[9] We combined the channel number data and permeability data using stopped-flow light scattering technique (Fig. 2B) and obtained the single channel water permeability in PB_{23}-PEO_{16} vesicles ($1.4\pm0.4\times10^8$ H_2O molecules per second). This value is close to that in lipid system ($3.5\pm1.0\times10^8$ H_2O molecules per second, Fig. 2C).[1]

In the future, we will systematically test the PAP channels in different BCPs systems, visually confirm the incorporation in giant unilamellar vesicles and discuss the interaction of different BCPs and the PAP channels. We will also plan to use fluorescence recovery after photobleaching (FRAP) technique to determine the diffusion coefficient of the PAP channels in lipid and BCPs bilayers and expect these results can provide insights into to maximizing the channel's packing density in BCP membranes.

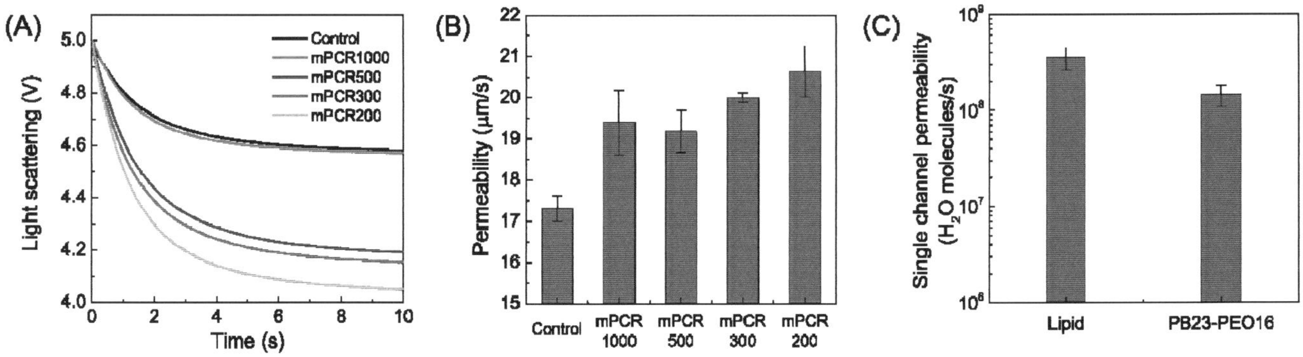

Fig. 2 Functional analysis of peptide-appended pillar[5]arene (PAP) artificial water channel in block copolymer vesicles. (A) The light scattering traces of PB_{23}-PEO_{16} polymersomes with different molar polymer-to-channel ratio (mPCR) after a rapid exposure to a hypotonic solution (−100 mM PEG600) on a stopped-flow instrument. We normalized all the traces to the same PMT value. (B) The water permeability of PB_{23}-PEO_{16} polymersomes with different mPCRs. (C) Single-channel water permeability of the PAP channel in lipid and block copolymer systems. The value in lipid vesicles was obtained from previous PNAS paper and the value in polymer system was calculated from the vesicles formed at an mPCR of 200 (Figure 2B) and fluorescence correlation spectroscopy (FCS) data. Data shown are the average of triplicates with error bars representing standard deviation.

00Reference

1 Shen, Y.-x. *et al.* Highly permeable artificial water channels that can self-assemble into two-dimensional arrays. *Proc. Natl. Acad. Sci. U.S.A.* **112**, 9810-9815, (2015).

2 Discher, D. E. & Eisenberg, A. Polymer Vesicles. *Science* **297**, 967-973, (2002).

3 Discher, B. M. *et al.* Polymersomes: Tough Vesicles Made from Diblock Copolymers. *Science* **284**, 1143-1146, (1999).

4 Kumar, M. *et al.* Highly permeable polymeric membranes based on the incorporation of the functional water channel protein Aquaporin Z. *Proc. Natl. Acad. Sci. U.S.A.* **104**, 20719-20724, (2007).

5 Rakhmatullina, E. & Meier, W. Solid-Supported Block Copolymer Membranes through Interfacial Adsorption of Charged Block Copolymer Vesicles. *Langmuir* **24**, 6254-6261, (2008).

6 Kumar, M. *et al.* High-Density Reconstitution of Functional Water Channels into Vesicular and Planar Block Copolymer Membranes. *J. Am. Chem. Soc.* **134**, 18631-18637, (2012).

7 Kowal, J. Ł. *et al.* Functional surface engineering by nucleotide-modulated potassium channel insertion into polymer membranes attached to solid supports. *Biomaterials* **35**, 7286-7294, (2014).

8 Zhang, X. *et al.* Natural channel protein inserts and functions in a completely artificial, solid-supported bilayer membrane. *Sci. Rep.* **3**, (2013).

9 Erbakan, M. *et al.* Molecular Cloning, Overexpression and Characterization of a Novel Water Channel Protein from *Rhodobacter sphaeroides*. *PLoS ONE* **9**, e86830, (2014).

Sorption Enhanced Mixed Matrix Membranes for High Temperature H_2/CO_2 Separation

Lingxiang Zhu, Deqiang Yin, Shailesh Ganesh Konda, Mark T. Swihart and Haiqing Lin*

Department of Chemical and Biological Engineering, University at Buffalo, The State University of New York, Buffalo, NY 14260, USA

*Corresponding author. Tel: +1-716-645-1856, Email: haiqingl@buffalo.edu (H. Lin)

Membrane technology has been widely explored for high temperature H_2 purification and CO_2 capture in the integrated gasification combined cycle (IGCC) processes. Polymeric membranes permeating H_2 and rejecting CO_2 at 100 - 200 °C have demonstrated promise for economical CO_2 capture. Conventional approaches are focused on rigid polymers with strong size sieving ability for H_2/CO_2 separation. In contrast, we investigate mixed matrix membranes (MMMs) containing palladium (Pd) nanoparticles, which have strong affinity towards H_2 and thus exhibit H_2/CO_2 solubility selectivity as high as 840 at 50 - 200 °C. Poly[2,2'-(m-phenylene)-5,5'-bisbenzimidazole] (m-PBI) is selected as a model polymer matrix. The m-PBI can also be crosslinked to improve the size-sieving ability of polymer phase. Uniform Pd nanoparticles (~10 nm) are prepared via hot-injection approaches, and effect of Pd loading (0 – 30 wt%) on the H_2/CO_2 separation properties is thoroughly evaluated. The loading of Pd nanoparticles increases H_2 permeability and H_2/CO_2 selectivity at temperature of 50 – 200 °C. For example, addition of 10 wt% Pd in the m-PBI increases the H_2 permeability from 2.3 Barrers to 3.5 Barrers, and H_2/CO_2 selectivity from 12 to 18 at 50 °C. This presentation will discuss pure-gas (H_2, N_2 and CO_2) permeation properties of the MMMs in details, and H_2 and CO_2 solubility in the MMMs to reveal the influence of H_2-Pd interaction on H_2/CO_2 separation performance.

Keywords: Mixed matrix membranes; Palladium nanoparticles; Polybenzimidazole; H_2/CO_2 separation; CO_2 capture

2016 North American Membrane Society Meeting

Fabricating Nanoporous Metallic Films as Components of Composite Membranes

Michael Detisch[1], Dibakar Bhattacharyya[1], and T. John Balk[1]

[1]University of Kentucky, Chemical and Materials Engineering, Lexington, KY

Magnetron sputtering is a physical vapor deposition method that allows the deposition of thin films of different materials on a variety of substrate materials. Sputtering allows fine control of the film thickness and composition through co-sputtering from multiple target materials. As part of this study thin films have been sputtered on top of microfiltration polysulfone membrane substrates. The resulting composite membranes have remained permeable under testing with deionized water. The base Nanostone PS35 membrane showed permeability of 1941 LMH at 1.7 bar, while the membrane-film composite had a permeability of 163 LMH at 1.7 bar.

Thin films of metallic alloys deposited in this way can be made nanoporous through a process called dealloying. The process involves the removal of the less noble component of an alloy by an etchant creating an open nanoporous structure. The pores created by this method commonly vary from a few nanometers to a few hundred nanometers.

This research focuses on using magnetron sputtering to deposit precursor metallic alloy films from 100 to 250nm thick on top of porous membrane substrates. These dense precursor films are then dealloyed to produce pore/ligament structures of approximately 10nm characteristic size. In these studies iron and palladium were chosen as a precursor alloy. A portion of the iron is etched away with sulfuric acid to generate a nanoporous structure. Fe/Pd nanoparticles have been used with success to dechlorinate various chlorinated organic compounds (COCs) for wastewater treatment purposes. Nanoporous Fe/Pd shows promise to be similarly reactive with COCs due to its high surface area and curved structure. Taken together this means the composite membrane produced by fabricating a high surface area, porous Fe/Pd film on top of a microfiltration or nanofiltration membrane substrate shows promise both as a catalyst and as a platform for separations.

This project is funded by NIH-NIEHS-SRC and by NSF KY EPSCOR at the University of Kentucky.

Contact author:

Michael J Detisch
University of Kentucky
Department of Chemical and Materials Engineering
177 F. Paul Anderson Tower
Lexington, KY 40506-0046
Telephone: (859)663-7681
Email: mjdeti2@uky.com

Tuning PIM-PI-OH/Z-MOF-Based Mixed-Matrix Membranes for Highly Efficient C_3H_6/C_3H_8 Separation

Ramy J. Swaidan, Xiaohua Ma, Ingo Pinnau*

King Abdullah University of Science and Technology (KAUST), Advanced Membranes and Porous Materials Center (AMPMC), Division of Physical Sciences and Engineering, Chemical and Biological Engineering Program, Thuwal 23955-6900, Saudi Arabia

*ingo.pinnau@kaust.edu.sa

Energy-efficient separation of propylene (C_3H_6) from propane (C_3H_8) is vital to the chemical process industry where C_3H_6 is used as feedstock for the production of a variety of important chemicals and plastics. A potentially significant marketplace exists for membrane technology in this separation if it can demonstrate the necessary performance. Currently, the required C_3H_6 purity is achieved by extremely energy-intensive cryogenic distillation. Recently, we reported the use of hydroxyl-functionalized polyimides of intrinsic microporosity (PIM-PIs) for C_3H_6/C_3H_8 separation. The best PIM-PI to date (PIM-6FDA-OH) had a mixed-gas C_3H_6/C_3H_8 selectivity of 15 combined with a C_3H_6 permeability of 2.5 Barrer at a feed pressure of 4 bar. One potential route to improving the permeability *and* selectivity is development of high-performance polymer-nanoparticle hybrid materials, also known as mixed-matrix membranes (MMM). For the first time, we report the fabrication and exceptional pure- and mixed-gas C_3H_6/C_3H_8 gas permeation properties of a PIM-PI-based MMM. Using PIM-6FDA-OH as the polymer matrix, we integrated a zeolitic metal-organic framework (ZMOF), ZIF-8, at various loadings ranging from 20% to 65% by weight. The membranes were tested with a 50:50 C_3H_6/C_3H_8 feed mixture at pressures up to 7 bar (p/p_{sat} ~0.8). Not only did the membranes demonstrate exceptional pure-gas properties (α >35, $P(C_3H_6)$ >10 Barrer), but the high performance was also strongly preserved in mixed-gas conditions. Exceptional C_3H_6/C_3H_8 mixed-gas selectivities ranging from 20 at 30% loading to 30 at 65% loading were achieved. We also observed minimal onset of plasticization, highlighting the strong potential of PIM-6FDA-OH/ZIF-8 mixed-matrix membranes for C_3H_6/C_3H_8 separations.

Carbon Molecular Sieve Membranes for Mixed Olefins/Paraffins Separations

Yu-Han Chu, William J. Koros

School of Chemical and Biomolecular Engineering
Georgia Institute of Technology, Atlanta, GA

Light olefins, such as ethylene (C_2H_4) and propylene (C_3H_6), are key building blocks for the petrochemical industry. The separation of olefins from paraffins is typically carried out by energy-intensive distillation. It is, therefore, appealing to develop olefins-selective membranes for use in a hybrid membrane-distillation process to retrofit the current distillation separation processes to lower the overall energy usage.

Carbon molecular sieve (CMS) membranes are able to exceed the limitation of the trade-off between productivity and efficiency of traditional polymeric membranes. In this work, starting materials and processing conditions were investigated in order to explore the impact on CMS membranes separation properties for mixed olefins/paraffins. CMS membranes were made having useful olefins/paraffins separation performance, which will be reported.

Abstract ID # 447907

High-throughput hollow-fiber (HT-HF) membrane module – a miniaturized, parallel format approach ideal for the accelerated design of environmental separation processes

There is growing interest in the use of microscale, high-throughput (HT) techniques due to their inherent advantages (increased speed of design, reduction in material requirements) over conventional filtration devices. All of the previous works in this area have focused on flat sheet membranes[1–6]. The hollow fiber (HF) membrane is a common format for environmental separation processes due to the larger membrane packing density, the cross-flow filtration regime, and the inherent ability to backflush the membrane. However, to our knowledge, no studies to date have developed a HT technique for HF membranes.

In this work, we developed the first ever high-throughput hollow fiber (HT-HF) membrane module; in its current format it can be used to run up to 6 filtration tests in parallel. Each test uses a single HF membrane which can be easily varied based on the length of the plastic tubing that connects the two compartments. Additionally, a set of three-way valves allows the tests to be run in either the 'outside-in' or 'inside-out' configurations and to backflush the membrane for cleaning. The initial experiments were performed using PVDF membranes (0.04 micron pore size), approximately 12 cm in length in the 'outside-in' flow configuration at a constant filtrate flux. Preliminary measurements of the hydraulic permeability demonstrated the reproducibility and usefulness of the HT-HF membrane module. A design-of-experiments approach was used to evaluate the effects of water source, filtration conditions, and physico-chemical cleaning conditions on membrane performance. The objective of this work is two-fold: to develop a detailed quantitative understanding of the fouling behaviour of HF membranes over a large design of filtration conditions, and to use appropriate theoretical frameworks to analyze the experimental results in terms of the underlying physico-chemical properties.

This technique has numerous applications in the field of environmental separations including the following: optimization of the filtration process used in wastewater plants (e.g. effect of operating conditions and addition of different coagulants to wastewater on filtration performance), fouling studies for natural water sources, and determining the fouling propensity in membrane bioreactors.

References

(1) Jackson, N.; Liddell, J.; Lye, G. *J. Membr. Sci.* **2006**, *276* (1-2), 31–41.

(2) Chandler, M.; Zydney, A. *J. Membr. Sci.* **2004**, *237* (1), 181–188.

(3) Kazemi, A. S.; Latulippe, D. R. *J. Membr. Sci.* **2014**, *470*, 30–39.

(4) Vandezande, P.; Gevers, L.; Paul, J.; Vankelecom, I.; Jacobs, P. *J. Membr. Sci.* **2005**, *250* (1-2), 305–310.

(5) Kong, S.; Aucamp, J.; Titchener-Hooker, N. J. *J. Membr. Sci.* **2010**, *353* (1-2), 144–150.

(6) Rayat, A. C. M. E.; Lye, G. J.; Micheletti, M. *J. Membr. Sci.* **2014**, *452*, 284–293.

Physicochemical and Photocatalytic Properties of Carbonaceous Char and Titania Composite Hollow Fibers for Wastewater Treatment

David K. Wang[a*], Muthia Elma[a,b], Julius Motuzas[a], Wen-Che Hou[c], Diego R. Schmeda-Lopez[a], Xiwang Zhang[d]

[a] *The University of Queensland, FIM²Lab - Functional and Interfacial Materials and Membrane Laboratory, School of Chemical Engineering, Brisbane Qld 4072, Australia. *Corresponding author: Fax: +61-7-3365 4199; Tel: +61 7 3365 6970; E-mail: d.wang1@uq.edu.au*

[b] *Chemical Engineering Department, Lambung Mangkurat University, Jl. A. Yani KM 36 Banjarbaru, South Kalimantan, 70714 Indonesia*

[c] Department of Environmental Engineering, National Cheng Kung University, Tainan City 70101, Taiwan

[d] *Department of Chemical Engineering, Faculty of Engineering, Monash University, Clayton, Victoria 3800, Australia*

Abstract

Advanced membrane technology that can treat polluted water and separate fresh water simultaneously is of serious research and industrial interests. In this work, we report the synthesis and characterization of two series of photocatalytic hollow fiber membranes consist of carbon char and titania with excellent control of physicochemical-mechanical-performance properties for the photo-degradation of organic dye, and simultaneously, separate fresh water. One of the major breakthroughs is that we were able to overcome the trade-off between the mechanical strength and photocatalytic efficiency of the hollow fibers by optimizing the amount of carbonaceous char and the anatase titania nanoparticles exposed on the hollow fiber surface for maximum mechanical strength and photocatalysis, respectively. As seen by Fig. 1, the optimized hollow fiber membranes with a very high proportion of the anatase TiO_2 phase (79%) reached a 90.4% photo-degradation of acid orange 7 with a bending modulus of 52 MPa. These membranes were found to be produced within a critical window of temperature (550–575 °C for 8 h) and time (6–8 h for 550 °C). Therefore, we believe this study contributes to the advancement of composite hollow fibers with predictable physicochemical and photocatalytic properties, and the potential to significantly reduce the processing and energy requirements for membrane fabrication and water treatment applications.

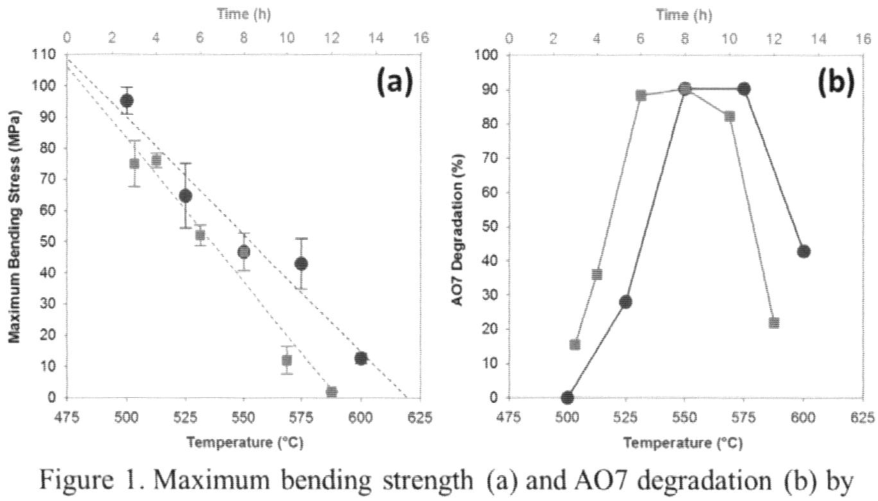

Figure 1. Maximum bending strength (a) and AO7 degradation (b) by carbon TiO_2 composite hollow fibers after 1 h under UV irradiation as a function of pyrolysis temperature and time.

***Keywords*:** titania; carbon; hollow fiber; membrane; photocatalysis; water treatment

Evaluation of Different Membrane Separation Processes for Treatment of Acid Mine Drainage and Mine Effluents.

Membrane separation has, in the recent years, gained a higher degree of acceptance by the mining industry. In the past decade there have been a number of examples of larger scale applications of membrane separation for treatment of complex mine effluent and discharge streams. This is an indication that with proper design, membrane separation is an economically viable water management option in the mining sector. Additionally in recent years there has been an increasing pressure, on mining companies, to implement membrane separation in order to meet increasingly stringent water discharge limits.

The objective of this work was to assess the use of pressure driven polymeric and ceramic membrane processes and electrodialysis at bench and pilot scale; for the treatment of Acid Mine Drainage (AMD) streams from different mine sites. Polymeric and ceramic membranes were assessed and compared for performance and fouling with a number of samples from different sites with a focus on metals and sulphate removal. The application of Electrodialysis was also evaluated as pre- and post-treatment to NF and RO, in addition to the recovery of acid and a concentrated salt stream from the AMD. A great deal of attention is paid to mitigating factors that impact membrane performance, process costs, and evaluation of strategies for treatment of the concentrate stream from the membrane separation process in the context of long-term mining operations and post closure site management.

In the assessment of ceramic ultrafiltration (UF) and NF membranes for the pretreatment and demineralization of AMD a comparison between surface modified and unmodified membranes is made with respect to resistance to fouling and rejection performance of the membranes.

Non-carbon adsorbent and oxidation catalyst for mercury removal from coal-fired flue gas

Wenqing Xu, Yinghong Wu, Yang Yang, Mingpan Shao, Tingyu Zhu*

Institute of Process Engineering, Chinese Academy of Sciences, Beijing, China.

Mercury emission from the coal-fired flue gas has become the precedent-controlled harmful pollutant because of its high volatility, long persistence, and bioaccumulation especially for Hg^0.The adsorption and catalytic oxidation methods are considered to be the most effective technique for controlling Hg^0. Among various sorts of non-carbon adsorbents and oxidation catalysts, minerals have been increasingly attracting scientists due to its porous, unique-structural, easily regenerated, environmental friendly, and cost-effective characteristics. Furthermore, the co-existing O_2, HCl, SO_2 and NO in the flue gas do have effect on the removal efficiency of Hg^0. Therefore in this work, we chose two kinds of minerals, HSZM-5 and montmorillonite, modifying them with $Fe(NO_3)_3$ and $Mn(NO_3)_2$ respectively, and discussed the effect of flue gas components on Hg^0 removal by adsorption and oxidation methods.

On the one hand, experimental results revealed that 3%Fe/ HSZM-5 sample was the optimal for Hg^0 removal within the temperature range of a SCR reaction as shown in Figure 1. In the presence of O_2 and HCl, the Hg^0 removal efficiency was 100%. And below 200°C, the products were adsorbed $HgCl_2$, but were both adsorbed and gaseous $HgCl_2$ when above 200°C. Without O_2, SO_2 inhibited the Hg^0 removal due to the competitive adsorption and formation of $Fe_2(SO_4)_3$. While when coexisting with O_2, SO_2 initially inhibited Hg^0 removal by competing the active sites, but subsequently promoted Hg^0 removal by producing active sites. NO could promote Hg^0 removal efficiency as the formation of the N-containing active species on the sample surface facilitated the Hg^0 oxidation to produce $Hg(NO_3)_2$ in the presence of NO and O_2. And

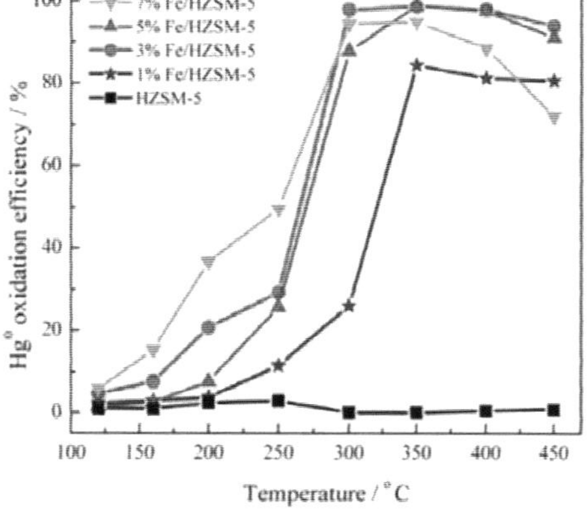

Figure 1: Hg^0 oxidation efficiency over Fe/HZSM-5 catalyst with different Fe contents at various temperatures. (8%O_2+balanced N_2)

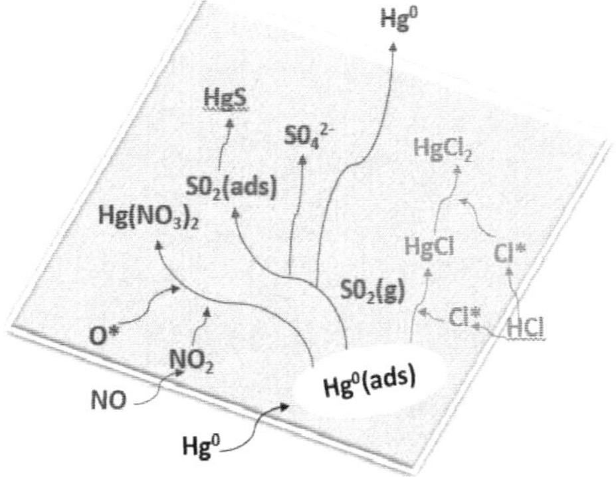

Figure 2: Proposed removal mechanism for Hg^0 over the Mn/MK10 sample.

then when the NO and O_2 were cut off, the Hg^0 concentration remained the same since the abundant NO-like species were adsorbed on the sample.

On the other hand, as for the modified montmorillonite(MK10), 4%Mn/MK10 sample showed the highest Hg^0 removal efficiency within the temperature range from 100℃ to 450℃. The proposed removal mechanism for various flue gas components could be seen in figure 2. In pure N_2, the Hg^0 removal efficiency was only 5% while rapidly increased to 99% when O_2 passed, indicating O_2 was the main active component for the Hg^0 oxidation. SO_2 appropriately promoted the Hg^0 removal when O_2 was non-existent, but might inhibited it when O_2 was existent. Because without O_2, SO_2 might absorb on the surface of the catalyst, transferred Hg^0 into HgS and therefore promoted Hg^0 removal. However, when O_2 existed, the formation of $MnSO_4$ competed for the active sites for mercury oxidation. HCl could remove 99% Hg^0 as Cl* oxidized Hg^0 into HgCl and $HgCl_2$. For NO, both MnOx and O_2 could adsorb and oxidize NO into NO_2, and further form $Hg(NO_3)_2$, thus 500 ppm NO promoted nearly 100% Hg^0 removal efficiency whether O_2 exist or not.

Towards Conductive Chemically-Active Membranes: Flavin-Functionalized Fibers and Electrodes Active in Photo-enhanced Oxygen Evolution

Madison J. Sloan[a], Rupam Sarma[b], Shankar Miller-Murthy and Anne-Frances Miller[a]*

Department of Chemistry, University of Kentucky[a]

Department of Chemical and Materials Engineering, University of Kentucky[b]

The fabrication of bio-electrode and fiber systems decorated with redox-active biomolecules, flavins, is demonstrated. Exploiting the photochemistry and electrochemistry of flavins, we explored the photo-electrochemical activity of flavin-functionalized electrode systems to assess their potential utility for sustainable energy production. Flavins were also grafted onto multi-walled carbon nanotubes to be assembled into membranes. As model systems, lumiflavin and flavin adenine dinucleotide were immobilized on carbon electrodes by dropcasting and covalent grafting techniques. Activity of these bio-electrodes towards generation of O_2 from H_2O was demonstrated. Irradiation of the electrode system with visible light led to increased activity of the electrodes with a 3-fold enhancement of oxidation of H_2O. Thus, we demonstrate functionalization and chemical activity of carbon materials that can be incorporated into membranes.

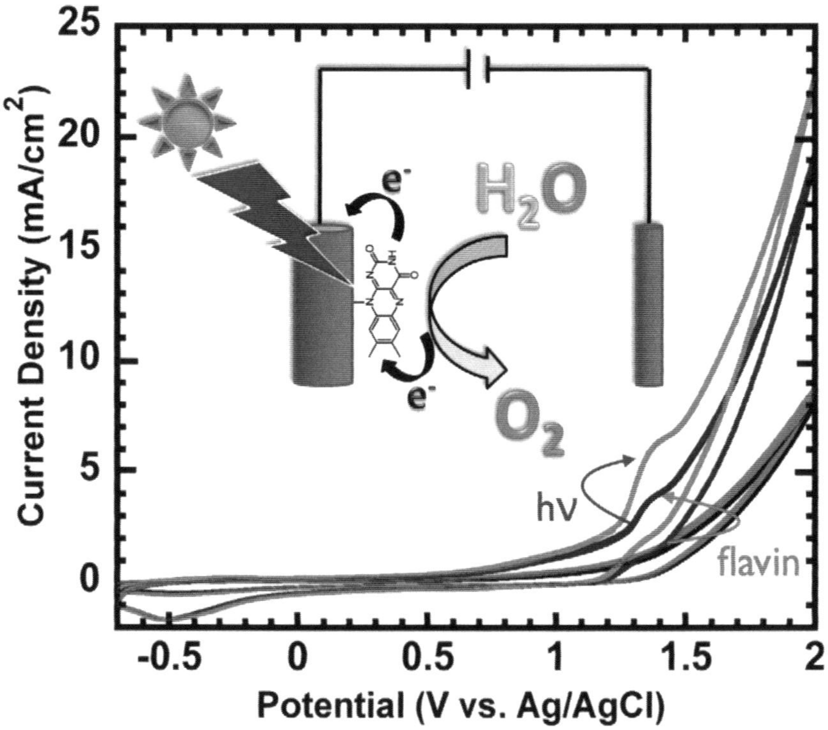

Characterization of biofouling on forward osmosis membranes used for long-term produced water treatment

Kevin Chan and Elizabeth Bell
Advisors: Prof. John Spear and Prof. Tzahi Cath
Department of Civil and Environmental Engineering
Colorado School of Mines, Golden, CO, USA

Oil and gas (O&G) produced water is a significant by-product of the upstream sector of the petroleum industry, generated at billions of gallons per day in the United States. Membrane processes such as forward osmosis (FO) can provide high quality permeate for beneficial reuse of produced water within the industry. The current research focuses on FO membrane systems utilizing cellulose triacetate (CTA) and polyamide (PA) thin film composite (TFC) membranes operated continuously for three weeks. Produced water from the Denver-Julesburg basin was used as a weekly batch feed solution while a 1 M sodium chloride (NaCl) solution served as the draw solution. On a daily basis, deionized water was added to the feed solution to replace permeated water and a 300 g/L NaCl solution was dosed into the draw solution to maintain consistent osmotic pressure. The feed and draw solutions flowed co-currently on opposite sides of the membranes at rates of 1.58 and 0.82 L/min, respectively. Membranes were subject to weekly chemically enhanced osmotic backwash cleaning with ethylenediamine-tetraacetic acid (EDTA) to investigate changes in treatment performance and fouling layer composition. After system deconstruction, biofilms on each membrane type were characterized with fluorescent and scanning electron microscopy (SEM) and 16S rRNA gene sequencing via the Illumina MiSeq platform. SEM images reveal similar biofilm buildup on both CTA and PA-TFC membranes after three weeks of continuous operation. Quantitative Insights Into Microbial Ecology (QIIME v1.9) bioinformatics analysis reveal significant differences in microbial community compositions between CTA and PA-TFC membranes. Analyses of feed and draw solutions reflect a greater breakthrough of microorganisms in PA-TFC treatment than in CTA. By addressing microbial community analysis, performance, and operation, we found the CTA membrane outperforms the PA-TFC membrane for the treatment of O&G wastewaters. Conclusions from this research provide a greater understanding of membrane biofouling that is critical for the development of effective treatment solutions for produced water.

Green Modification of Outer Selective P84 Nanofiltration (NF) Hollow Fiber Membranes for Cadmium Removal

Jie Gao, Shi-Peng Sun, Wen-Ping Zhu, Tai-Shung Chung*

Department of Chemical and Biomolecular Engineering
National University of Singapore, Singapore 117576

*Corresponding author
Tel: +65-65166645; fax: +65-67791936; Email: chencts@nus.edu.sg

Abstract:

Nanofiltration (NF) is a membrane filtration process with a membrane pore size in the sub-nano scale. Due to its special rejection mechanisms: size exclusion and charge repulsion, NF is able to offer good rejections to multi-valent heavy metal ions without much compromising water fluxes as compared to those of reverse osmosis (RO) membranes.

There is a growing interest in thin-film composite (TFC) NF hollow fiber membranes. The TFC layer can be formed on both inner and outer surfaces of hollow fiber membranes. The outer selective TFC hollow fiber membranes may have less mass transfer resistance and lower chances being blocked by foulants compared to the inner ones. Thus, the outer selective TFC fibers can be fabricated into smaller fiber dimensions with a higher package density in modules. Despite these advantages, the molecular design of an ideal TFC layer on the outer surface of hollow fiber membranes remains a challenge, making the mass production difficult. Since most TFC layers are made from interfacial polymerization of m-phenylenediamine (MPD) and trimesoyl chloride (TMC), removing excess solution after the MPD coating is the major obstacle in the process. Vacuum assist interfacial polymerization or wipe drying are required to make a defect free polyamide layer. Another obstacle is that a large amount of alkane solvents has to be used during interfacial polymerization, which brings extra costs and environmental issues. Eliminating alkane solvents is critical to lower manufacture costs and yield a green production process.

In this study, green methods to prepare the outer selective TFC hollow fiber membranes were explored by firstly modifying the membrane substrate with polyethyleneimine (PEI) and then by water soluble small molecules such as glutaraldehyde (GA) and epichlorohydrin (ECH). Using P84 polyimide as the substrate, not only do these modifications decrease substrate's pore size, but also vary surface charge by making the membranes less positively charged. As a result, the resultant membranes have higher rejections against salts such as Na_2SO_4, NaCl and $MgSO_4$. The PEI and then GA modified membrane has the best separation performance with a NaCl rejection over 90 % and a pure water permeability of 1.74 ± 0.01 $Lm^{-2}bar^{-1}h^{-1}$. It also shows an impressive rejection of 94 % for $CdCl_2$ removal during long-term stability tests. The $CdCl_2$ rejection remains higher than 90% when testing it from 5 to 60 °C. This study may provide useful insights for green manufacturing of next-generation NF hollow fiber membranes.

Acknowledgements:

The authors would like to thank Chinese Gansu province for the research project entitled "Membrane development to removal heavy metals" with the grant number of R-279-000-360-597 and GSK-EDB Trust Fund for the project entitled "New membrane development to facilitate solvent recovery and pharmaceutical separation in pharmaceutical syntheses" with the grant number of R-706-000-019-592. The authors also acknowledge Mr. Z.L. Cheng, Dr.

X. Li, Ms, Y. Zhang, Dr. Z.Z. Zhou and Dr. Y.K. Ong for all their helps and suggestions on the experimental work. Thanks are also due to Mr. K.P. Ng for his help on the NF set-up.

High performances of poly(arylene ether sulfone) with tetra(quaternary ammonium) group for solid alkaline exchange membrane fuel cell applications.

Jieun Choi [a,b], *So Young Lee* [a], *Jong Hyun Jang* [a], *Dirk Henkensmeier* [a], *Sung Jong Yoo* [a], *Yung-Eun Sung* [b], *Hyoung-Juhn Kim* [a]*

[a] Fuel Cell Research Center, Korea Institute of Science and Technology (KIST), 39-1 Hawolgok-dong, Seongbuk-gu, Seoul 136-791, Republic of Korea

[b] Department of Chemical and Biological Engineering, Seoul National University, 599 Gwanak-ro, Gwanak-gu, Seoul, 151-742, Republic of Korea

Email: hjkim25@kist.re.kr

Solid alkaline exchange membrane fuel cells (SAEMFCs) are expected to overcome the limitations of proton exchange membrane fuel cells (PEMFCs). Oxygen reduction reactions at the cathode are facilitated under basic environment than acidic one. Thus it is enabled to use non-Pt catalysts in SAEMFCs. Furthermore, as hydroxyl ions migrate from the cathode to the anode, it does not allow crossover of methanol through the membrane.

Among the SAEMFC components, the AEM is a key element which determines the performance and durability of the fuel cell. In this study, poly(arylene ether sulfone)s with tetra(quaternary ammonium) hydroxide were prepared for solid alkaline exchange membrane fuel cells. Using the monomer with four tertiary amine groups, we synthesized poly(arylene ether sulfone)s which conduct hydroxyl ions. Tertiary amine groups were derivatized to quaternary ammonium functional groups. This synthetic path can precisely control the amount of quaternary ammonium hydroxide groups in the polymer structure. The synthetic polymers and prepared membrane were characterized by NMR, thermogravimetric analysis, water uptake, conductivity, and cell performance. We also evaluated the long-term durability of the membrane after several alkali treatment condition.

Investigation of Ultrafiltration, Nanofiltration, Biofiltration and Ozonation for the Removal of Microcystin-LR from Water

On August 2, 2014, the greater Toledo area woke up to a Do Not Drink or Boil Water Advisory. The advisory was due to the presence of a cyanotoxin (algal toxin) produced by cyanobacteria in Lake Erie called microcystin-LR in the drinking water supply that has a WHO provisional guideline of 1 mg/L. Upon entering the City of Toledo Collins WTP crib at the Lake Erie intake, potassium permanganate is added to the water for mussel control. While potassium permanganate is needed to control mussels, it lyses cyanobacteria cells, releasing algal toxins to the water. The water is then pumped nearly three miles to the Low Service Station, where powdered activated carbon (PAC) is added to the water for taste and odor control, and the water is transported approximately six miles to the WTP (High Service Station). PAC is generally effective for removal of algal toxins through adsorption onto its surface. At High Service, alum, lime and soda ash are added to the water for coagulation-flocculation, softening, and removal of metals. The water is then sand filtered, carbonated and chlorinated before being sent to the distribution system. However, traditional physicochemical water treatment processes, such as coagulation-sedimentation-filtration, have been shown to only be partially effective for the removal of whole algal cells and not effective for the removal of algal toxins. Furthermore, chlorination is effective for oxidizing algal toxins at relatively high free chlorine concentration as long as the pH is below 8; however, for corrosion control, Toledo water is kept at a pH above 9. Therefore, the treatment process was not enough of a barrier to prevent microcystin-LR from entering the drinking water supply. By August 4, the water treatment plant increased its PAC dosage by nearly four times, and while the toxin was removed, a significant amount of PAC sludge was produced and the cost of PAC addition was unsustainable. The objective of this project is to study alternative water treatment processes, namely membrane separations, biofiltration and ozonation, for the effective removal of algal toxins.

Fundamental Studies of Ion Transport in Ion-Exchange Membranes

Jovan Kamcev, Benny D. Freeman*, Donald R. Paul

McKetta Department of Chemical Engineering, The University of Texas at Austin, USA

**freeman@che.utexas.edu*

Fundamental understanding of ion sorption and transport in ion-exchange polymers is critical for developing high performance membranes for water purification and energy generation applications. Despite the long history of literature on the topic, the current state of understanding is not sufficient to build the appropriate connections between polymer structure and transport properties. Experimental techniques for characterizing ion sorption and transport in charged membranes have been established, however, a simple theoretical framework to describe the experimental findings is missing. In this work, a framework for interpreting ion sorption and diffusion in ion-exchange membranes has been formulated and tested against experimental data. For the membranes used in this study, the framework could describe, and in some cases predict, concentration gradient (i.e., salt permeability) and electric field (i.e., ionic conductivity) driven ion transport in charged membranes.

Equilibrium portioning of ions between the membrane and solution phases was modeled using the classical Donnan theory. Previous studies have often overlooked the importance of including ion activity coefficients (i.e., non-ideal effects) in this model, mainly because no reliable models or experimental techniques were available for obtaining these values in the membrane phase. In this study, Manning's counter-ion condensation theory, originally developed for polyelectrolytes, was extended to ion-exchange membranes and used to calculate membrane ion activity coefficients. Solution ion activity coefficients were obtained by the Pitzer model. This framework was validated using experimental data for ion sorption of various electrolytes in commercial ion exchange membranes. Remarkably good agreement between theory and experiment was observed without using adjustable parameters.

Ion diffusion coefficients in charged membranes are affected by at least two factors: (1) the geometric impedance by the polymer chains and (2) electrostatic effects due to the fixed charge groups. The geometric factor was modeled using an approach developed by Mackie and Meares, which presumes ion diffusion occurs in the solvent in the swollen polymer matrix and the polymer chains act as impenetrable obstructions. This model has no adjustable parameters. The electrostatic influence of the fixed charge groups on the diffusing mobile ions was modeled using an approach set forth by Manning. Manning's model on diffusion is a direct extension of his counter-ion condensation theory used to predict ion activity coefficients. The framework for modeling ion sorption and diffusion in charged polymers developed in this study was validated using experimental salt permeability and ionic conductivity data for commercial charged membranes. Remarkably, this approach was able to predict salt permeability coefficients in charged membranes with no adjustable parameters and accurately describe ionic conductivity data using only one adjustable parameter. The generality and shortcomings of this framework are discussed.

Correlating Nafion Membrane Microstructure with Vanadium Redox Flow Battery Performance

Wei Wang, Zimin Nie, Xiaoliang Wei, Bin Li, Vijayakumar Murugesan, Vincent Sprenkle

Pacific Northwest National Laboratory, 902 Battelle Blvd, Richland, WA 99354, USA

wei.wang@pnnl.gov

This poster describes the correlation established between the Nafion membrane microstructure and the vanadium redox flow battery performance. The impact of the equivalent weight on the microstructure of the membrane and thus the performance of VRBs is systematically studied. By tailoring the equivalent weight and membrane thickness, the new membrane contains optimal pore geometry with extremely low vanadium-ion permeability was developed. In addition, its area resistance is comparable to, and its cost is significantly less than, the widely used Nafion 115 membrane. Excellent VRB single-cell performance (89.3% energy efficiency at $50mA{\cdot}cm^{-2}$) was achieved along with a stable cyclical capacity over prolonged cycling (>200 cycles). The chemical composition (Figure 1), pore geometry, transport properties and flow cell performance will be reported.

Acknowledgements

The authors would like to acknowledge financial support from the U.S. Department of Energy's (DOE's) Office of Electricity Delivery and Energy Reliability (OE) (under Contract No. 57558). Pacific Northwest National Laboratory is a multi-program national laboratory operated by Battelle for DOE under Contract DE-AC05-76RL01830.

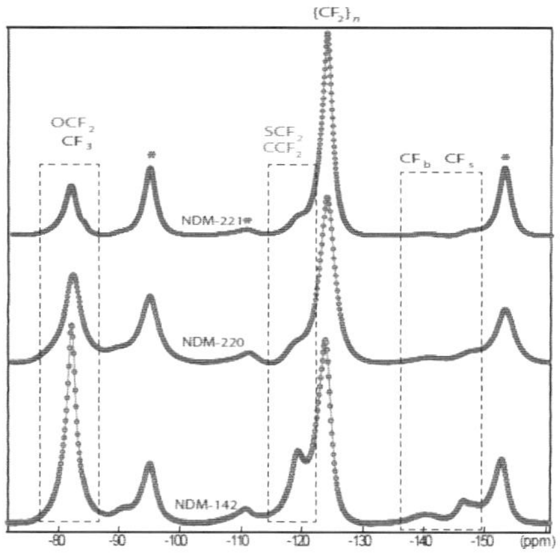

Figure 1. ^{19}F MAS-NMR spectra measured at an 11.7 T magnetic field with a spinning speed of 13.5 kHz for NDM142, NDM220, and NDM221 membranes. The circles and red line represents the experimental spectra and the fitted curve with deconvolution analysis, respectively (see text for details). The asterisk (*) indicates MAS spinning side bands.

Maximizing Yield, Cost Savings and Environmental Benefits Using Osn Membrane Technology for Efficient Purifications.

Abstract

Organic solvent nanofiltration (OSN) is a molecular separation technology based on membranes. It allows the separation of molecules dissolved in an organic solvent on basis of size caused by shape and charge of the molecules.

Compared to other unit operations OSN is still regarded as a new unit operation with its first commercial successes in the early 90s. Since then a number of new applications and processes were developed that prove the potential of organic solvent nanofiltration as a resource efficient non-thermal separation technology and enabler of new processes and even new products. Applications of organic solvent nanofiltration are found in the oil- and petrochemical industry, in the food and bioproduct industry, in the bulk chemistry as well as in the fine chemistry and pharmaceutical industry. Examples include purification of raw materials, concentration of temperature sensitive products, solvent swap of temperature sensitive products, product recovery from mother-liquors, color impurity removal, homogeneous catalyst removal, recovery and reuse, concentration of chromatography streams etc. For a lot of these applications OSN is in an ecological and economical way a strong alternative to thermal separation processes.

Recent successes of OSN in technical processes mark the break-through of this technology. This ascent is further driven by global mega trends as OSN serves material efficiency, energy efficiency and the demand for high quality products.

OSN also allows new processes as its characteristics are unique since OSN combines non-thermal and continuous separation with no need for thermal energy or for additives and recycle streams in most cases. Therefore, OSN competes with and also complements conventional separation techniques like distillation, evaporation, adsorption, absorption, chromatography, crystallization, etc.

To corroborate the technical relevance of the unique advantages of OSN and at a stage of break-through this lecture will give an overview about recent applications of OSN and highlight emerging trends. Benefits such as mass efficiency via increase of product yields, reduction of process time, and reduced use of high-energy consuming technologies will be presented. An outlook about ways to a customizing toolbox of OSN membrane properties will finalize the presentation (fig. 1).

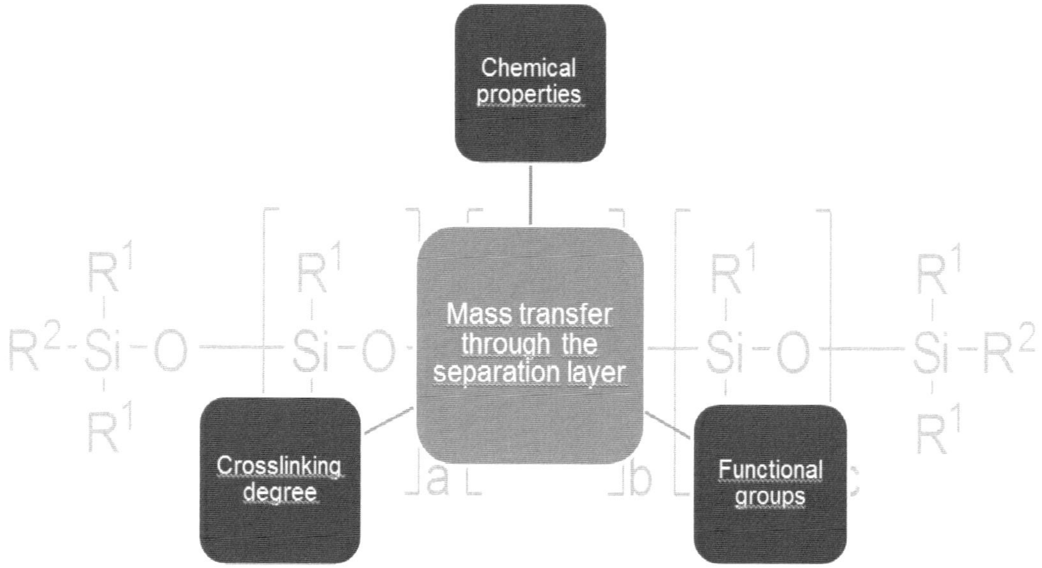

Figure 1: Levers to influence mass transfer through a silicone based membrane separation layer on a polymer basis

Comparison of Fouling Mechanisms of Four Nanofiltration Membranes: Optimal Membrane Selection for Stephenfield Regional WTP, MB, Canada

I. Moran*, B. Gorczyca
University of Manitoba Department of Civil and Environmental Engineering, University of Manitoba, 15 Gillson St. Winnipeg, Manitoba, Canada, R3T 5V6.

Corresponding Author* Email: ummorani@umanitoba.ca

Abstract: Dual membrane filtration plants, incorporating microfiltration (MF) and nanofiltration (NF) have become widely used in potable water treatment as a method that can meet increasingly stringent THMs standards. Due to the high soluble material content in Manitoba and surrounding areas, a major concern for plant optimization is NF fouling. Over operational time, the foulant layer acts as an inhibitor of permeate flux and can significantly decrease membrane element longevity and increase plant operational cost. The extent of membrane fouling is determined by the specific interaction between the membrane and soluble material in the feed water. A detailed understanding of NF fouling mechanisms is necessary to optimize plant operation. Physical and chemical analyses were carried out on four different membranes (DOW FILMTEC NF90-400/34i, DOW FILMTEC NF270-400/34i, DOW FILMTEC BW30XFR-400/34i, DOW FILMTEC XFRLE-400/34i) to determine the optimal membrane for installation in the Stephenfield Regional WTP in southern Manitoba. Post-MF water from the Stephenfield Regional WTP was run in cross flow at 10bar (145psi) across each membrane at different operational time steps (1, 4, 12, 24, 48 hours). Virgin membrane surface roughness appears to be the governing factor for the degree and mechanism of fouling. Atomic force microscopy showed the initially rough NF90 and BW30 membranes experience a higher initial rate of fouling due to the higher surface area to attach and surface contours for soluble material to grab. The non-fouling resistant NF90 and NF270 membranes experienced a flux restoration from fouling over time. This phenomenon is predicted to be the dual result of membrane compaction and scouring. The flattening process of membrane peaks pulls down the overbearing layer of foulant causing an increase in surface roughness with greater susceptibility to the effects of cross flow scouring. The BW30 and XLE, coated with a fouling-resistant polyether block amide compound, experienced a lesser degree of fouling. The retained mass of foulant on the NF90 was three times greater than the other tested membranes due to the high virgin membrane surface roughness and lack of fouling resistant coating. This understanding of fouling mechanisms, as well as permeate quality, was used to select the NF270 as the optimal element for longevity and operational cost.

Keywords: Membrane Fouling; Plant Optimization; High Soluble Material Content

INTRODUCTION

Motivation

In a water treatment plant (WTP), membrane filtration is used as a water treatment process to remove particulate and dissolved material in water to improve the quality of water prior to distribution. Over operational time, this particulate and dissolved material builds up on the surface of a membrane and can have important implications regarding flux (flow rate) and material removal efficiency, among others. An understanding of fouling mechanisms and behavior can be used to predict the short- and long-term consequences of fouling or to make an informed decision as to which membrane, within a set of membranes, is most capable of withstanding the effects of fouling for a particular raw water source. For a water treatment plant, this can be important to meet both the demands of the client and federal and provincial water quality standards. This thesis investigates the relationship between four nanofiltration membranes and raw water from the Stephenfield Regional WTP in southern Manitoba and to make an informed decision as to which membrane is most suitable for installation in the Stephenfield Regional WTP.

Literature Review

The membrane selection process is fundamental to long-term plant cost and operation. A membrane that experiences a faster rate of fouling due to surface morphology, hydrophobicity or other key properties may not be the optimal choice for installation. Vrijenhoek et al. discuss the influence of membrane surface properties on the initial rate of colloidal fouling of reverse osmosis and nanofiltration membranes. The conducted research compares the fouling mechanisms of two nanofiltration membranes (Osmonics HL, Dow-FilmTec NF-70) and two reverse osmosis membranes (Trisep X-20, Hydkronautics LFC-1) [Vrijenhoek et al. (2001)].

The four tested membrane showed vastly different virgin membrane surface roughness. Compared to the relatively smooth HL membrane, the X-20, NF-70 and LFC-1 had surface roughnesses approximately three, four and five times greater than the HL membrane, respectively. The experiment conducted by Vrijenhoek et al. compared the decline in flux of each of these membranes to the relative surface roughness and a 0.99 positive correlation coefficient was determined between these two variables. The strong positive correlation between initial surface roughness and flux decline is a result of what the authors describe as “valley clogging” which is further supported by three-dimensional surface scans taken by atomic force microscopy. In its context, valley clogging refers to the rapid deposition of material in the valleys of an initially rough membrane which causes a rapid increase in foulant mass and subsequent decline in flux. This phenomenon is less prominent in membranes with low initial roughness as there are fewer valleys to get caught in and less surface area to attach. The reasonability of the concept of valley clogging, supported by a strong correlation between it and early flux decline, may be important in the membrane selection process. This data puts an asterisk on the “time zero” permeate flux and may shift our attention to the permeate flux after sufficient valley clogging has occurred.

Virgin membrane zeta potential was considered as a proponent of initial fouling. Zeta potential is likely only a significant factor when parts of the membrane surface are exposed. In other words, zeta potential will have a stronger impact regarding foulant-on-membrane fouling rather than foulant-on-foulant fouling. Flux decline was normalized to the membrane with the smallest decline in flux, the HL membrane, and compared with zeta potential. All other variables held constant, the correlation coefficient between these two variables was 0.05. This small correlation coefficient suggests a weak correlation between zeta potential and the four tested membranes in the study. It is important to note than none of the membranes in this study are coated with a polyether block amide compound – any one of several commercially available amide coatings which alters the electrostatic properties of the uppermost layer of the membrane resulting in the repulsion of common fouling material. This experiment has demonstrated the lack of correlation between virgin membrane zeta potential and the two reverse osmosis and two nanofiltration membrane in the study. Further investigation may be necessary to determine any correlation between zeta potential and membranes coated with a polyether block amide compound.

Salt rejection is a measure of the passage of dissolved salts past the surface of a membrane. The four tested membranes were subjected to a silica dose of 200mg/L in operation for the two hours and the initial and final salt concentrations were compared. The resultant correlation coefficient between flux decline and salt rejection was 0.87. In this experiment, the X-20 and LFC-1 membranes had the same salt rejection, but a nearly twofold difference in normalized flux decline. Due to the similarities in salt rejection between these two membranes, Vrijenhoek et al. suggest a coincidental relationship between salt rejection and flux decline and rather focuses on the relationship between salt rejection and virgin membrane surface roughness; the rough initial surface of the membrane causes rapid deposition of the water containing a high concentration of the soluble salt particles. While it is not directly stated in this report, it is expected that the major proponent of salt rejection is relative pore size or molecular weight cut-off. This is supported by the manufacturer-specified salt rejection of the LFC-1 and X-20 having the same value of 99.0% and experimentally rejecting the same amount of salt. The relationship between salt rejection and surface roughness is explained in more detail in Section 1.2.1 of this report.

The experiments conducted by Vrijenhoek et al. provide readers with a basic understanding of reverse osmosis and nanofiltration membrane fouling mechanisms. The report emphasizes the impact of virgin membrane surface roughness in the early stages of fouling – fouling is more rapid for membranes with high virgin surface roughness due to high friction between the membrane and soluble material by what the author describes as "valley clogging". This idea is supported by data showing a strong correlation between surface roughness and flux decline. The report shows an insignificant correlation between membrane hydrophobicity and flux decline and what the author describes as a coincidental relationship between salt rejection and flux decline.

While this report effectively shows the relationship, or lack thereof, between flux decline and the aforementioned parameters, there are still properties that either require further investigation or that are not covered. The membranes used in this research were made of different material, but were not coated with a polyether block amide compound – a developing membrane property generally used against waters of high soluble material content as a way to avoid fouling [Jiraratananon et al. (2002)]. The effects of these polyether block amide compounds, found commercially through different membrane manufacturers, may be of particular interest. In addition, the experiment isolated the effects of membrane compaction by subjecting the membrane sample to 24 hours at normal operating pressures for 24 hours prior to testing. While membrane compaction is not part of the scope of this study, it may be important in developing a more complete understanding of the membrane properties over time; there may be a relationship between membrane compaction and early membrane fouling under typical operating conditions.

Effects of Membrane Fouling

Over operational time, the amount of foulant retained on the surface of a membrane increases and prevents the passage of permeate across the membrane by "flux inhibition" [Mallevialle et al. (1996)]. This mechanism can be seen in Figure 1. Controlling the permeate flux can be of great importance for WTP operators in order to meet the demand of customers. Increasing pressure in the filtration system can increase the permeate output, but can have negative energy consumption (cost) implications. Finding the optimal balance between these two parameters, supplemented by the cost and effectiveness of pre-treatment, is imperative for treatment plants to meet their cost goals [Schäfer et al. (2001)]. Depending on the classification of membrane, treatment plants can either use a backwashing cycle to remove the attached foulant or use chemical cleaning agents to restore flux. Membrane flux may not be fully restored to its original condition due to irreversible fouling – fouling that is strongly bound to the membrane and cannot be removed by backwashing or chemical cleaning [Kimura et al. (2004)] – but can help increase permeate output and membrane longevity.

Figure 1: The typical relationship between degree of fouling and flux over operational time for regular and fouling resistant membranes.

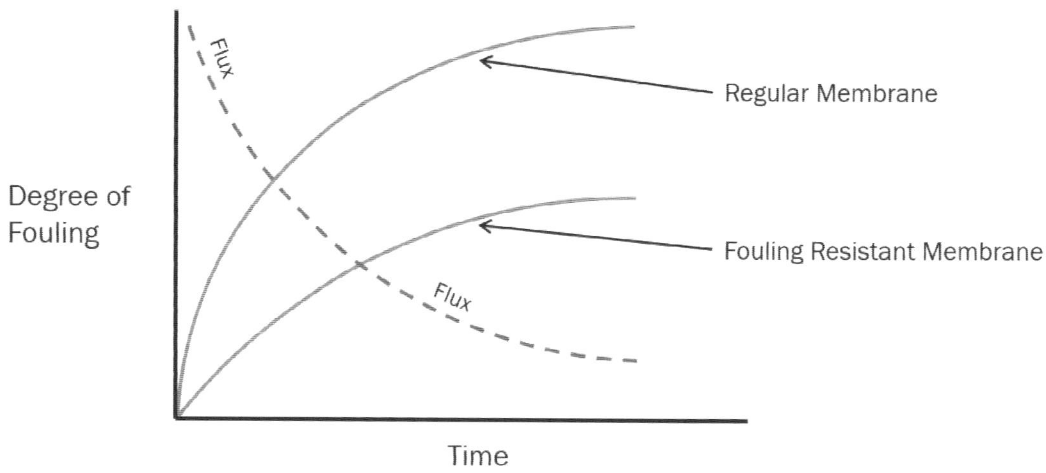

Effects of Membrane Compaction

Membrane compaction is a naturally occurring process in membrane filtration in which the surface contours or peaks are flattened by an applied pressure. This is an irreversible process that occurs in the early service life of a membrane and the effects quickly level off to a relative maximum. The degree of compaction is proportional to the applied pressure. The permanent flattening of peaks results in a permanent reduction in surface area which, in turn, causes an overall reduction in flux [Persson et al. (1995)]. An illustration of membrane compaction can be seen in Figure 2.

Figure 2: a) A virgin membrane with high initial surface roughness and b) the same membrane after an applied pressure for a given amount of time.

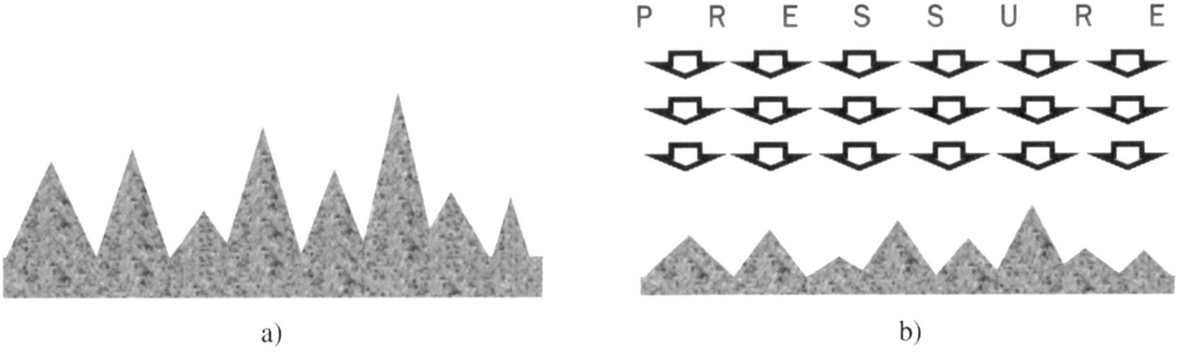

Fouling Resistant Membranes

Some membranes are coated with a polyether block amide compound, known commercially under the trademark name PEBAX®. Many commercially available nanofiltration membranes from DOW Chemical and Process are labeled as 'fouling resistant' as a result of this extra coating. Louie et al. showed that membranes coated in PEBAX® 1657 exhibited a decrease in surface roughness by 35-63%, a reduction of permeate flux by 29-81% and no significant change in hydrophilicity [Louis et al. (2006)]. Membranes with this attribute are often used in treatment of water with high salt content, such as brackish or sea water, and generally have a higher rejection of dissolved material. A simple comparison in fouling between a regular membrane and one coated with a polyether block amide compound can be seen in Figure 1.

METHODOLOGY

Membrane Selection Considerations

The Stephenfield Regional WTP in southern Manitoba is upgrading their water treatment system from chemical coagulation to a dual microfiltration and nanofiltration membrane filtration system. The need for this upgrade is due to the inefficiency of the chemical coagulation to remove DOC. This high DOC has resulted in failure of the effluent to meet the provincial and federal drinking water guidelines for trihalomethanes (THMs). It is expected that replacing the existing system with dual membrane filtration will enhance Stephenfield's removal efficiency of DOC and reduce the trihalomethane formation potential (THMFP).

Stephenfield Regional WTP Pilot-Scale Set-Up

The pilot-scale membrane filtration system at the Stephenfield Regional WTP is to simulate the performance of membranes to be installed in the full-scale operation. A schematic of the pilot-scale plant can be seen in Figure 3. Surface water is collected from the Boyne river and transferred to a chemical precipitation contact tank. The water is then pumped into the microfiltration system containing a single microfiltration element. Concentrate water from the microfiltration system is discarded into the drainage system. Post-microfiltration water is pumped into a secondary transfer tank before pumped into the reverse osmosis/nanofiltration system. The nanofiltration system contains three nanofiltration elements (Stage 1, Stage 2, Stage 3) ran in series. Permeate is collected at each stage and blended together to produce the plant's final effluent. Concentrate water from the nanofiltration system is discarded in the drainage system. The microfiltration and nanofiltration elements can be swapped out for new elements at the discretion of the plant operators.

Experimental Design Basis

Four nanofiltration membranes were chosen for small-scale testing. These membranes include the Dow FilmTec NF90-400/34i, Dow FilmTec NF270-400/34i, Dow FilmTec BW30XFR-400/34i and Dow FilmTec XFRLE-400/34i. Supplier-provided characteristics of these membranes can be seen in Table 1.

Table 1: Manufacturer-specified membrane attributes.

	NF90	NF270	BW30	XLE
Composition	Polyamide Thin-Film Composite	Polypiperazine Thin-Film Composite	Polyamide Thin-Film Composite	Polyamide Thin-Film Composite
Salt Rejection	Moderate	Moderate-High	High	High
Other Attributes	High Productivity	High Productivity	Brackish Water Fouling Resistant	Brackish Water Fouling Resistant Extra Low Energy

Bench-Top Membrane Filtration System

A custom bench-top membrane filtration system was used to carry out the small-scale testing. 75L of post-microfiltration water from the Stephenfield Regional WTP and placed in a 125L Equinox feed tank. Feed water was pumped through a WEG Hydra-Cell constant flow pump forcing 7L/min through the system. Water was pumped into a Sterlitech CF042 membrane filtration cell and ran in crossflow across the surface of the encased membrane. Depending on the needs of the experiment, permeate is either collected for analysis or recycled back into the feed tank. A flow diagram of the experimental setup can be seen in Figure 4.

Figure 3: Flow chart for the pilot-scale membrane filtration system at the Stephenfield Regional WTP, MB, Canada.

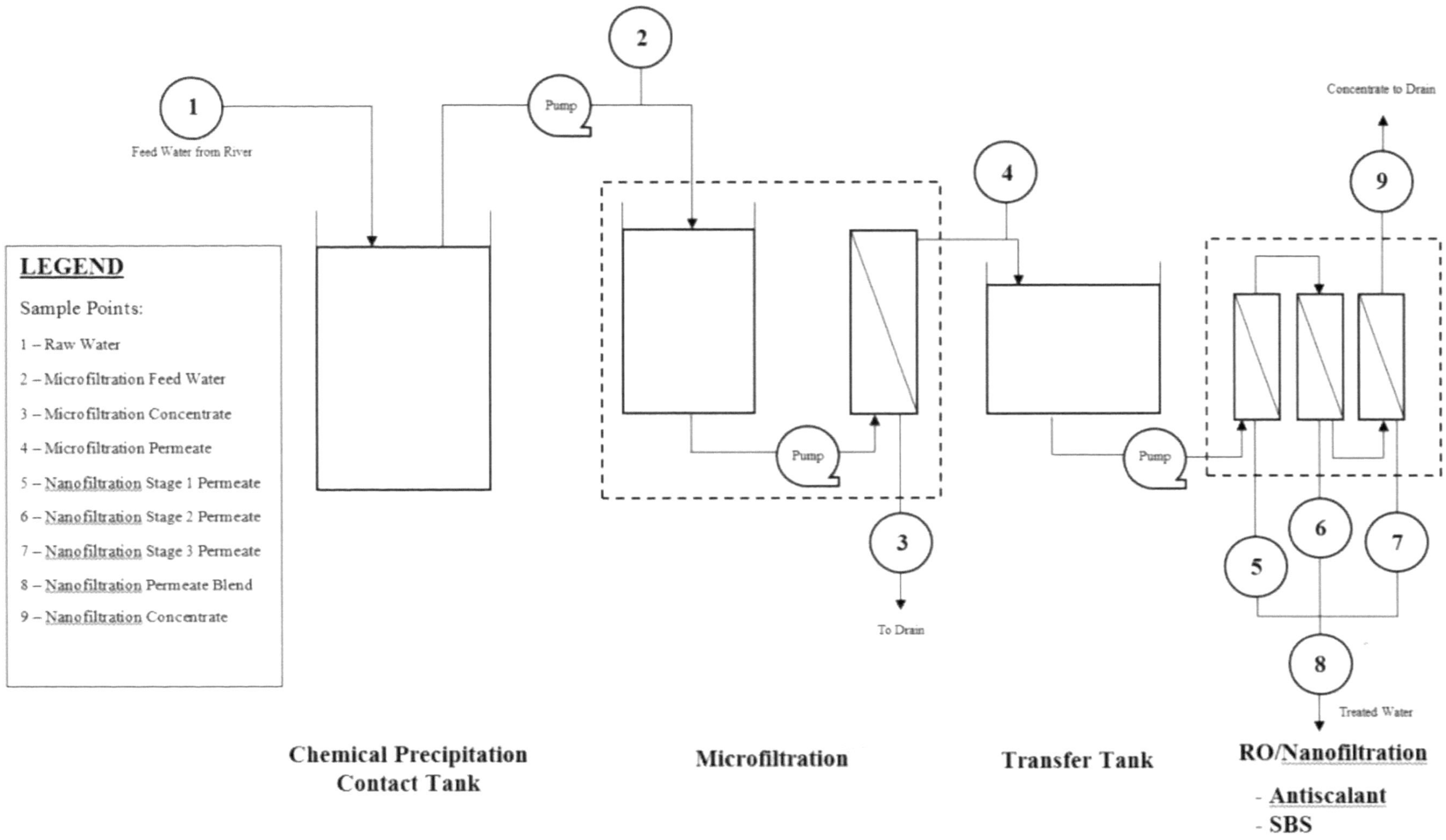

Figure 4: Flow chart for the bench-top membrane filtration system at the University of Manitoba.

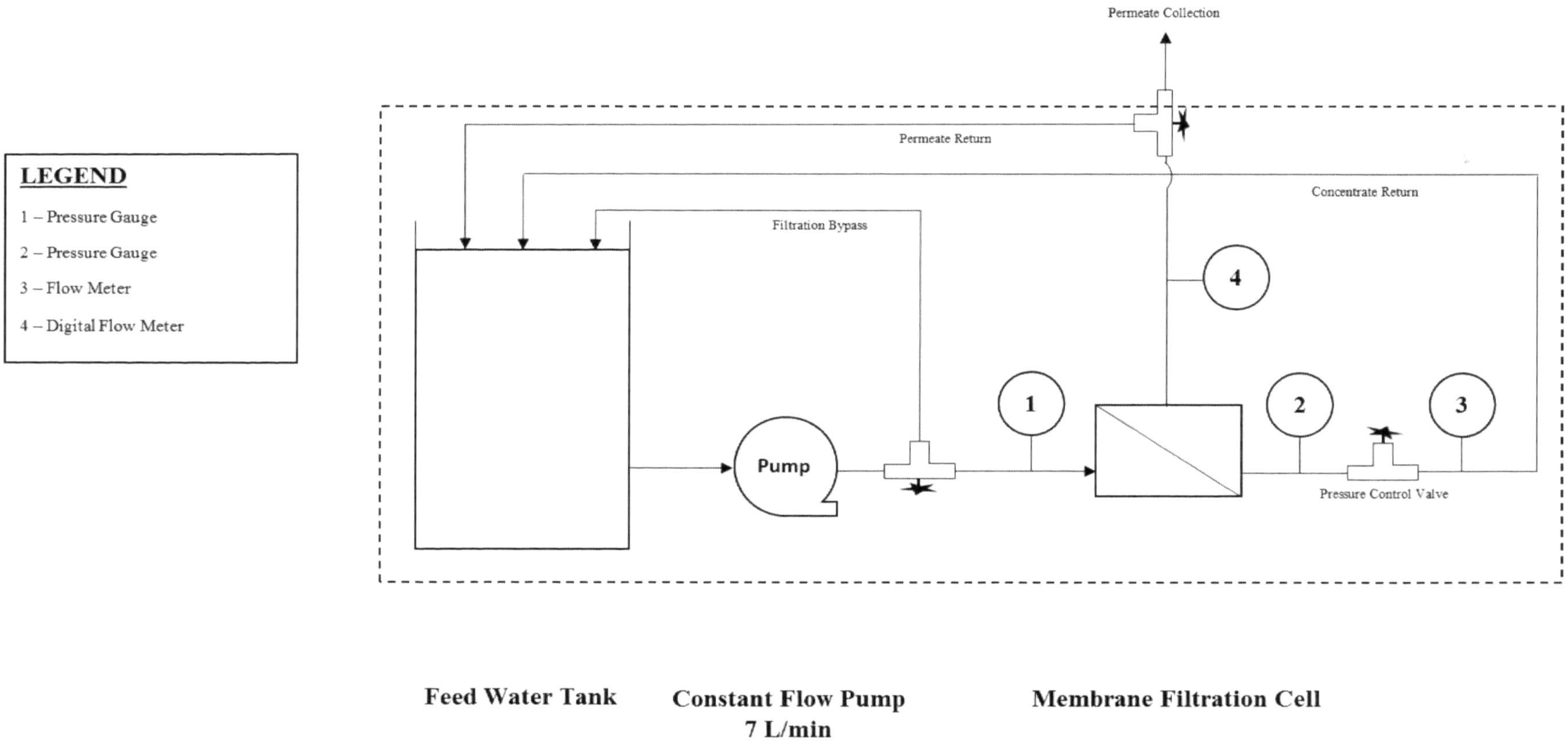

Operating Conditions and Measurements

Membrane partitions were cut from a flat sheet membrane sample to an approximate size of 4.25x9cm and placed in a 10% methanol-in-water solution for 24 hours prior to filtration. Each NF membrane was ran in the bench-top membrane filtration system for 1, 4, 12 and 24 hours. Pressure was held constant at 10bar (145psi) for all trials. Flux measurements were taken at the intervals specified in Table 2. Membranes were removed from the filtration cell and placed in a sealed beaker containing DI water until subsequent testing.

Table 2: Flux measurement recording schedule.

Time [min]	Interval [min]
0 - 10	1
10 - 20	2
20 - 60	5

Removal Efficiency of Raw Water Characteristics

Characteristics of the feed water will be compared to the permeate characteristics to determine the removal efficiency of dissolved constituents. Testing for manganese will be carried out in accordance to ASTM D858-12 Standard Test Methods for Manganese in Water, testing for iron in accordance with ASTM D1068-10 Standard Test Methods for Iron in Water, testing for alkalinity in accordance with ASTM D1067-11 Standard Test Methods for Acidity or Alkalinity in Water and testing for hardness in accordance with ASTM D1126-12 Standard Test Method for Hardness in Water. Measurements for DOC were taken using a Teledyne Tekmar TOC Fusion TOC Analyzer.

Characteristics of Membrane and Foulant by Chemical and Oxidative Testing

Chemical and oxidative testing was carried out to show the change in membrane and foulant characteristics over time. Three drops of methylene blue dye were placed on the active surface of the fouled membrane and left for ten seconds. The dye was rinsed off with deionized water and dye absorbance was noted. On a separate partition, three drops of 0.1N hydrochloric acid was placed on the active surface of the membrane. The presence of any effervescence indicates the presence of carbonates in the foulant. The Fujiwara test was performed on a third membrane partition. A solution was prepared with 2mL of 0.1N sodium hydroxide and 1mL of pyridine. The membrane partition was placed in the solution and gently heated over a standard Bunsen burner. A change in color to pink indicated the presence of oxidizing halogens in the foulant.

Surface Complexity

The change in surface complexity over time was measured by two distinct methods. The effects of membrane compaction were monitored by subjecting each membrane to the same tests prescribed above, but filtered instead with deionized water in crossflow across the surface of the membrane. As membrane compaction is a direct result of applied pressure, the decline in flux is directly proportional to the decline in surface area.

A Veeco Dimension 3100 atomic force microscope with a NanoScope IVa Controller was used to show the change in foulant complexity over time. A 100μm^2 surface area was plotted in contact mode at a speed of 10μm/s. Relative surface roughness and three-dimensional surface images were collected for analysis.

RESULTS AND DISCUSSION

Filtration Influent Water Quality

Table 3: Post-microfiltration / nanofiltration feed water from the Stephenfield Regional WTP. Collection date: July 8th, 2015.

Analyte	Concentration
Alkalinity (mg/L $CaCO_3$)	501
pH	8.13
Hardness (mg/L $CaCO_3$)	360
Total Organic Carbon (mg/L)	18.6
Manganese (mg/L)	1.807
Iron (mg/L)	0.78
Conductivity (mS/cm)	1.152

Removal Efficiency of Raw Water Characteristics

Manitoba surface waters are characterized by very high levels of DOC, making its removal an important consideration to reduce the THM formation potential. Removal efficiencies of DOC< hardness, alkalinity and manganese can be seen in Figures 5 through 8. All membranes effectively removed over 98% DOC by the late stage. The NF270 was the only membrane to leave a significant residual of hardness, alkalinity and manganese at roughly 20% each. This residual hardness and alkalinity may be desirable to avoid post-treatment addition of hardness, but the residual manganese may cause the plant to fail to meet its target residual manganese. This residual is likely a reflection of the molecular weight cut-off (relative indication of pore size), but may also be attributed to the different membrane material; the NF270 is a polypiperazine thin-film composite. Further investigation as to the impact of different membrane material will be completed at a later date.

Figure 5: The lab-simulated removal efficiency of DOC at stage 1 and stage 3.

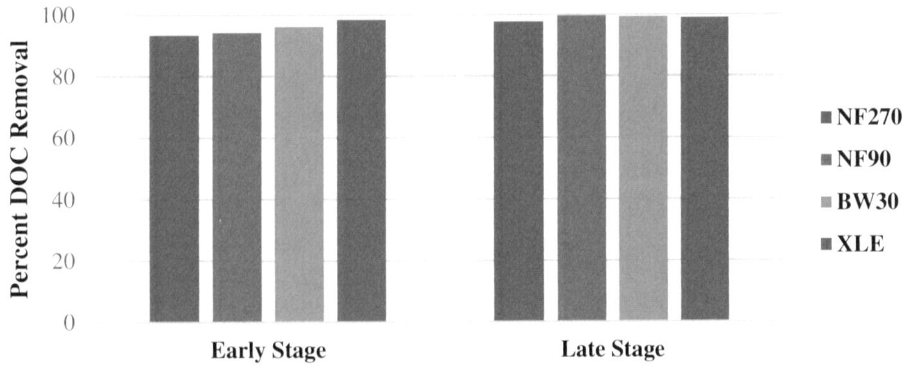

Figure 6: The lab-simulated removal efficiency of hardness at stage 1 and stage 3.

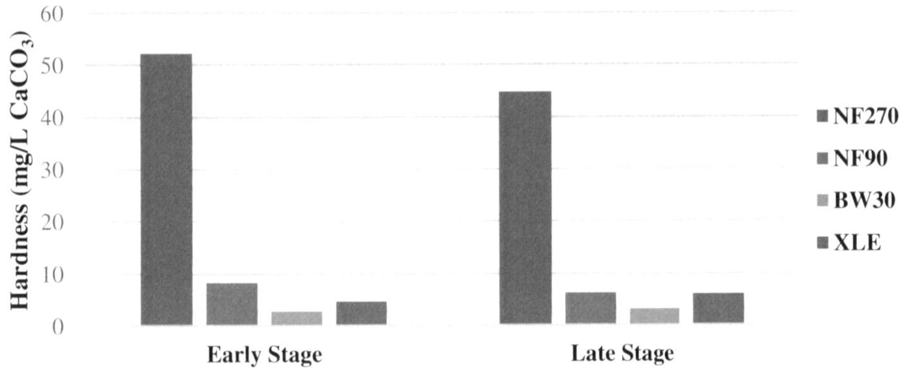

Figure 7: The lab-simulated removal efficiency of alkalinity at stage 1 and stage 3.

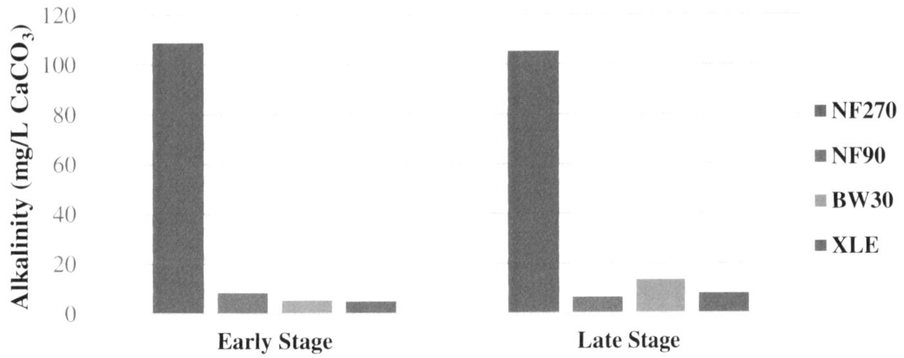

Figure 8: The lab-simulated removal efficiency of manganese at stage 1 and stage 3.

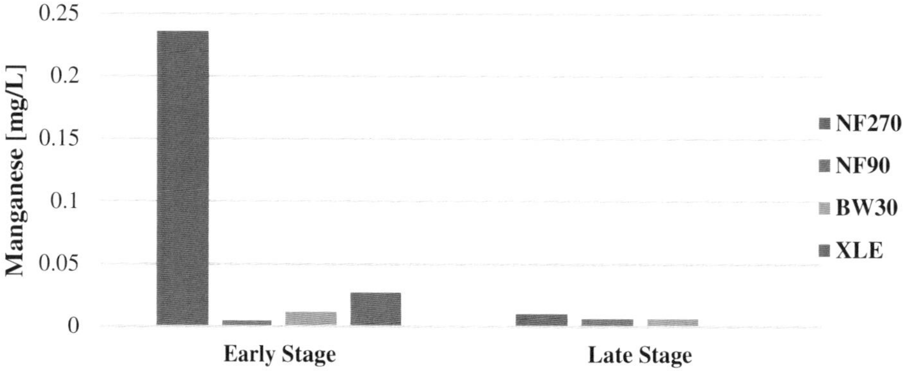

Flux Measurements

A high permeate flux is desirable for Stephenfield to meet the demand of their clients. The high flux of the NF270 makes it a preferable choice while the BW30 had a relatively low flux as observed in Figure 9. However, permeate flux is often inversely proportional to the removal efficiency of soluble material. This idea is supported by Figure 6 through Figure 8 in which the NF270 allowed the significantly larger passage of hardness, alkalinity and DOC than the other three tested membranes. Flux decline due to fouling and/or compaction is another major consideration. The large declines in the NF90 and XLE are explained later.

Figure 9: Permeate flux over time.

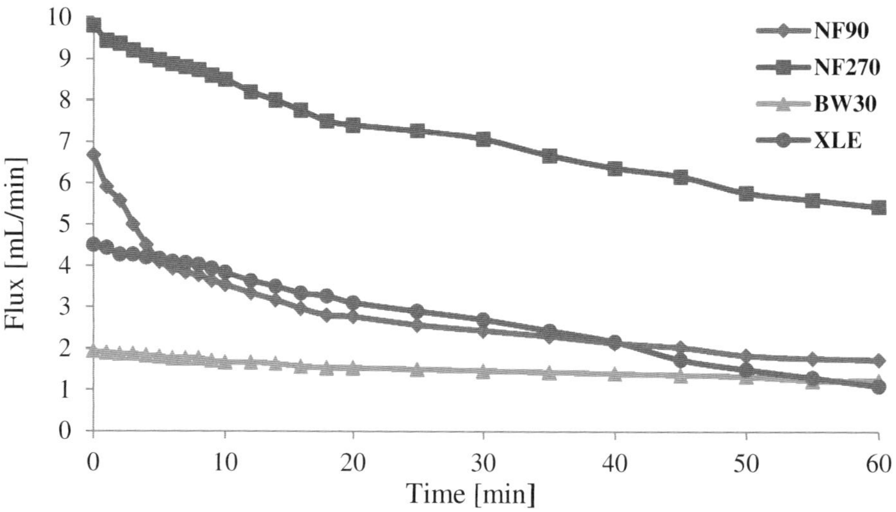

Flux Decline Due to Fouling and Compaction

The effects of membrane compaction were differentiated by subjecting each of the membranes to separate trials containing nanofiltration influent from Stephenfield and deionized water. In all cases, the net effect of flux decline increased over time. A longer test would likely indicate a

Figure 10: The effects of fouling, membrane compaction and the sum of these two over time for the a) NF90, b) NF270, c) BW30 and d) XLE membranes.

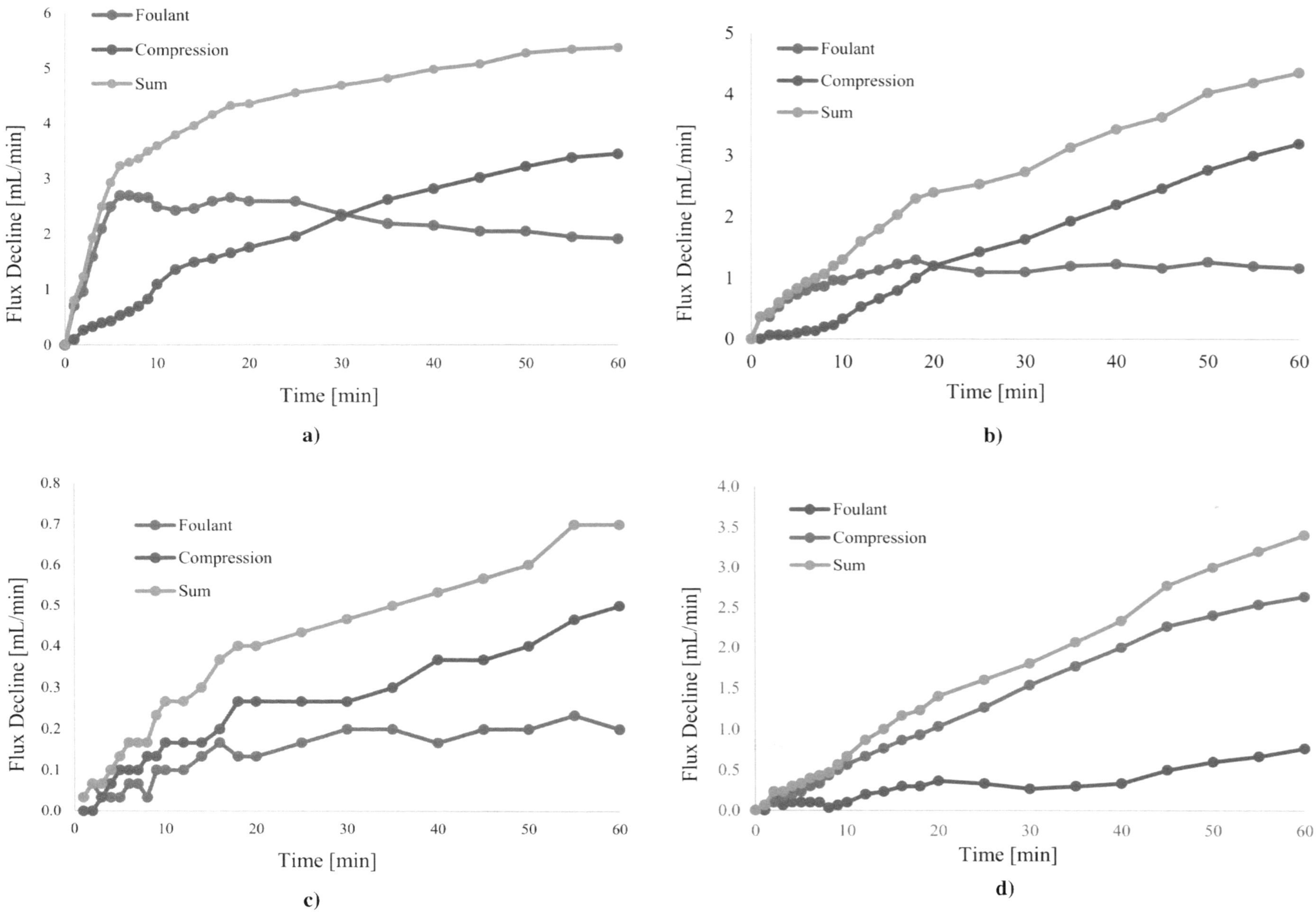

relative maximum. As expected, the effects of fouling were significantly less for the fouling resistant BW30 and XLE membranes; the effects of fouling were approximately three and six times greater for the NF270 and NF90. While the XLE had a relatively high decline in flux, a large part of the decline in flux can be attributed to compaction. This is because the "Xtra Low Energy" membrane is not suitable for high pressures as with the 10bar applied pressure in this experiment.

Unexpectedly, the effects of fouling decreased over time for the NF90 and NF270. This suggests either a change in foulant structure or the physical removal of the attached foulant resulting in the observed flux restoration due to fouling. These results were consistently supported through triplicate trials. A change in foulant structure that would cause the observed flux restoration would be a result of a decrease in foulant density. As the membrane is under a constant applied pressure with overbearing foulant, this relationship is less likely. Focus then shifts to the physical removal of foulant. Scouring is the physical removal of material from a surface due to an applied horizontal shear. The effects of scouring can be observed in the three-dimensional surface scan of the BW30 membrane after 12 hours in Figure 13. The rate of scouring is greater for surfaces with a higher roughness. The proposed relationship between scouring and membrane compaction will be described later.

Retained Mass of Foulant

The NF90 exhibited nearly three times the retained mass of foulant as the other three tested membranes. This is shown graphically in Figure 11. The lack of retained foulant on the BW30 and XLE membranes can be explained by their polyether block amide (commercially sold as PEBAX©) coating. Further investigation regarding the interaction between the polyether block amide and the soluble material in the feed water is necessary. While the NF270 is not coated with the polyether block amide, the lower mass of foulant can be attributed to its relatively smooth surface area which will be discussed in detail in the latter sections of this report.

Figure 11: Retained mass of foulant on the surface of the nanofiltration membrane over the course of 24 hours of operation.

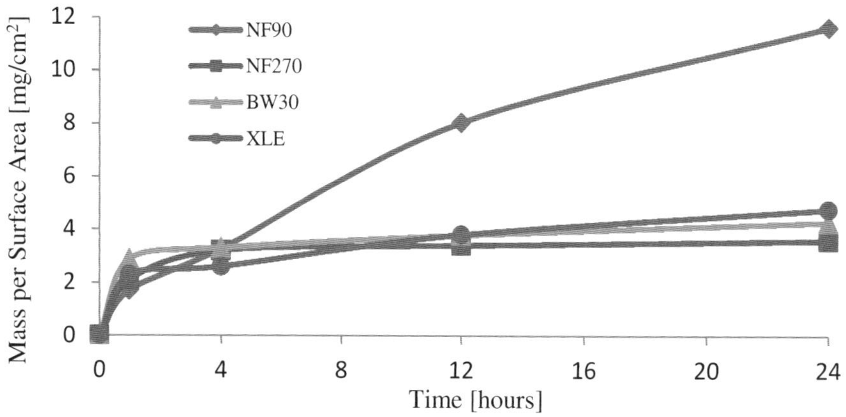

Chemical and Oxidative Testing

Chemical and oxidative testing was carried out to determine the post-filtration condition of the nanofiltration membranes and to make conclusions about specific foulant characteristics. Three drops of methylene blue dye were placed on each of the four tested membranes at each of the operational times listed in the experimental methods section of this report. As discussed the fouling and compaction section of this report, the XLE membrane is not suitable at the 10bar applied pressure in the experiment which made it the likeliest candidate for post-filtration imperfections. However, in all cases the rinsed dye left no residue on the membrane indicating no visible damage to the active surface of the membranes.

The Fujiwara and acid tests were carried out to determine whether the foulant contained oxidizing halogens and carbonates, respectively. In all cases, these tests failed to show the presence of oxidizing halogens or carbonates.

Surface Complexity

The membrane surface roughness shown in Figure 12 was derived from the three-dimensional surface profiles seen in Figure 13. It is important to note that membrane samples maintained their moisture during scans so as to give the most representative scan of the membrane at the particular operational time.

Figure 12: The change in surface roughness for each of the four tested membranes over time.

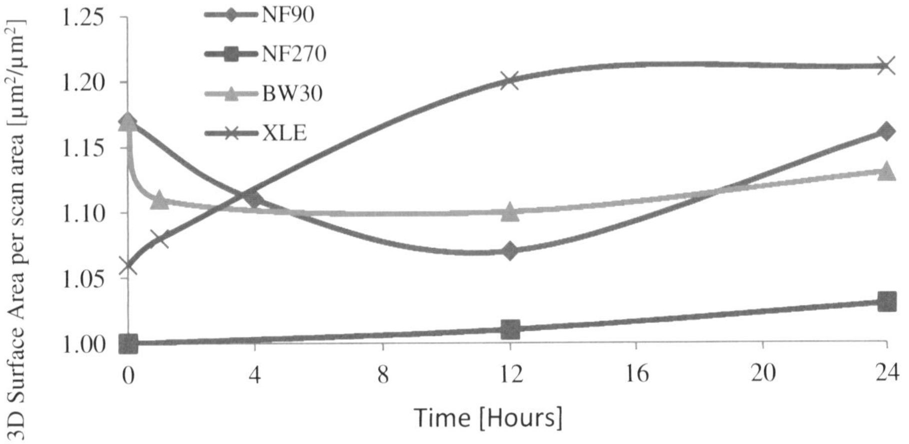

Virgin membrane surface roughness appears to be one of the governing factors for the degree of fouling and fouling mechanisms. The initially smooth NF270 and XLE membranes performed as expected: the foulant had accumulated on the surface of the membrane creating steeper peaks and deeper valleys. The initially rough NF90 and BW30 membranes followed the phenomenon labeled "valley clogging" by Vrijenhoek et al. The soluble material in the feed water initially filled in the contours of the membrane which effectively smoothed out the outer surface. It is expected that the rate of membrane fouling is greater during this time period as there is a larger amount of membrane surface area for the foulant to attach and more grab due to the peaks and valleys. This notion of a high rate of fouling is supported by Figure 10 showing the effects of

Figure 13: The Atomic Force Microscopy (AFM) 3-dimensional surface scans for each of the four tested membranes and each of the operational times.

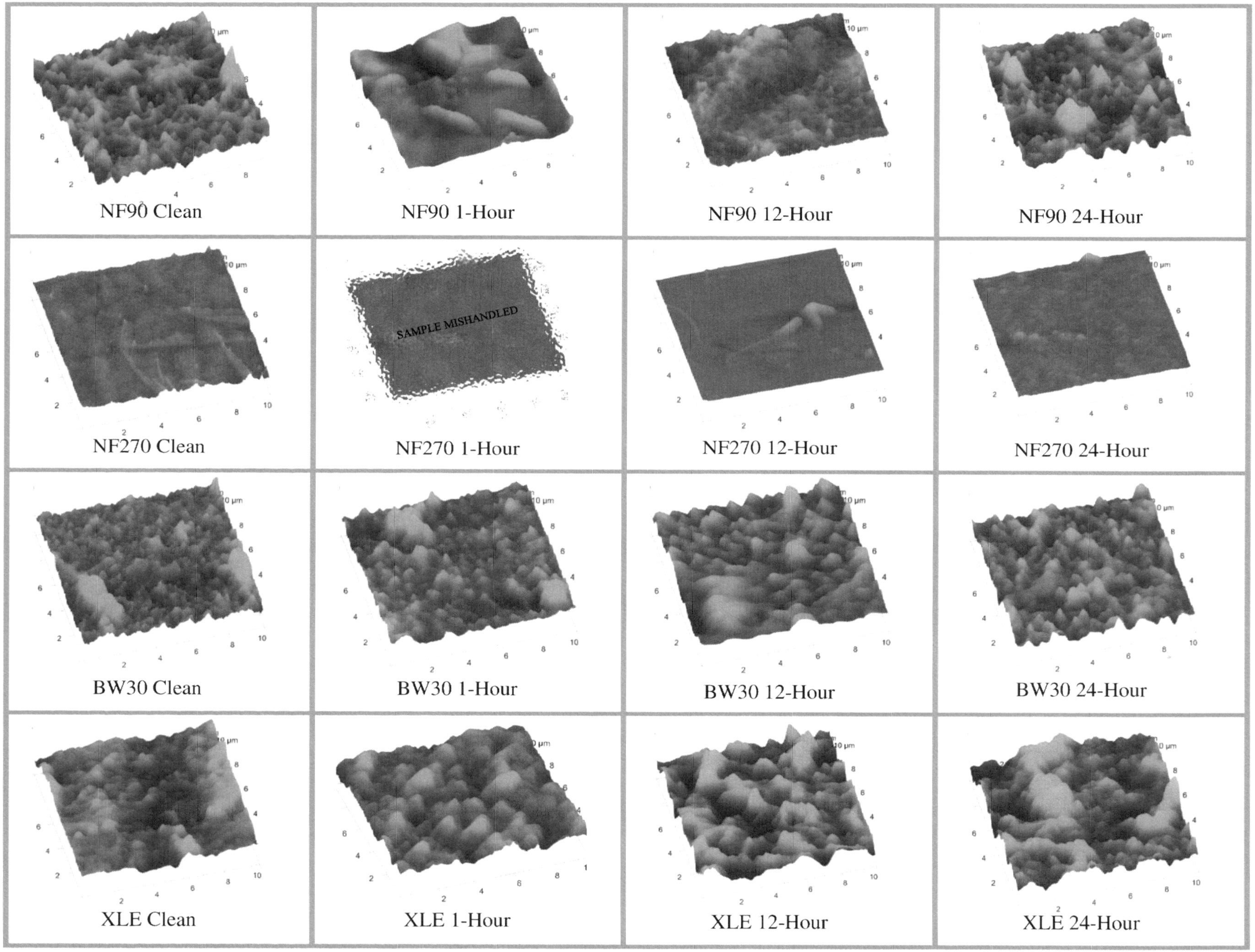

fouling on the decline in flux. These figures show a greater decrease in flux due to fouling in the early stages of the experiment before showing indications of levelling off after a sufficient period of time. This early decline in flux due to fouling is particularly high for the non-fouling resistant, initially rough NF90 membrane. From this point, the foulant complexity had increased in a similar manner to the NF90 and XLE membranes.

Another crossflow membrane filtration phenomenon, known as scouring, can be observed in the 12-hour AFM scan for the BW30 membrane as seen in Figure 13. Under the crossflow conditions of the experiment, horizontal shear is given by the water under the applied 10bar pressure. The effects of scouring can be used to explain the decrease in the effects of fouling for the NF90 and NF270 in Figure 9a and Figure 9b. As foulant is scraped off the surface, some of the lost flux due to fouling is restored.

Scouring is a potential explanation for the flux restoration due to fouling, but what would occur if the experiment was to continue? Would the effect of fouling on flux for the NF90 and NF270 eventually reach zero? It can be predicted that, in the long run, the addition of foulant by filtration and the removal of foulant by scouring will reach equilibrium. A more important consideration is to when this equilibrium will occur. Given all other variables held constant during the experiment, a possible explanation is a relationship between compaction and fouling. Consider the rough surface of the NF90 covered by a thin layer of foulant making the surface entirely smooth. A schematic of this can be seen in Figure 14a. At a later point in the experiment after membrane compaction has occurred, the foulant complexity has changed. This can be seen in Figure 14b. In both cases, the mass of foulant has been conserved and distance from the surface of the membrane to the surface of the foulant has also been conserved.

Figure 14: The predicted method for the relationship between flux decline due to fouling and flux decline due to compaction. a) A membrane with a high virgin surface roughness with smooth layer of foulant and b) the same membrane after compaction has occurred.

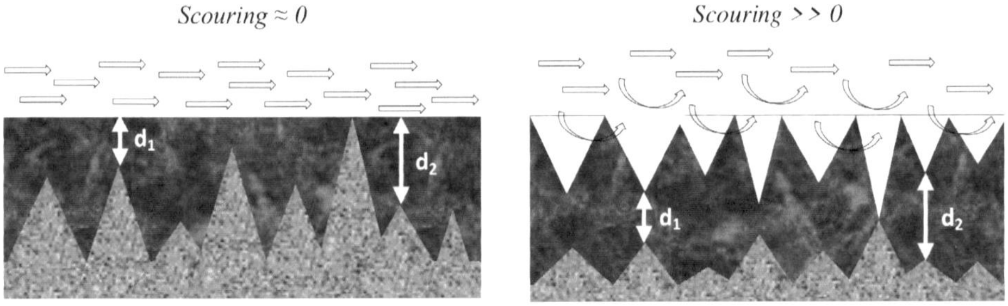

a) b)

The reduction of peak height of the membrane has effectively pulled in the layer of foulant causing a change in the surface complexity. The resultant surface has a greater roughness to which scouring is more likely to occur. The predicted long-term relationship between the effects of fouling and compaction can be seen in Figure 15. Further investigation regarding this claim will be done at a later date.

Figure 15: The predicted long-term flux decline over time due to fouling and membrane compaction.

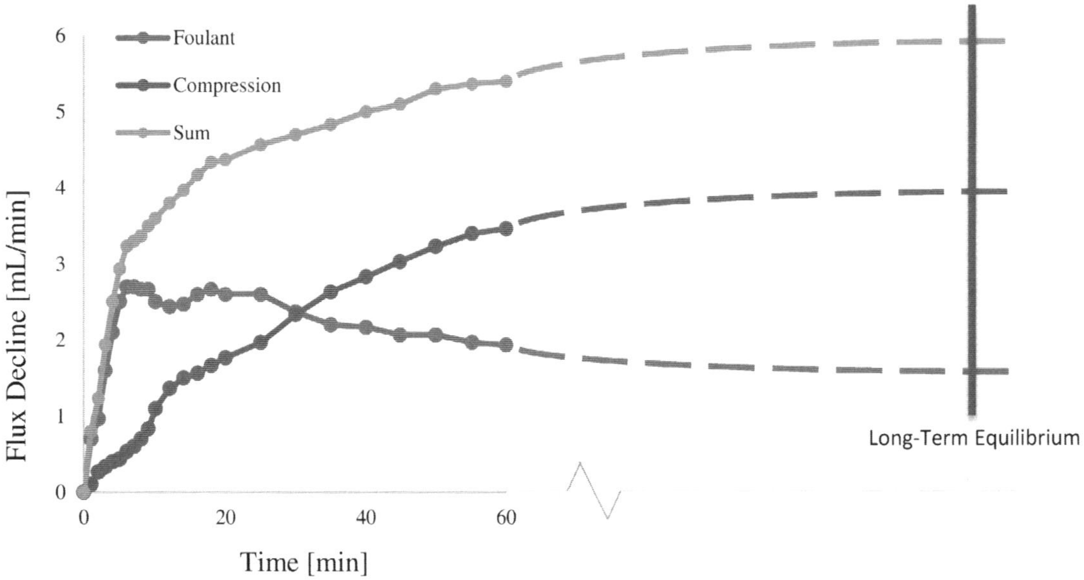

Conclusions

Based on the results as a whole, it is clear to see that membrane performance is heavily influenced by the characteristics of the feed water and the properties of the membrane. The NF270 had the lowest removal efficiency of hardness, alkalinity and manganese, but the highest overall flux. The BW30 showed the lowest flux over time and the NF90 experienced the largest degree of flux decline. Three-dimensional surface profiling of the tested membranes over time suggests a unique relationship between initial surface complexity and fouling patterns. The latter parts of this section will provide an explanation for the observed changes in membrane and foulant characteristics over time.

Each membrane was tested for their removal efficiencies of DOC, hardness, alkalinity and manganese. As expected, the membrane with the greatest manufacturer-specified pore size, the NF270, allowed a significantly greater passage of ionic species and manganese. The suffix -270 in the NF270 is a relative indication of pore size. By the same description, the pores of the NF90 and BW30 are three and nine times smaller than the NF270. The removal efficiencies of DOC are roughly the same between all membranes. The passage of potentially large and complex DOC is expected to be lower than the relatively small ionic species. Subsequent testing will be conducted at a later date to determine exact pore size; this will be done by subjecting the membrane to different solutions containing a single, known molecular weight and observing the passage of this carbon (otherwise known as molecular weight cutoff testing). That being said, water treatment plant effluent for general consumer purposes maintains a residual hardness and alkalinity. At roughly 50mg/L $CaCO_3$ for hardness and 100mg/L $CaCO_3$ for alkalinity, the NF270 would not require post-treatment addition of hardness and alkalinity as the other three membranes. However, this comes at the cost of excessive manganese residual in the effluent – an important consideration for membrane selection of the plant operators at Stephenfield due to the relatively high levels of manganese in the surface water.

The observed fouling mechanism for membranes with low virgin membrane surface roughness is very different than the observed fouling mechanism for membranes with high virgin surface roughness. In this experiment, low virgin surface roughness membranes are epitomized by the NF270 and XLE membranes and high virgin surface roughness membranes are epitomized by the NF90 and XLE. Roughness measurements were derived from the AFM scans by taking the ratio of three-dimensional surface area to two-dimensional scan area. In the order of increasing virgin membrane surface roughness: the accumulation of foulant on the NF270 was low due to the nearly-flat surface of the membrane. This smoothness left the foulant with less surface area to attach to or ridges to get caught in. The accumulation of foulant on the XLE membrane followed a similar trend as the NF270 membrane, but increased in surface roughness much faster. This steeper incline in surface roughness can be attributed to the foulant having more surface area to attach to and more valleys to get caught in. The NF90 and BW30 membranes performed very differently. Instead of solely increasing in surface complexity, these membranes began the experiment with high virgin surface roughness, declined in roughness from 0 to 12 hours and experienced an increase in roughness after 12 hours. This is a result of what Vrijenhoek et al. described as valley clogging – the deposition of foulant in the cracks and valleys of the membrane and often leads to a rapid decline in flux. Once the valleys have been sufficiently smoothed out, the surface complexity from further deposition of foulant will increase. Degree of fouling and subsequent flux inhibition can have an important impact on the decline in flux.

The retained mass of foulant appears to be influenced by presence of a fouling-resistant polyether block amide compound. The BW30 and XLE membranes, coated with one of these compounds, exhibited a much lower flux decline due to fouling. The XLE experienced twice as much flux decline due to fouling as the BW30, but roughly one third and one sixth that of the NF270 and NF90, respectively.

Given the information above, as well as other plant operator considerations, the NF270 membrane was selected for pilot-scale testing in September 2015. This membrane was selected for its high permeate flux, low degree of fouling and residual hardness and alkalinity. The performance of the NF270 was monitored for the following two months and failed to meet the effluent goals for manganese content. At the discretion of the plant operators, the stage 2 and stage 3 membranes were swapped out for NF90 membranes to improve the overall removal efficiency of manganese. Pilot-scale performance with the new membrane configuration is under constant monitoring.

Future Projects

Although the experimental results explain some of the key components of nanofiltration membrane fouling, further research is required in order to develop a more complete understanding. One of the tested membranes was a polypiperazine thin-film composite, the NF270, unlike the polyamide thin-film composition of the other three. This membrane featured a smoother surface, higher flux and lower removal efficiency of soluble material. Further research will be conducted to compare the performance of the polyamide and polypiperazine thin-film composites. Further to this membrane property, virgin membrane hydrophobicity and zeta potential will be tested and compared between the membranes with and without the fouling-resistant polyether block amide compound to show any trends between these and fouling patterns. The conducted experiment gave an indication of the short-term effects of both fouling and compaction, but failed to prove the long-term effects. A longer experiment will be conducted to determine the relative long-term equilibrium point between fouling and compaction and observe any notable relationship between fouling and compaction. Further to this, membrane samples will be subjected to 1, 4, 12 and 24 hours of filtration with DI water and analyzed with atomic force microscopy to show the isolated effects of membrane compaction over time. A more thorough analysis of the foulant will be conducted to identify the composition and further analyze the specific relationship between membrane and foulant. Techniques include, but are not limited to, Fourier transform infrared spectroscopy, fractal analyses, small-angle neutron scattering, cross-section splicing and scanning/transmitting electron microscopy. Plant monitoring and optimization of the developing Stephenfield Regional WTP is expected to continue until project completion.

REFERENCES

Vrijenhoek, Eric M. and Seungkwan Hong and Menachem Elimelech (2001), “Influence of membrane surface properties on initial rate of colloidal fouling of reverse osmosis and nanofiltration membranes”, *Journal of Membrane Science Volume 188*, February, pp.115-128.

Jiraratananon, R. and D. Sampranpiboon and R.Y.M. Huang (2002), “Pervaporation separation and mass transport of ethylbutanoate solution by polyether block amide (PEBA) membranes:, *Journal of Membrane Science Volume 210 Issue 2*, pp389-409.

Malvialle, Joël and Peter E. Odendaal and Mark R. Wiesner (1996), “Water Treatment Membrane Processes”, American Water Works Association, ed. New York, NY, McGraw-Hill.

Schäfer, A.I, A.G Fane, T.D Waite (2001), “Cost factors and chemical pretreatment effects in the membrane filtration process of waters containing natural organic matter”, *Water Research Volume 35 Issue 6*, pp.1509-1517.

Katsuki, Kimura et al. (2004), “Irreversible membrane fouling during ultrafiltration of surface water”, *Water Research Volume 38 Issure 14-15*, pp.3431-3441.

Persson, Kenneth M. and Vassilis Gekas and Gun Trägårdh (1995), “Study of membrane compaction and its influence on ultrafiltration water permeability”, *Journal of Membrane Science Volume 100 Issue 2*, pp.155-162.

Louis, Jennifer S. et al. (2006), “Effects of polyether-polyamide block copolymer coating on performance and fouling of reverse osmosis membranes”, *Journal of Membrane Science Volume 280 Issue 1-2*, pp.762-770.

Facile One-Step Modification of Membrane Surface Using Hydrophilic Polymers for Wastewater Treatment

Nima Shahkaramipour, Sankara Narayanan Ramanan, Cheng Kee Lai, Chong Cheng, and Haiqing Lin*

Department of Chemical and Biological Engineering, University at Buffalo, The State University of New York, Buffalo, NY 14260, USA

* Corresponding author. Tel: +1-716-645-1856, Email: haiqingl@buffalo.edu (H. Lin)

Abstract

Polymeric membranes have been widely used for wastewater treatment. However, the membranes can be fouled by suspended solids and dissolved organic matters presented in the water, which block the membrane pores on the surface or inside the supports and thus decrease the water flux. Surface modification to enhance the hydrophilicity has been touted as a promising approach to reduce the fouling. In this study, we have modified ultrafiltration (UF) membranes using superhydrophilic zwitterionic polymers by facile one-step coating on the surface, aiming to reduce the fouling. More specifically, three commercial UF membranes, PES-2, PAN-50, and PS-100 are coated with a solution containing polydopamine (PDA) and sulfobetaine methacrylate (SBMA). The PDA and SBMA react and form a thin film on the surface, while PDA provides adhesion to the membrane surface and SBMA provides superhydrophilic properties. The coating enhances the surface hydrophilicity, which is confirmed by contact angle measurement. The effect of coating time and composition of the coating solutions on the water flux in the membranes is investigated. The membranes coated with zwitterionic materials (PDA+SBMA) demonstrate better stability and higher water flux than those of PDA coated and uncoated ones, when tested with wastewater containing bovine serum albumin (BSA) using a crossflow system. For instance, the modified PES-2, PAN-50, and PS-100 with PDA and SBMA, increased the water permeance by 65%, 81%, and 61%, respectively, compared to analogous uncoated membranes, after 3 hour filtration using a 3g/l BSA solution.

The relation between structure-performance of thin film composite membranes and the tools used for their fabrication method

Kelly Briceño [1], Irakli Javakhishvili [2], Haofei Guo [3], Knud Villy Christensen [1], Birgir Norddahl [1], Søren Hvilsed [2], Frank Lipnizki [3]

[1] Department of Chemical Engineering, Biotechnology and Environmental Technology, University of Southern Denmark, Campusvej 55, 5230 Odense M, Denmark

[2] Danish Polymer Centre, Department of Chemical and Biochemical Engineering, Technical University of Denmark, Building 227, 2800 Kongens Lyngby, Denmark

[3] R&D department, Business Center Membrane, Alfa Laval A/S, Stavangervej 10, 4900, Nakskov, Denmark

Keywords: Polyamide, thin film composite membranes, fabrication variables, interfacial polymerization.

For more than 30 years polyimides (PA) have been one of the main polymers for the fabrication of thin film composite membranes. Several researchers have assessed the main fabrication variables that influence the final structure of the polyamide layers including monomer concentration, solvents, additives in aqueous and organic phases and type of support to mention a few [1,2]. However, few studies have reported the influence of the fabrication tools used in the preparation of the PA layer. Each research group report their own recipe and fabrication method according to the general criteria of avoiding bubble formation during fabrication as that would later compromise the function of the membrane, but each process step must be well controlled to obtain reliable and consistent membrane performance. In general, the fabrication of the polyamide membrane layer is carried out in two steps. A polymeric support is initially brought in contact with the aqueous phase containing *m*-phenylene diamine (MPD) monomer and then with the organic phase containing the trimesoly chloride (TMC) monomer in order to promote PA formation through interfacial polymerization. The critical step occurs immediately after the support has been in contact with the aqueous phase, because formation of irregular drops could be initiation points for defects during the second step. Several tools have been used to eliminate the drop formation including air-knifes, rubber wipers, rubber rollers, glass rollers, or for that matter the absence of any tool using only water evaporation. In this work different methods of avoiding drop formation during the membrane preparation are tested to evaluate how the preparation methods influence the membrane structure and the final membrane properties. Understanding the membrane formation and consequently the defect formation will help to control and reproduce membrane preparation both in laboratory and industrial scale [3]. Specifically this work focuses on the preparation of a polyamide membrane layer supported by a polysulfone support after immersion in MPD/aqueous solution followed by immersion in TMC/heptane solution. The polysulfone support is in contact with the MPD/aqueous phase using immersion or pipetting. Further the use of a rubber wiper or absence of any tool to eliminate the aqueous solution droplets is evaluated. Both NaCl rejection and water and NaCl fluxes are evaluated, in order to relate them with the structural characteristics of the membranes using SEM, contact angle and streaming potential.

[1] Asim K. Ghosh, Byeong-Heon Jeong, Xiaofei Huang, Eric M.V. Hoek , Impacts of reaction and curing conditions on polyamide composite reverse osmosis membrane properties, Journal of Membrane Science 311 (2008) 34–45

[2] Liwei H, McCutcheon J, Impact of support layer pore size on performance of thin film composite membranes for forward osmosis, Journal of Membrane Science 483(2015)25–33

[3] R. Rangarajan, N.V. Desai, S.L. Daga, S.V. Joshi, A. Prakash Rao, V.J. Shah, J.J. Trivedi, C.V. Devmurari, K. Singh, P.S. Bapat, H.D. Raval, S.K. Jewrajka, N.K. Saha, A. Bhattacharya, P.S. Singh, Paramita Ray, G.S. Trivedi, N. Pathak, A.V.R. Reddy, Thin film composite reverse osmosis membrane development and scale up at CSMCRI, Bhavnagar, Desalination 282 (2011) 68–77

Facile One-Step Modification of Membrane Surface Using Zwitterionic Polymers for Wastewater Treatment

Nima Shahkaramipour, Sankara Narayanan Ramanan, Cheng Kee Lai, Chong Cheng, and Haiqing Lin*

Department of Chemical and Biological Engineering, University at Buffalo, The State University of New York, Buffalo, NY 14260, USA

* Corresponding author. Tel: +1-716-645-1856, Email: haiqingl@buffalo.edu (H. Lin)

Abstract

Polymeric membranes have been widely used for wastewater treatment. However, the membranes can be fouled by suspended solids and dissolved organic matters presented in the water, which block the membrane pores on the surface or inside the supports and thus decrease the water flux. Surface modification to enhance the hydrophilicity has been touted as a promising approach to reduce the fouling. In this study, we have modified ultrafiltration (UF) membranes using superhydrophilic zwitterionic polymers by facile one-step coating on the surface, aiming to reduce the fouling. More specifically, three commercial UF membranes, PES-2, PAN-50, and PS-100 are coated with a solution containing polydopamine (PDA) and sulfobetaine methacrylate (SBMA). The PDA and SBMA react and form a thin film on the surface, while PDA provides adhesion to the membrane surface and SBMA provides superhydrophilic properties. The coating enhances the surface hydrophilicity, which is confirmed by contact angle measurement. The effect of coating time and composition of the coating solutions on the water flux in the membranes is investigated. The membranes coated with zwitterionic materials (PDA+SBMA) demonstrate better stability and higher water flux than those of PDA coated and uncoated ones, when tested with wastewater containing bovine serum albumin (BSA) using a crossflow system. For instance, the modified PES-2, PAN-50, and PS-100 with PDA and SBMA, increased the water permeance by 65%, 81%, and 61%, respectively, compared to analogous uncoated membranes, after 3 hour filtration using a 3g/l BSA solution.

Vibrational and X-ray Spectroscopy and Microscopy Reveal Irreversible Fouling Mechanisms of Reverse Osmosis Membranes during Water Reuse: Direct Evidence of Aluminum Coagulant Carryover

Mutiara Ayu Sari and Shankar Chellam

Department of Civil Engineering, Texas A&M University, College Station, TX 77843-3136

Water reuse is becoming increasingly important to quench the thirst of a growing population. Reusing municipal wastewater necessitates advanced technologies such as reverse osmosis (RO) to remove the myriad contaminants present even after tertiary treatment to produce potable water. Since water reuse is a relatively new application for RO, factors contributing to membrane fouling are not yet well understood.

The objective of this research is to identify dominant foulants and RO fouling mechanisms during water reuse. Membrane morphology, organic functionalities and elemental composition were analyzed using scanning electron microscopy-energy dispersive spectra (SEM-EDS), atomic force microscopy (AFM), Fourier Transform Infrared (FTIR) spectroscopy, and X-ray Photoelectron Spectroscopy (XPS). Membranes were chemically cleaned using different chelating agents at low and high pH depending on the nature of the foulants revealed from the surface analysis. Autopsies were performed on RO membranes after chemical cleaning and compared with the virgin membrane. Two 4-inch diameter thin-film composite spiral wound modules (KOCH HR-4040) were extracted from an actual two-stage pilot-plant implemented in support of design of municipal water reuse for this research.

SEM-EDS showed higher relative amount of aluminum in the 1st stage suggesting that fouling was dominated by the carryover of amorphous Al precipitates from pretreatment. The Al concentration was lower in the 2nd stage suggesting the carryover was throughout the 2-stage pilot-plant or that Al was precipitating *in situ*. Silicon was not detected in all samples suggesting the dominance of Al (hydr)oxides rather than aluminosilicates. Al, Si, Cl, P, and Ca were detected in XPS scans of fouled membranes but were absent in virgin membranes confirming the importance of inorganic fouling.

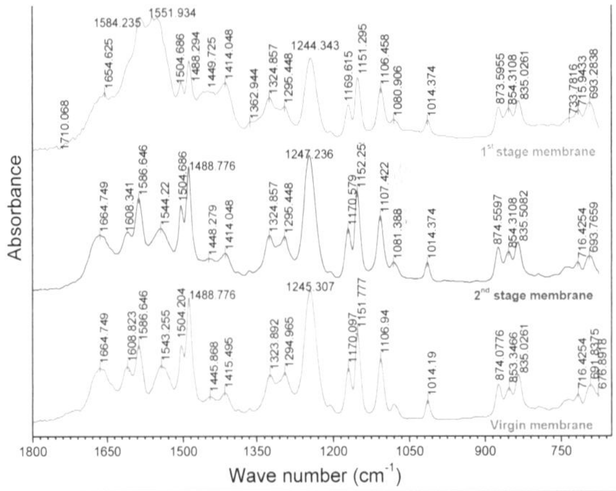

Figure 1. *ATR-FTIR spectra of clean and irreversibly fouled membranes exhibited changes in peak size and locations.*

Proteinaceous compounds were relatively insignificant as the amide I peak (1653 cm^{-1}) did not intensify significantly compared to the amide II (1540 cm^{-1}) band despite shifting towards a wave number that indicates bacterial origin (i.e. around 1654 cm^{-1}). Carbohydrate-like compounds that usually appear as a broad band in the region of 900-1100 cm^{-1} (bacterial fingerprints region) were also not observed (Figure 1). Hence, natural organic matter of biological origin was not a major foulant.

Efficient Al removal during EDTA cleaning and the higher Al concentrations measured in spent EDTA cleaning solution demonstrated that it was more efficient than citric acid. XPS did not identify any Ca after both citric acid and EDTA cleaning even though analysis of spent cleaning solutions suggests that some Ca should remain on the membrane after EDTA cleaning. Therefore, fouling in large-scale systems appears to be heterogeneous where the small spot size of X-ray analysis may not detect all the foulants. FTIR spectra showing reduction of signals from metal complexation by chemical cleaning and effects of fouling and cleaning on membrane surface morphology will be discussed in detail during the presentation.

In conclusion, prominent inorganic fouling was evidenced by SEM-EDS, XPS, and FTIR. Al, Ca, and P were found to be the most abundant elemental (inorganic) foulants. The 1st stage membrane was fouled to a higher degree compared to the 2nd stage due to larger Al deposition attributed to carryover from coagulation pretreatment. FTIR further confirmed lower organic fouling due to weak signals from protein-like and carbohydrate-like moieties from biological constituents. EDTA better removed Al whereas citric acid was efficient in removing Ca. Interestingly, high resolution XPS scans result for regenerated membranes still showed residues from the chemical cleaning agents despite thorough rinsing by ultrapure water after cleaning. Since EDS or XPS survey spectra could not easily detect small amount of residues it is very easy to overlook their presence. Nonetheless, since high flux recovery was obtained by single step cleaning, the impact of having chemical residuals from cleaning may be insignificant. In any case, a longer pure water rinsing duration is recommended after cleaning to better remove all traces of the cleaning agents and avoid their build up over multiple cleaning cycles.

The influence of surface morphology on the performance of reverse osmosis membrane from the perspective of water reclamation

Jianzhong Xia

The water environment in China is challenged by three main threats, water pollution, water scarcity and water insecurity. About 70% of the cities are facing a water shortage, and the overall deficiency of water supply across the whole country is about 50 billion m^3 each year. One effective way to solve the aforementioned problem is to use MBR/RO(NF) hybrid process to generate new water source by reclaiming the municipal wastewater. However, conventional NF/RO membrane is not designed for this purpose- the energy cost, antifouling property, and salt rejection are not fully optimized for this application. To overcome the energy cost issue, high productivity membrane is needed. Due to the self-limiting film growth theory, it is hard to increase the flux by reducing the polyamide thickness. However, it is practical to increase the apparent flux by controlling the surface morphology of fully aromatic polyamide NF/RO membrane. This surface morphology is usually described as "ridge and valley", which can actually increase the surface area in three dimensions. There are lots of ways to control the surface morphology and the morphology indeed varies in a wide range. In this presentation, several different kinds of "ridge and valley" structures will be presented. The correlation between the membrane performance and morphology will be discussed in detail. Besides the lab test and characterizations, pilot study in our water reclamation plant is also conducted. Overall, a balance among flux-rejection-fouling is a must for a practical engineering project.

Comparative Study of NF/RO Membranes for Sustainable Desalination of Produced Water and Frac Flowback Wastewater Following Biological Pretreatment

Stephanie M. Riley, Tzahi Y. Cath

Colorado School of Mines, Department of Civil and Environmental Engineering

Development of unconventional oil and gas (O&G) resources is reshaping the energy landscape of the United States. Technological advances in hydraulic fracturing and horizontal drilling have made previously unrecoverable resources economically viable. However, extraction from these formations requires hundreds of thousands of gallons of clean water per well and generates substantial (and unavoidable) volumes of wastewater. These wastewaters (e.g., hydraulic fracturing flowback and produced water (PW)) contain varying concentrations of total dissolved solids (TDS), ranging from 10,000 mg/L to more than 200,000 mg/L, oil, grease, hydrocarbons, and various organic and inorganic matter. Consequently, a multi-barrier approach is required to effectively treat O&G waste streams for beneficial reuse.

Biological and membrane processes are promising methods to treat O&G wastewaters. Biological technologies have proven effective at removing organic material from a variety of waste streams including domestic waste streams, landfill leachate, and oily wastewaters; tight/dense membranes are most established in desalination of seawater and brackish waters. For sustainable membrane treatment, optimizing water recovery, and reducing fouling propensity, pretreatment targeting removal of organic constituents and suspended solids is critical. Integration of biologically-active filtration (BAF), ultrafiltration (UF), and nanofiltration/reverse osmosis (NF/RO) results in comprehensive removal and degradation of contaminants, reducing membrane fouling propensity due to BAF pretreatment, and producing a high quality permeate suitable for reuse applications such as hydraulic fracturing fluid and irrigation.

The current study evaluates the performance of NF/RO membranes following BAF and UF pretreatment of Denver-Julesburg (DJ) basin PW and frac flowback wastewater. The main objectives include (a) optimizing BAF and UF pretreatment for maximum removal of organic compounds, (b) evaluating NF/RO fouling propensity and membrane performance based on solids, inorganic, and organic rejection, and water flux, and (c) identifying favorable NF/RO membranes for treatment of PW and flowback, based on membrane characterization and performance. Bench-scale experiments with BAF have been conducted to acclimate a biofilm and optimize degradation of organic matter. In addition, experiments with UF were conducted to further reduce the turbidity of the BAF effluent and prepare it for NF/RO desalination. Ongoing NF/RO experiments compare the performance of various membranes (e.g., NF90, NF270, BW30 (DOW Chemical Company, Midland, MI), and NFX (Synder Filtration, Vacaville, CA)). Following rejection tests, membrane properties for each NF/RO membrane are being investigated using scanning electron microscopy (SEM) and atomic force microscopy (AFM) (morphology), contact angle (hydrophobicity), zeta potential (surface charge), and Attenuated Total Reflectance Fourier Transform Infrared Spectroscopy (ATR-FTIR) (surface chemistry) to evaluate variations in fouling.

A Fouling comparison study of algal, bacterial, and humic organic matters in seawater desalination pretreatment using UF ceramic membranes

Mohammed A. Namazi [a,b]*, Sheng Li [a], Noreddine Ghafour [a], Gary L. Amy [a], TorOve Leiknes [a]

[a] *Water Desalination and Reuse Center, 4700 King Abdullah University of Science and Technology, Thuwal 23955-6900, Saudi Arabia*

[b] *Desalination Technologies Research Institute (DTRI), Saline Water Conversion Corporation (SWCC), Jubail 31951, Saudi Arabia*

Abstract

This paper investigates three types of organic matters, namely algal organic matter (AOM), bacterial organic matter (BOM), and humic organic matter (HOM). These organics are different in properties and chemical composition. Therefore, they were systematically studied to understand whether they behave differently or similarly in reference to organic fouling of ceramic ultrafiltration (UF) membranes. (AOM), (BOM) and (HOM) were compared in terms of organic content, fouling behavior, and removal efficiency. UF experiments have been conducted at constant flux mode using (5 KDa) and (50)kDa ceramic membranes. Six filtration cycles were tested in duplicates, with a cycle consisting of thirty minutes operation followed by hydraulic backwashing for one minute. AOM and BOM were extracted from marine algae, *Chaetoceros affins* (*CA*) and marine bacteria, *Pseudidiomarina atlantica* (*P. atlantica*) respectively in the lab, while commonly used (Suwannee River) humic acid was used as HOM. These organics have been spiked separately into synthetic seawater feed with a total TOC of (0.5 mg/L). Results showed that more TEP/organics were removed by the 5 kDa membranes compared to the 50 kDa membrane, which is accounted for by lower MWCO. The UF 5 kDa membrane also showed low fouling formation than 50 kDa membrane for all of three types of organic matter tested. Analysis of the fouled membranes by SEM images showed that fouling was dominated by cake layer formation for the 5 kDa membrane while pore blockage followed by cake layer formation is apparent for the 50 kDa membrane. The SEM images clearly show that a cake layer is formed for all types of organics tested; the flux was stable over the filtration periods for AOM and BOM, most probably due to the high porosity of the cake layer. On the other hand, HOM showed more fouling due to the combination of pore blockage and cake layer formation. AOM and BOM organics exhibited a similar fouling behavior and mechanism. However, AOM was higher than BOM in terms of TEP concentrations. This is probably attributed to high polysaccharides concentration in AOM whereas BOM was higher in protein content than AOM. HOM exhibited a different organic content, fouling behavior and mechanism compared to AOM and BOM. This is most likely due to the difference in size of HOM compared to the other organics. Furthermore, the divalent cations such as calcium and magnesium showed a significant impact on membrane fouling. That's probably because of the presence of calcium and magnesium ions made the membranes and organic matter less negatively charged and easier to deposit on membranes, thus, enhancing the membrane fouling significantly.

Keywords: SWRO pretreatment; UF Ceramic membranes; Fouling, AOM, TEP.

Fouling behaviour of forward osmosis TFC membranes during wastewater reclamation using feed streams mimicking treated effluent

Machawe M Motsa[a*], Bhekie B. Mamba[a], Arne R.D Verliefde[b]

[a]Nanotechnology and Water Sustainability (NanoWS) Research Unit, College of Science, Engineering and Technology, University of South Africa, Florida Campus, P.O Box 392, Pretoria 003, South Africa.

[b]Particle and Interfacial Technology Group (PaInT), Department of Applied Analytical and Physical Chemistry, University of Gent, Coupure links 653 B-9000 Ghent, Belgium

**Corresponding author. Tel.: +27 11 559 6148, Fax: + 27 11 559 6425, E-mail address: machawemmvulane@gmail.com*

In this work forward osmosis (FO) membranes were used to treat feed water simulating secondary wastewater. Particular emphasis was made on determining the fouling characteristics of the different fractions of effluent organic matter (EfOM) using alginate, bovine serum albumin (BSA), humic acid and octanoic acid as model foulants. These were representative of carbohydrates, proteins, humic substances and fatty acids, respectively, in treated wastewater effluents. The results revealed that the FO process had a potential in treating wastewater effluent. The work also demonstrated that under our experimental conditions humic acid and octanoic acid had no significant influence on permeate flux loss. However, alginate and bovine serum albumin resulted in significant total flux loss, due to alginate-calcium complexation that led to gel-layer formation on the membrane surface. The protein molecules in BSA were largely neutralized resulting in multiple layer adsorption onto the membrane surface. Foulant-foulant interactions were examined by performing experiments with the different foulant mixtures. It was observed that the mixture of proteins and polysaccharides caused a further decline in permeate flux due to synergistic interactions. Sequential fouling experiments demonstrated that proteins dominated the fouling process in the presence of other co-foulants. Membrane surface topology and charge were found to influence foulant attachment. The fouled membranes were then cleaned through osmotic backwash using ultrapure water. It was revealed that fouling reversibility was a function of permeate rate and shear forces induced by cross-flow velocity sweeping across the membrane surface.

Keywords: *Bovine serum albumin, Complexation, Effluent organic matter, Flux decline, Thin-film composite.*

PROCESS OPTIMIZATION ON CLEAN WATER RECOVERY FROM REAL REACTIVE DYEING EFFLUENTS

B. Kocer[1], M. Erkanli[1], R.D. Zaf[1], L. Yilmaz[2], U. Yetis[1], Z. Culfaz-Emecen[2]

[1]Department of Environmental Engineering, [2]Department of Chemical Engineering

Middle East Technical University, 06800 Ankara, Turkey

Of all textile dyeing processes, reactive dyeing occupies a more prominent position for the coloration of cellulosic, and also wool and nylon fibers. The degree of fixation is low for reactive dyes compared to other types of dyestuff, as the functional group also binds to water. The effluents from reactive dyeing are therefore heavily colored, contain high concentrations of organic matter with a low BOD to COD ratio, and also salt. The aim of this study is to investigate the optimal process for clean water recovery from reactive dyeing wastewater(RDW). RDW sample were taken from the last three steps of reactive dyeing line of a cotton textile factory that comprises of four sequential baths of dyeing, warm rinsing, washing with soap, and lukewarm rinsing. Waste dyeing bath solution was deliberately excluded due to its extremely high salt and color contents and the mixed RDW from the last three steps was treated by nanofiltration (NF) with and without microfiltration (MF) and ultrafiltration (UF) pretreatment as shown by Figure 1. The treatment performance was assessed with respect to color, organic matter, turbidity, and conductivity removals, and also filtration flux and membrane fouling characteristics. Firstly, 0.45 µm MF and 150 kDa UF PES membranes were tested using a 400 mL Amicon Stirred-Cell. With MF at 0.2 bar, TOC and turbidity removals were 20% and 90%. Better TOC and turbidity removal efficiencies of 27% and 93% were achieved with UF at 0.5 bar, with no reduction in conductivity. Filtration fluxes were similar for both membranes. Secondly, direct NF studies were conducted with DOW NF90 membrane using SEPA CF membrane system at 2 different cross-flow velocities (CFV) of 58.9 and 103.2 mL/s and 3 different trans-membrane pressures (TMP) of 4, 10 and 14 bar (Table 1). At all CFVs and TMPs, more than 99% color removal (at 436, 525 and 620 nm) was reached. Highest TOC and conductivity removals of 92% and 98% were achieved at both 10 and 14 bar TMP and CFVs of 58.9 and 103.2 mL/s. Permeate conductivity was below 1 mS/cm at 10 and 14 bar. At all experimental conditions, fouling resistances were similar. Permeate flux did not increase with an increase in CFV. But, as expectedly it was found to increase with an increase in TMP and measured as 0.35, 10.24, and 22.65 L/m^2.h, at 4, 10 and 14 bar, respectively. The overall results indicated that sufficiently good NF performance can be achieved at the TMP of 14 bar and CFV of 58.9 mL/s.

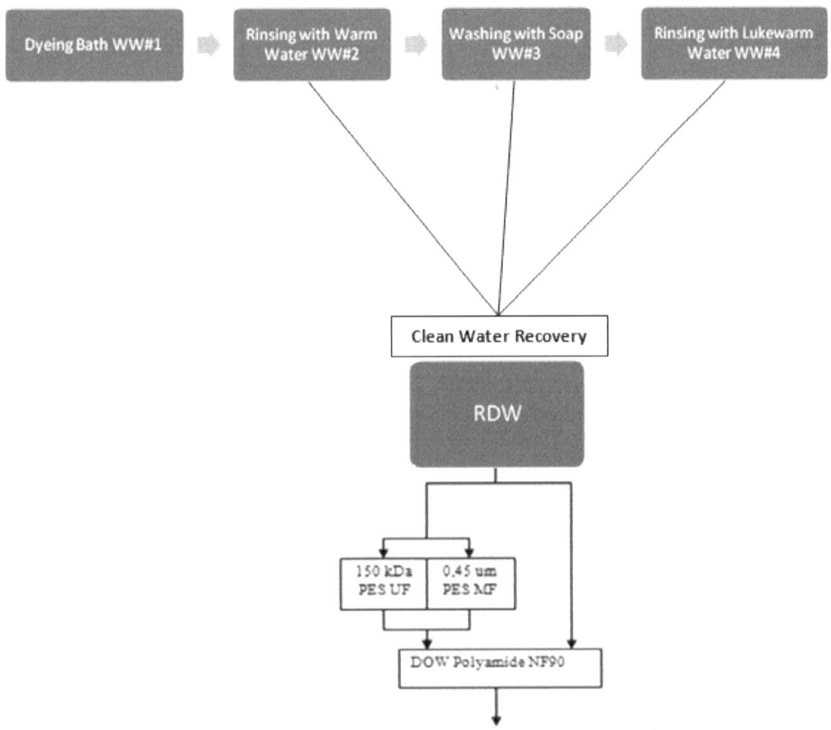

Figure 1: Experimental Diagram

Table 1: Feed and Permeate TOC, Turbidity, Conductivity and Absorbance Values

Parameter	CFV (mL/s)	58.9	103.2	58.9	103.2	58.9	103.2
	FP (bar)	4		10		14	
TOC (mg/L)	Feed	514.00	473.40	568.00	450.36	475.27	448.32
	Permeate	42.93	36.81	9.53	7.57	6.07	6.49
Turbidity (NTU)	Feed	49.43	44.1	41.40	38.13	38.37	44.13
	Permeate	1.67	2.99	0.99	1.61	2.08	3.28
Conductivity (mS/cm)	Feed	14.64	14.31	14.27	14.27	14.27	14.31
	Permeate	3.64	2.95	0.88	0.94	0.66	0.59
Absorbance (at 436 nm)	Feed	1.89	1.31	2.34	1.36	1.53	1.53
	Permeate	0.00	0.01	0.00	0.00	0.01	0.01
Absorbance (at 525 nm)	Feed	2.41	1.33	3.16	1.42	1.70	1.67
	Permeate	0.01	0.01	0.00	0.00	0.01	0.01
Absorbance (at 620 nm)	Feed	0.41	0.30	0.52	0.29	0.31	0.33
	Permeate	0.00	0.01	0.00	0.00	0.01	0.00

RECOVERY AND REUSE OF REACTIVE DYEING WASTEWATER BY ULTRAFILTRATION

M. Erkanli[1], B. Kocer[1], R.D. Zaf[1], L. Yilmaz[2], Z. Culfaz-Emecen[2], U. Yetis[1]

[1]Department of Environmental Engineering, [2]Department of Chemical Engineering

Middle East Technical University, 06800 Ankara, Turkey

Reactive dyeing process employs salt for increasing the affinity of the dyestuff towards the fiber and produces a wastewater that has high organic and salt load, and high color content. In this study, the treatment of reactive dyeing wastewater (RDW) by ultrafiltration (UF) in order to recover and reuse the sodium chloride and the water while leaving the spent dyes in the reject stream is considered. Two UF membranes were tested for the treatment and recovery of the RDW resulting from dyeing and proceeding rinsing and washing baths. The RDW samples taken from a textile industry were firstly strained and then subjected to UF using a cross-flow GE Osmonics SEPA lab-scale membrane filtration system. The UF membranes tested were 5 kDa PES and 2 kDa Thin Film GE Osmonics. Three scenarios presented in Figure 1 were considered for two stage applications of 5 kDa and 2 kDa after observing removal efficiencies of both membranes with raw RDW. Filtration through 5 kDa UF provided TOC and color removal rates (absorbance at 525 nm) of 74.6±2.0% and 87.7±0.3%, respectively. Salt (conductivity) removal which we desire to keep low, was below 2%. Treatment of raw RDW with 2 kDa UF resulted in 82.7±1.0%, 97.5±0.1% and 9.3±2.7% removal rates for TOC, color and conductivity, respectively. In order to achieve further TOC and color removal, second stage UF experiments are conducted for permeates of 2 kDa and 5 kDa UF membranes. As a result, second stage 5 kDa application for 5 kDa permeate did not have much change in removal rates. Second stage 2 kDa UF treatments to first stage permeates caused a significant increase in permeate quality removing the color more (Table 1). During these experiments filtration fluxes for 5 kDa and 2 kDa with raw RDW was observed to be 10.5±1.8 lmh and 4.8±0.1 lmh, respectively. Among three scenarios, 5 kDa + 2 kDa serial application having higher filtration fluxes at first stage and 2 kDa + 2 kDa achieving better permeate quality in terms of TOC and color content were chosen for further studies. Permeates of two scenarios (5 kDa + 2 kDa and 2 kDa + 2 kDa) were used to dye a piece of cotton fabric in dyeing tests carried out in the laboratory of the textile factory that the RDW samples were taken from. Same steps were followed mimicking the dyeing process of the factory but with the treated wastewater this time instead of tap water. Both permeates were successful in cotton dyeing with dark red. Permeate of 5 kDa + 2 kDa exhibited a satisfactory performance in coloring middle and dark shade green, while permeate from 2 kDa + 2 kDa showed similar results with light and middle shade green. When the effect of trans-membrane pressure was investigated for 5 kDa + 2 kDa UF application, it was seen that increasing pressure results in an increase in starting filtration flux for both membranes. A sharper decrease compared to lower pressure filtration was observed in filtration flux of 5 kDa. Volume reduction factor (VRF) was increased to see the change in permeate quality with decreasing concentrate ratio, and it was seen that a VRF value of 10 caused a decrease of 2-7% for color, TOC and turbidity removal rates for 2 kDa only.

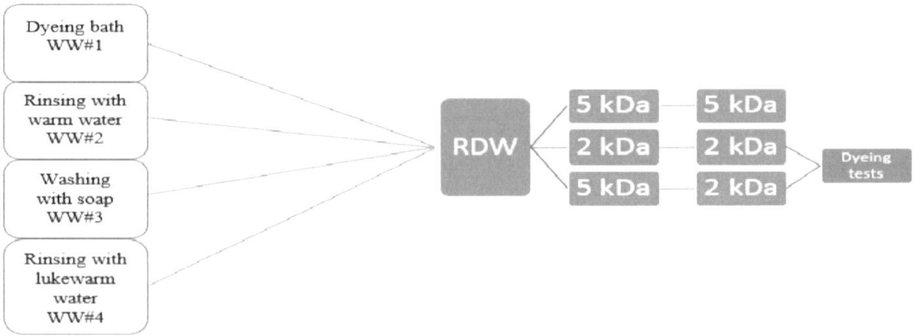

Figure 1 : UF treatment scenarios

Table 1: Feed and permeate characteristics with removal efficiencies

	Feed for 5 kDa	Permeate of 1st stage 5 kDa	Permeate of two stage 5 kDa	Rejection of 5 kDa + 5 kDa (%)	Feed for 2 kDa & 5 kDa	Permeate of 1st stage 2 kDa	Permeate of two stage 2 kDa	Rejection of 2 kDa + 2 kDa (%)	Permeate of 1st stage 5 kDa	Permeate of 5 kDa + 2 kDa	Rejection of 5 kDa + 2 kDa (%)
TOC (mg/L)	248.75±3.43	59.58±5.68	49.92±5.66	79.9	228.90±4.63	44.22±3.31	30.57±5.31	86.6	61.35±1.60	43.65±2.35	80.9
Turbidity (NTU)	54.5±1.4	3.1±0.3	1.3±0.1	97.6	35.2±0.5	1.0±0.0	0.5±0.1	98.5	1.3±0.0	1.0±0.1	97.3
Conductivity (mS/cm)	27.97±0.12	27.17±0.01	27.63±0.02	1.4	28.09±0.03	26.01±0.06	25.48±0.07	9.3	28.23±0.01	26.54±0.07	5.6
Absorbance at 525 nm	1.9319± 0.0001	0.2421± 0.0000	0.1372± 0.0001	92.9	1.8905± 0.0001	0.0444± 0.0000	0.0114± 0.0001	99.4	0.2290± 0.0001	0.0240± 0.0000	98.7

Sustainable Antibiofouling Properties of Thin Film Composite Forward Osmosis Membrane with Rechargeable Silver Nanoparticles Loading

Zhongyun Liu, Yunxia Hu*

Key Laboratory of Coastal Environmental Processes and Ecological Remediation;

Research Center for Coastal Environmental Engineering and Technology of Shandong Province;

Yantai Institute of Coastal Zone Research, Chinese Academy of Sciences,

Yantai, Shandong Province 264003, P.R. China

*Corresponding author, Tel: +86-535-2109236; E-mail: yunxiahu@yic.ac.cn

ABSTRACT

Microbial attachment and biofilm formation on filtration membrane can greatly compromise its flux and separation efficiency. Here, a simple and facile approach has been developed to in situ generate silver nanoparticles (Ag NPs) on the thin film composite forward osmosis (TFC FO) membrane for sustainable antibiofouling performances. Mussel-inspired dopamine chemistry was applied to grow polydopamine coating on both surfaces of FO membranes, followed by the generation of Ag NPs upon a simple dip coating in silver nitrate aqueous solution. Furthermore, the Ag NPs deposited membranes had a long-term silver release profile with rechargeability for multiple times upon their depletion, and exhibited strong sustainable bactericidal efficacy against Gram-negative bacteria and Gram-positive bacteria. The Ag NPs had controllable effect on the membrane performances including the water flux and reverse salt flux in the FO test mode. Our easily applied antibacterial strategy may apply to other types of filtration membranes with diverse material surfaces and may pave a new way to achieve the sustainable membrane antibiofouling performance in a large scale.

Second interfacial polymerization of thin-film composite polyamide membrane with polyethyleneimine for forward osmosis applications

Liang Shen, Yan Wang*

School of Chemistry and Chemical Engineering, Huazhong University of Science & Technology, Wuhan, 430074, P.R. China

In this work, a fast and simple surface modification-second interfacial polymerization (SIP) of thin-film composite (TFC) membranes using polyethyleneimine (PEI) of various molecular weights, was applied to improve the separation performance for forward osmosis (FO) applications. Compared to the control TFC membranes, PEI - modified membranes exhibit higher water flux and lower fouling propensity. The effects of PEI molecular weight and PEI concentration on the membrane morphology and separation performance were investigated systematically with various characterizations. Post-treatment to surface modified membranes was carried out using four aqueous solutions of different pH values. Meanwhile, the FO performance, intrinsic separation properties and antifouling properties of these post-treated membranes were also studied. In comparison with the surface-modified TFC membranes with PEI, the post-treated TFC membranes show higher water flux and improved anti-fouling properties, especially those post-treated with acidic or alkaline aqueous solutions.

Simulation of a Full-Scale Forward Osmosis-Nanofiltration Hybrid System through mass-balance

Sherub Phuntsho[*], Jung Eun Kim, Ho Kyong Shon**

Centre for Technology in Water and Wastewater, School of Civil and Environmental Engineering, University of Technology, Sydney (UTS), Broadway, NSW 2007, Australia

Corresponding authors: Email: *sherub.phuntsho@uts.edu.au, **Hokyong.Shon-1@uts.edu.au

1. Background

Water scarcity is becoming a significant issue worldwide and, desalination and water reuse are two of the reliable solutions. Desalination is however currently not cost-effective for irrigation. Recent efforts have been to develop less capital and energy intensive desalination technologies such as forward osmosis (FO) that drives based on natural osmotic process unlike the conventional reverse osmosis (RO) desalination process that uses lot of energy. Based on recent studies, FO hybrid systems could be more effective when the water treatment involves highly challenging waters [1, 2]. Fertiliser driven FO or FDFO process is seen as one of the potentials for practical application of FO process [3, 4]. The diluted fertiliser DS from the FO process after dilution contains nutrients essential for the plants and hence can be used for fertigation of crops if the nutrient concentration meets the required standard. Recently, FDFO-NF system has been studied at a pilot-scale level by this group [5].

The main objective of this study is to conduct a full-scale FDFO-NF simulation of how the various process parameters in the hybrid system affect each other when operated in a continuous closed-loop system based on the simple mass balance of the volumetric flows, draw solutes and the feed solutes. This study provides an enhanced understanding of some of the options and limitations of operating full-scale FDFO-NF hybrid system in a continuous closed-loop system. The result of this study is also applicable to any closed loop membrane FO-RO and FO-NF hybrid system for any applications.

2. Methodology

The simulation is entirely based on the mass balance in the closed FDFO-NF hybrid system and hence the process performances including performance mechanisms are not taken into considerations. The mass balance covered the following aspects of the hybrid system: volumetric solution flows, draw solutes and feed solutes. The simulation was carried out assuming $(NH_4)_2SO_4$ or SOA as DS, NaCl solution as FS. Some of the other important simulation parameters used are specific reverse solute flux or $SRSF_D$ = 0.5 gL^{-1} SOA based on our pilot-scale FDFO-NF operation in the field recently for CTA FO membrane [5]. The NaCl rejection rates of FO and NF membranes and SOA rejection rate of NF membrane is assume at 90% for simulation study unless stated otherwise.

3. Discussion and conclusions

The simulation results show how varying the process parameters could have profound implications on the plant performance and plant design. Following are the summary of the findings from this simulation study:

- Unlike in a lab-scale FO studies, the initial DS flow rate and its concentration in a full-scale FO-NF hybrid system with fixed plant capacity cannot be set at an arbitrary values as they directly influence the final diluted DS concentration including the plant capacity as shown in Fig 1a.

- For a fixed FO-NF plant capacity, the initial DS flow rate and its initial net concentration are inversely related (Fig 2b). A net DS mass loading rate (product of initial concentration difference between the DS and FS and the initial DS flow rate) is one of the most important parameters for the design of the continuous closed-loop FDFO-NF system. The net mass-loading rate is constant for a fixed plant capacity irrespective of what the initial DS concentration or the DS flow rate is used but its value increases with the increase in the plant capacity and the feed concentration.
- The reverse diffusion of draw solute through the FO membrane could add a significant cost to the FO-NF plant due to need for its replenishment and also due to the need for special brine treatment and management showing the need to develop FO membranes with high reverse flux selectivity for commercial applications.
- Depending on the rejection rates of the FO and NF membranes, accumulation of feed solutes could occur in the closed-loop FDFO-NF system which increases rapidly with the number of DS recycling but reaching a steady peak concentration. The accumulated feed solute concentration in the recycled DS could however provide enhanced driving force of the DS as long as it does not affect the final product water quality in terms of Na^+ or Cl^- concentrations for irrigation. This could be a challenge for FS with higher TDS such as seawater unless FO membrane with high rejection is used.
- This study demonstrates limitations of the FO-NF hybrid process and the options available for improving the operation of a closed-loop hybrid system.

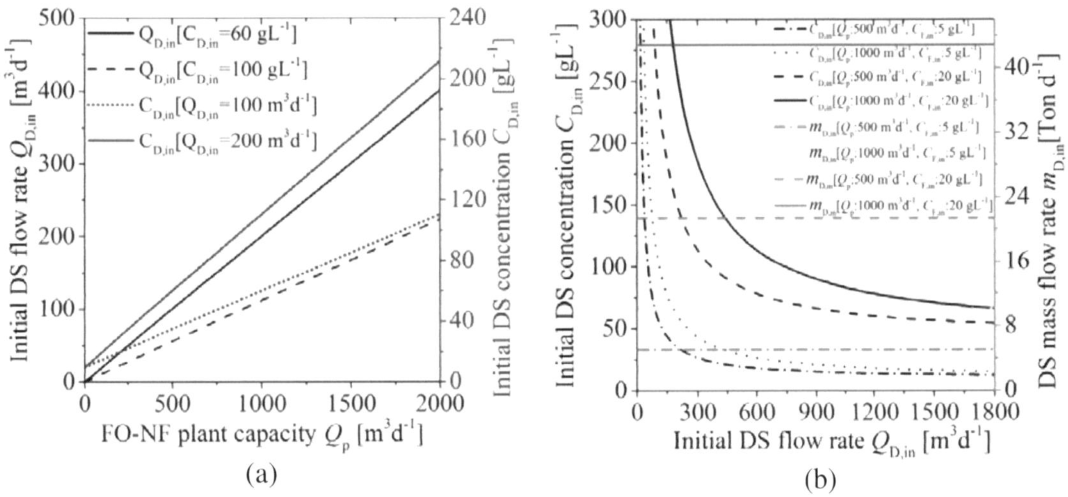

Fig 1: Influence of each process parameters over the other process parameters. (a) FO-NF plant capacity on the required initial DS flow rates and DS concentrations and (b) initial DS flow rate on the required initial DS concentration and the net mass flow or loading rates. $C_{D,in}$ & $Q_{D,in}$: initial DS concentrations & flow rates, Q_p: FDFO-NF plant capacity, $m_{D,in}$: DS net mass loading rate.

References

1. Chen, G., et al., Desalination, **366**: p. 113-120.
2. Coday, B.D., et al., Desalination, **333**(1): p. 23-35.
3. Hoover, L.A., et al., Environmental Science & Technology, **45**(23): p. 9824–9830.
4. Phuntsho, S., et al., Reviews in Environmental Science and Bio/Technology, **11**(2): p. 147-168.
5. Phuntsho, S., et al., Journal of Membrane Science, **508**: p. 22-31.

Modified microporous polypropylene flat sheet membrane via TIPS for membrane distillation

WANG Yujie[a,b]*, ZHAO Zhi ping[b], LUAN Jinyi[a], XI Zhenyu[a], YANG Yongqiang[a]

a,Environmental Protection Research Institute, BRICI, SINOPEC, Beijing, 100013, CHINA.

b,Beijing Institute Technology, Chemical Engineering & Environment, Beijing, 100081, CHINA.

* correspondence,wangyuj.bjhy@sinopec.com

It has become more and more important to replace toxic solvents using greener alternatives during the membrane production process. In this work, environmental friendly binary diluents composed of Carnauba wax as latent solvent were employed for dissolving polypropylene to form homogenous solution and prepare microporous polypropylene flat sheet membrane via thermally induced phase separation process (TIPS).

The effect of binary diluents on polypropylene membrane structures and performances was investigated. The phase diagram of the PP/binary diluents system as a function of polymer weight fraction, and the phase diagram as a function of Carnauba wax fraction in the binary diluents were investigated as well. We also tried to figure out how the existence of a latent diluent influenced the mechanical properties, pore size distribution. In order to evaluate the practical aspects of latent diluent, analytical techniques were used including: scanning electron microscopy (SEM) for investigating morphology across the membrane and differential scanning calorimetry (DSC) to investigate the crystallization behavior of polymer and diluents.

It was found that the morphology of the membranes was strongly affected by binary diluents. Indeed, physicochemical characteristics of diluents influenced the interactions between binary diluents and polypropylene. Finally, the interconnected spongy structure inside microporous membrane was achieved successfully, the mechanical properties, mean pore size and pore size distribution were further controlled as well.

In the desalination test of vacuum membrane distillation, the water permeability

changed widely depending on different binary diluents, however, the salt rejection remained almost more than 99.0%.
Key words: polypropylene flat sheet membrane; thermally induced phase separation; membrane distillation, green solvent

Development of Anti-Fouling, Anti-Microbial Membranes for Wastewater Treatment

Steven Weinman[a], Maria Bass[b], Viatcheslav Freger[b], Moshe Herzberg[c], Scott Husson[a]

[a]Department of Chemical & Biomolecular Engineering, Clemson University, Clemson, SC 29634
[b]Wolfson Department of Chemical Engineering, Technion - Israel Institute of Technology, Haifa, 32000 Israel
[c]Zuckerberg Institute for Water Research, Ben Gurion University of the Negev, Sede-Boqer Campus, Midreshet Ben Gurion, 84990 Israel

Treatment of impaired waters for beneficial use exposes the membranes to feed waters containing biological and abiotic species, which leads to fouling and loss of membrane productivity over time. Since reduction in flux due to fouling is one of the largest costs associated with membrane processes in water treatment, new coatings that limit and reverse fouling would have significant economic and societal impacts. Developing these advanced coatings is the focus of our work.

Prior studies to control biofouling largely have focused either on preventing microorganisms from attaching or killing the microorganisms once they do attach. Based on considerable literature and experience, we know that each of these control strategies alone can be effective but not sufficient for controlling biofouling. Our hypothesis is that membranes coated with dual-mode polymer nanolayers can limit and reverse membrane biofouling by switching reversibly between passive (antifouling) and active (antimicrobial) states of biofouling control. This strategy differs fundamentally from most other surface modification strategies that rely solely on passive control (i.e., using coatings designed only to weaken foulant adhesion and, thus, limit their accumulation) or active control (e.g., adding biocides).

In this poster, we will introduce the methodologies used to coat surfaces with carboxybetaine zwitterion dual-mode polymer nanolayers designed to combat biofouling. We will explain the chemical switching mechanism and use FTIR and streaming potential measurements to support the reversible switching of the polymer nanolayer between its antifouling and antimicrobial states. Subsequently, the feasibility of the approach will be tested in bacterial deposition and membrane biofouling experiments using appropriate model systems. Initial bacterial deposition results show a similarly low bacteria initial attachment rate as compared to other common zwitterion coatings.

Highly permeable peptide-appended pillar[5]arene artificial water channel in block copolymer membranes

Yue-xiao Shen, Tingwei Ren, Daniel Tsai, Ian T. Sines, Manish Kumar*
Department of Chemical Engineering, Pennsylvania State University, University Park, PA, 16802, USA
*Corresponding author: 155 Fenske Laboratory, Pennsylvania State University, University Park, PA, 16802, USA, (814) 865-7519, manish.kumar@psu.edu

000We have reported that peptide-appended pillar[5]arene (PAP) artificial water channels (Fig. 1) combine the advantages of biological water channels Aquaporins (AQPs) and carbon nanotubes (CNTs).[1] It can be synthesized using simple chemistry at gram scale in the lab and provide stable structure and good solvent compatibility. The water conductance of PAP is in the range of AQPs. It can also be vertically aligned in lipid membranes, with much higher packing density than current synthetic CNTs-based membranes.

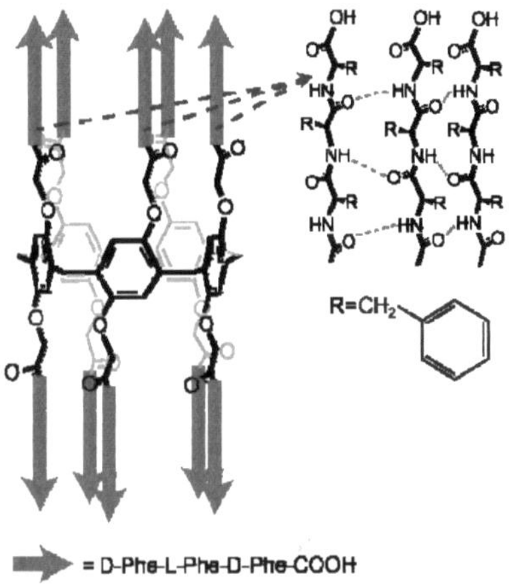

Fig. 1 Structure of the peptide-appended pillar[5]arene (PAP) channel.

Although the initial testing of the PAP channels was conducted in lipid systems, this is not an ideal format for commercial separation materials. We continue to investigate the functionality of these channels in block copolymer (BCP) membranes. BCPs, with lipid-like structures, are most commonly synthesized for biomimetic applications because of the advantages of BCPs membranes including high mechanical and chemical stability,[2,3] low water and gas permeability[4] and customizable properties (e.g., a larger range of membrane thickness[2] and end groups[5]). We are testing a series of poly(butadiene)-*b*-poly(ethylene oxide) (PB-PEO) diblock copolymers and polymethyloxazoline-polydimethylsiloxane (PMOXA-PDMS-PMOXA) triblock copolymers. These two kinds of BCPs have different hydrophobicities and thicknesses, and been shown to functionally incorporate water channel proteins aquaporin 0,[6] potassium channel[7] and α-hemolysin.[8]

We incorporated fluorescently labeled channels into polymersomes using the film rehydration method. The number of channels per vesicle was measured based on a fluorescence correlation spectroscopy (FCS) technique.[9] We combined the channel number data and permeability data using stopped-flow light scattering technique (Fig. 2B) and obtained the single channel water permeability in PB_{23}-PEO_{16} vesicles ($1.4\pm0.4\times10^8$ H_2O molecules per second). This value is close to that in lipid system ($3.5\pm1.0\times10^8$ H_2O molecules per second, Fig. 2C).[1]

In the future, we will systematically test the PAP channels in different BCPs systems, visually confirm the incorporation in giant unilamellar vesicles and discuss the interaction of different BCPs and the PAP channels. We will also plan to use fluorescence recovery after photobleaching (FRAP) technique to determine the diffusion coefficient of the PAP channels in lipid and BCPs bilayers and expect these results can provide insights into to maximizing the channel's packing density in BCP membranes.

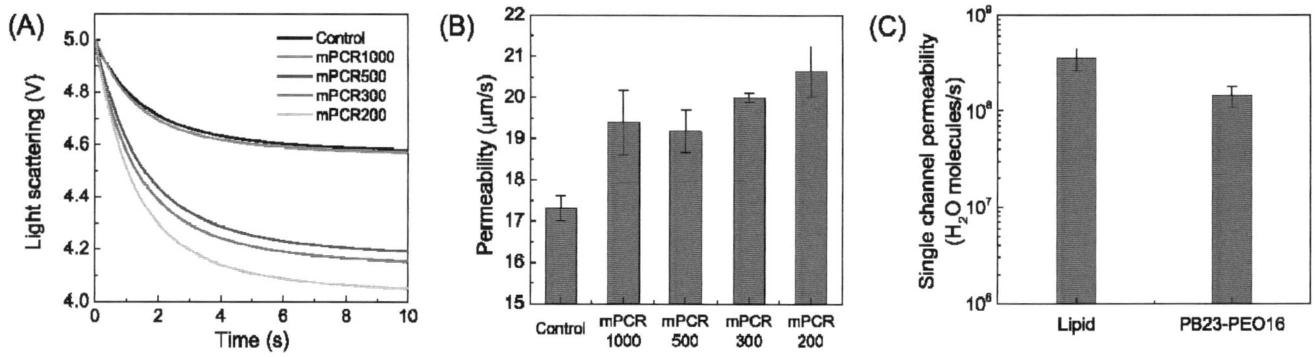

Fig. 2 Functional analysis of peptide-appended pillar[5]arene (PAP) artificial water channel in block copolymer vesicles. (A) The light scattering traces of PB_{23}-PEO_{16} polymersomes with different molar polymer-to-channel ratio (mPCR) after a rapid exposure to a hypotonic solution (−100 mM PEG600) on a stopped-flow instrument. We normalized all the traces to the same PMT value. (B) The water permeability of PB_{23}-PEO_{16} polymersomes with different mPCRs. (C) Single-channel water permeability of the PAP channel in lipid and block copolymer systems. The value in lipid vesicles was obtained from previous PNAS paper and the value in polymer system was calculated from the vesicles formed at an mPCR of 200 (Figure 2B) and fluorescence correlation spectroscopy (FCS) data. Data shown are the average of triplicates with error bars representing standard deviation.

00**Reference**

1 Shen, Y.-x. *et al.* Highly permeable artificial water channels that can self-assemble into two-dimensional arrays. *Proc. Natl. Acad. Sci. U.S.A.* **112**, 9810-9815, (2015).

2 Discher, D. E. & Eisenberg, A. Polymer Vesicles. *Science* **297**, 967-973, (2002).

3 Discher, B. M. *et al.* Polymersomes: Tough Vesicles Made from Diblock Copolymers. *Science* **284**, 1143-1146, (1999).

4 Kumar, M. *et al.* Highly permeable polymeric membranes based on the incorporation of the functional water channel protein Aquaporin Z. *Proc. Natl. Acad. Sci. U.S.A.* **104**, 20719-20724, (2007).

5 Rakhmatullina, E. & Meier, W. Solid-Supported Block Copolymer Membranes through Interfacial Adsorption of Charged Block Copolymer Vesicles. *Langmuir* **24**, 6254-6261, (2008).

6 Kumar, M. *et al.* High-Density Reconstitution of Functional Water Channels into Vesicular and Planar Block Copolymer Membranes. *J. Am. Chem. Soc.* **134**, 18631-18637, (2012).

7 Kowal, J. Ł. *et al.* Functional surface engineering by nucleotide-modulated potassium channel insertion into polymer membranes attached to solid supports. *Biomaterials* **35**, 7286-7294, (2014).

8 Zhang, X. *et al.* Natural channel protein inserts and functions in a completely artificial, solid-supported bilayer membrane. *Sci. Rep.* **3**, (2013).

9 Erbakan, M. *et al.* Molecular Cloning, Overexpression and Characterization of a Novel Water Channel Protein from *Rhodobacter sphaeroides*. *PLoS ONE* **9**, e86830, (2014).

Modeling Nanofiltration: The Fundamentals and its Application in Softening and Pretreatment

Omar Labban[a], Chang Liu[b;c], Tzyy Haur Chong[b;c], John H. Lienhard V[a];

[a]Rohsenow Kendall Heat Transfer Laboratory, Department of Mechanical Engineering, Massachusetts Institute of Technology, Cambridge MA 02139-4307, USA

[b]School of Civil and Environmental Engineering, Nanyang Technological University, Singapore 639798, Singapore

[c]Singapore Membrane Technology Centre, Nanyang Environment and Water Research Institute, Nanyang Technological University, Singapore 637141, Singapore

Abstract

The high selectivity of nanofiltration (NF) for multivalent ions has made it very attractive for scale-preventive softening and pretreatment applications. A better predictive capability of the separation efficacy of NF membranes, and what features govern this performance, however, is needed to facilitate predictive design for large-scale. Modeling NF membranes often poses a serious challenge given the variety of separation and transport mechanisms that influence the process. In this study, an overview of the fundamentals underlying the calibration of NF membranes models, based on the Donnan Steric Pore Model with Dielectric Exclusion (DSPME-DE), will be discussed. Experiments run on a recently developed low-pressure cross-linked layer-by-layer (LbL) NF membrane, and the results obtained from this first model of the membrane will be presented. Additional experiments run on hard water and artificial seawater will be considered to assess the membrane's potential for low-pressure softening and pretreatment.

Xiaobo Dong

Department of Chemical and Material Engineering, University of Kentucky, KY

177 F. Paul Anderson Tower xiaobodongmax@hotmail.com

Title: Investigation of environmentally friendly solvents for casting of water treatment membranes

Abstract: Membrane technology has been proven to be effective in recent years due to its benefits in membrane water treatment technology. Immersion precipitation is a conventional method to fabricate porous polymeric membranes. In this method, a dope solution is prepared by dissolving the polymer in a solvent, such as DMSO, DMF, and NMP. A membrane is then formed by the precipitation of the dope polymer in an anti-solvent bath, such as water. However, most of currently used solvents are volatile and toxic to the environment. Their disposal after use and the residue in the system has potential to cause damage to the environment. The aim of this study is to investigate 'green solvents' to cast membranes in order to minimize the environmental impacts. Polymer dope solutions will be prepared using several environmentally friendly solvents focusing on the polymer chemistry and thermodynamics of mixing. The membranes will be casted and characterized for their performance, and then compared with traditional membranes. Furthermore, the economic feasibility will also be considered. As a conclusion, this study will aim at making an environmentally friendly membrane with high performance in water treatment.

Title: Advancements in Switchable Polarity Solvent Forward Osmosis (SPS FO) Water Treatment Technology

Authors: Birendra Adhikari*, Christopher J. Orme, Michael G. Jones, Catherine M. Hrbac, Daniel S. Wendt, Aaron D. Wilson

Email: birendra adhikari @INL.gov

Switchable Polarity Solvent Forward Osmosis (SPS-FO) technology is an emerging technology in the area of high salinity natural and industrial water treatment. It involves forward osmosis treatment of high salinity feed water using novel sets of draw solutions called switchable polarity solvents (SPSs). Switchable polarity solvents are tertiary amines which dissolve in an aqueous phase by forming bicarbonate salts after reacting with carbon dioxide. Upon the application of heat, carbon dioxide is liberated and SPSs become insoluble. Our group at Idaho National Laboratory (INL) identified 1-cyclohexylpiperidine (CHP) as an SPS with smooth switchable behavior, good chemical compatibility and high osmotic potential of up to 570 atm that can conceivably cause saturated brine solution causing solid NaCl sedimentation. SPS-FO technology is attractive because low-cost heat energy can be used to eject carbon dioxide out of the bicarbonate and the ability to address high salinity feeds and recover a large fraction of the water in the feed solution. We are developing this technology for commercial deployment by designing and optimizing the various process units. These units include: the gas contactor, degasser, liquid-liquid mechanical separator, forward osmosis, double staged 'nanofiltration' and reverse osmosis unit and a stationary polishing step. This is leading to a scale up effort of this technology. We have concluded that this technology will significantly reduce the cost of managing the waste water in oil and gas industries and has potential in other industrial water management.

The water flux properties of negative and positive electric charged monodisperse particles using water treatment membrane

Seung Moon Woo, Dong Hee Choi[1], Sang Yong Nam*

Department of Materials Engineering and Convergence Technology, Engineering Research Institute, Gyeongsang National University, Jinju, 660-701, Korea

*walden@gnu.ac.kr

ABSTRACT

Water scarcity is a result of poor management of the sufficient available water resources and environmental pollution. It already affects every continent and around 3 billion people around the world. In other to resolve these problems, various alternatives were developed to meet the demands of water usage [1].

Recently, water treatment system using membrane has been considered as one method due to its economic feasibility and efficiency compared with another process [2]. In this study, monodispersed polystylene particles with micron size, positive and negative charges were prepared by emulsion polymerization to measure UF permeation characteristics. We selected Track-etched Polycarbonate (PC) membrane for evaluation of water treatment. Because PC membrane has hydrophilic property, uniformed pore size and superior mechanical properties. The prepared PS latex colloids were dispersed in feed solution and then this is used to evaluate of the commercial PC membrane. Particle size, chemical structure, thermal characteristics and surface charge of PS latex colloids were evaluated using FE-SEM, FT-IR, DSC and zeta potential measurement equipment respectively. Finally, we investigated water permeability and rejection of the membrane with PS particles to confirm application possibility in field of water treatment

REFERENCES

[1] Ermolina, I. and Morgan, H., *J. Colloid Interface Sci.*, **285**(1), 419-428 (2005)

[2] Reese, C.E., Guerrero, C.D., Weissman, J.M., Lee, K.T. and Asher, S.A., *J. Colloid. Interf. Sci.*, **232**(1), 76-80 (2000)

Exploring Water Reuse with a Novel FO-DCMD Hybrid Membrane Process

Gavin Zirkel, Jairo Luque Villanueva, and Andrea Achilli
Environmental Resources Engineering Department, Humboldt State University

The water and wastewater industry accounts for approximately 3% of the total energy consumption of the United States [1]. This energy demand is associated with substantial costs, both monetary and environmental. As existing water treatment facilities age and new regulations are introduced to water resources, there is a push to research innovative methods for water treatment. The purpose of this project was to examine a novel approach to wastewater treatment involving a hybrid membrane process capable of producing potable quality effluent.

The overall process examined in this project involved forward osmosis (FO), direct contact membrane distillation (DCMD), and biological processes to recover energy and nutrients (anaerobic membrane bioreactors (AnMBR), Sharon, Annamox, and struvite precipitation). The FO-DCDM hybrid membrane process directly concentrates influent wastewater, extracting potable quality effluent. During the FO step, osmotic pressure causes fresh water to permeate across a selectively permeable membrane into a draw solution of high ionic concentration. Utilizing a vapor pressure gradient across a hydrophobic microporous membrane, the DCMD process then extracts pure water vapor from a heated draw solution into a potable quality distillate. FO was selected as the concentrating process due to its potential to be less prone to fouling than pressure driven membrane processes [2]. DCMD was selected as the purifying process due to the potential availability of low quality waste heat at treatment facilities, which would drive the process with no additional energy input.

Throughout experimentation, HTI and Porifera® FO membranes and GE® MD membranes were used. NaCl draw solutions of several concentrations were tested for water flux and reverse solute flux characteristics. Varying temperature couples were applied to the DCMD process to evaluate fluxes of DCMD as well as the affect of various draw solution temperatures on the FO process. The developed bench-scale FO-DCDM system was automated using National Instruments™ LabVIEW software and NI-DAQ devices. The system was comprised of two membrane modules, four fluid tanks, two circulation pumps, an immersion heater, a proportional control valve, and a heat exchanger. Temperatures were monitored using resistance temperature detectors. Tank volumes were determined using pressure transducers, and ion concentrations were determined using conductivity probes. Total organic carbon (TOC) and total nitrogen (TN) were determined using a Shimadzu TOC-L analyzer. Proportional-integral-derivative (PID) controllers were used to control the immersion heater as well as the proportional control valve.

The poster will summarize the results from bench-scale FO-DCMD experiments. These results include carbon and nitrogen rejections from surrogate wastewater trials, as well as process specific water fluxes and reverse salt flux associated with the FO process. Two approaches for the heat utilization in MD will also be discussed. From completed FO-DCMD experiments, it was found that together, FO and DCMD are capable of rejecting 98% of both TOC and TN. This points to the hybrid process being very effective as a method of extracting nutrients and producing high quality effluent. It was also found that FO fluxes are more dependent upon draw solution temperature than on draw solution concentration. Higher draw solution temperatures led to higher FO fluxes while reducing reverse solute flux. Using a "percent on" heating approach rather than a PID for the draw solution led to more stable fluxes. This heating approach is more representative of an industrial scale-up, so the hybrid process is suited for such an application.

References:

[1] U.S. Environmental Protection Agency, Evaluation of Energy Conservation Measures for Wastewater Treatment Facilites, EPA 832-R-10-005, September 2010.

[2] Cath, Tsahi Y., Amy E. Childress, Menachem Elimelech. (2006). "Forward osmosis: Principles, applications, and recent development." *Journal of Membrane Science*. 281, 70-87.

Optimizing design parameters for thin film composite hollow fiber membranes and module for forward osmosis

Jian Ren, Jeffrey R. McCutcheon*

Department of Chemical and Biomolecular Engineering,
Center for Environmental Sciences and Engineering,
University of Connecticut,
191 Auditorium Rd. Unit 3222, Storrs, CT 06269-3222, USA
* jeff@engr.uconn.edu

Hollow fiber membranes have been considered a valuable platform for osmotic membrane processes because of their high packing density relative to flat sheet membranes. This study evaluates an approach to make thin film composite (TFC) hollow fiber membranes for forward osmosis (FO) by simply employing commercial hollow fiber ultrafiltration (UF) membranes as support material. A thin polyamide film with excellent selective properties was synthesized on the inner surface of hollow fibers via interfacial polymerization. Besides demonstrating the feasibility to be used as good TFC FO membrane supports, the commercial hollow fiber UF membranes also provide a systematic platform with consistent properties to study structure-performance relationship of FO hollow fiber membranes. A series of commercial hollow fiber membranes were used to evaluate how molecular weight cutoff (MWCO), fiber dimensions (inner and outer diameter) and module dimensions (length and width) impacted overall performance of the TFC membranes and module. With the availability of reproducible membranes and modules, we have developed a computational fluid dynamics (CFD) model with COMSOL Multiphysics in order to optimize hollow fiber module design for FO application. The comparison between the model and experimental results provided insight into both how more accurate models can be developed and how these models can be used to design better modules without costly experimental testing.

Fabrication and characterizations of hollow fiber membranes for removing virus and bacteria

Seung Moon Woo[1], Youn Suk Chung[2], Sang Yong Nam[1*]

[1]Department of Materials Engineering and Convergence Technology, Engineering Research Institute, Gyeongsang National University, Jinju 660-701 Korea
[2]Puremem Co. Ltd, Hwasung 445-909 Korea
[*]Corresponding author e-mail: walden@gnu.ac.kr

In present time, we need advanced water treatment technology for separation of virus and bacteria in effluent which occur epidemic and waterborne diseases. Water purification system is mainly devided into two categorization like Reverse osmosis (RO) and Ultrafltration (UF). Membrane used in these systems requires higher durability because of operating in harsh condition. Of these, the membrane using in UF system has many advantages like higher efficiency and lower energy consume for water treatment compared with RO system [1]. In many kinds of membrane, hollow fiber type membrane is possible to make easily and to get optimized property by control of various spinning conditions such as temperature of coagulation bath, concentration of polymer, addition of additive, air gap and internal coagulation [2]. In this study, polysulfone hollow fiber membrane was successfully prepared by phase inversion method for separation of virus and bacteria. When we prepare the hollow fiber membrane, we controlled various factors such as the polymer concentration, air gap and internal coagulation to investigate effect to membrane property. Morphology of surface and cross section of membrane were measured by field emission scanning electron microscope (FE-SEM). Water flux of membrane was measured using test modules. Mean pore diameter of membrane was calculated using rejection of polystyrene (PS) latex beads for separation of virus and bacteria. Flux and mean flow pore diameter of prepared membrane show 1.5 LPM, 0.03 μm at 1.0 kgf/cm^2. The bacteria and virus removal performance of prepared UF membranes were over 6 logs.

REFERENCES

[1] Y. Yang, "The Influence of Nano-sized TiO_2 Fillers on the Morphologies and Properties of PSF UF Membrane", Journal of Membrane Science, Vol. 288, pp 231-238, year 2007.
[2] L. Fiksdal, "The Effect of Coagulation with MF/UF Membrane Filtration for the Removal of Virus in Drinking Water", Journal of Membrane Science, Vol. 279, pp 364-371, year 2006.

The validation of a stepwise model of direct contact membrane distillation for large-scale applications

Joanna Murphy, Jacob Rowe and Andrea Achilli

Direct contact membrane distillation (DCMD), a thermally driven membrane separation process, has been shown to offer considerable promise as a method of producing high purity product water. The process offers many benefits over other membrane separation technologies and other configurations of membrane distillation for water treatment, including near complete rejection of solutes, the ability to operate using higher concentration feed solutions, the lower energy requirement of the process and the ability to utilize waste, solar or other low grade heat to offset some of the process energy input.

If DCMD is to be adopted as a commercially viable option for water treatment, three main obstacles to the large-scale implementation of the process must be overcome. The process must be shown to be effective on a large scale, the energy requirement of the process must be minimized, and membranes optimized for DCMD must become available commercially.

In an attempt to address the first two of these challenges, a stepwise model of DCMD was developed to predict outlet temperatures and water flux of DCMD systems [1]. The stepwise model uses an experimentally derived membrane distillation coefficient (MDC), a membrane-specific coefficient relating transmembrane vapor pressure difference to pure water flux, and known operating temperatures and flow rates to predict the system's performance. For this model to be considered a useful design tool, it must be validated at multiple scales.

To realize this objective, a counter-current flow lab-scale plate and frame DCMD system (Figure 1), intended to produce 6 L/hr of distillate water was built and tested to obtain experimental water flux and outlet temperature data. The model was applied to the lab-scale system, reflecting the specific geometry and operating conditions of this system. Predictions from the stepwise model were compared to experimental data to validate the model's utility in predicting flux, outlet temperatures and energy consumption for larger scale systems.

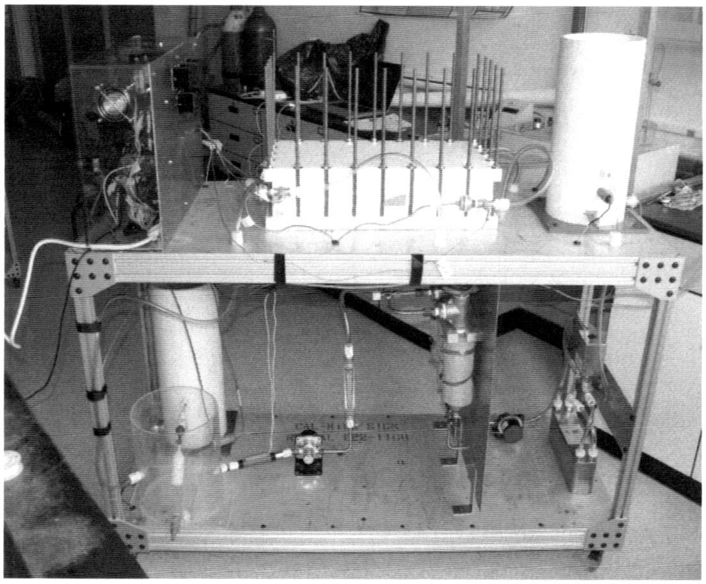

Figure 1: The lab scale membrane distillation system that was used to validate the model presented in Gustafson et al. [1].

Preliminary results indicate that the modeled water flux data are in good agreement with the experimentally obtained ones (Figure 2). This suggests that the model can be used to predict water flux and outlet temperatures of lab-scale DCMD system and therefore maximize water production while minimizing energy consumption.

In this presentation, the application of the model as a design tool to determine the ideal operating conditions for the lab- scale DCMD system will be discussed. Results of the model validation will be presented. The required membrane area to produce 6 L/hr under a variety of operational constraints, as determined from the stepwise model, will be discussed. Additionally, predicted energetic requirements for this system and the potential for heat recovery within the system predicted by the validated model will be presented. Such heat recovery would further reduce the energy requirements for DCMD, improving the economic feasibility of the process.

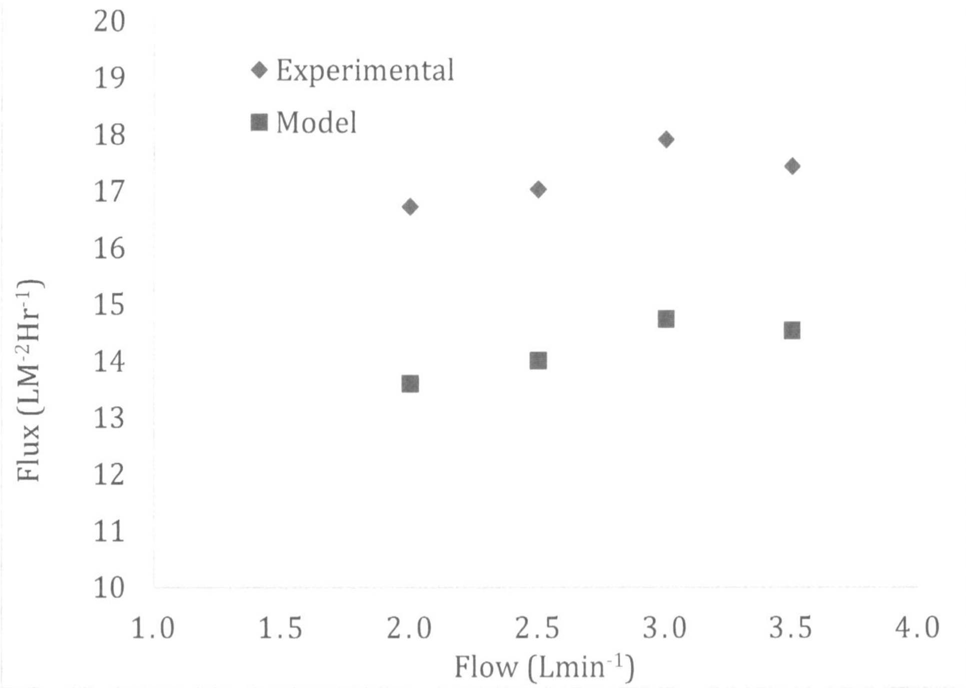

Figure 2: The model under-predicts distillate flux at several flow rates. However, it accurately reflects the relationship between flow and flux. All experiments shown were performed using distillate of 24 °C, 0.122 m^2 of membrane area and a feed solution of 50 g/L NaCl.

Reference

[1] Ryan D. Gustafson, Joanna R. Murphy, Andrea Achilli. 2016. "A stepwise model of direct contact membrane distillation for application to large-scale systems: Experimental results and model predictions." *Desalination* 378: 14-27.

Polyamide thin-film composite membranes for potential raw biogas purification: experiments, modelling and scale-up

P. Izak[1], M. Simcik[1], M.C. Ruzicka[1], M. Karaszova[1], Z. Sedlakova[1], J. Vejrazka[1], K. Friess[2]

[1] *Institute of Chemical Process Fundamentals, Czech Academy of Sciences, Rozvojova 135, 16502 Prague 6, Czech Republic*
[2] *Department of Physical Chemistry, University of Chemistry and Technology, Prague, Technicka 5, 166 28 Prague 6, Czech Republic*

Biogas, seems to be a very good candidate for the replacement of depleting fossil fuels. The raw biogas is created by the decomposition of waste through the process of anaerobic digestion and has to be purified in order to reach a fuel quality. Unwanted impurities such as CO_2 or acidic gases [1] have to be removed while the others impurities with concentrations lower than 1 vol. % are marginal for the biogas upgrading [2, 3]. The recent breakthrough in biogas upgrading represents the membrane separations with water-swollen thin polymer (skin) layer composite membranes [4-6] when the presence of water vapor in raw biogas assists in the effective gas separation. For this reason, pre-treatment such as water vapor removal is avoided and a raw biogas feed is directly in contact with the composite membrane.

Two swollen polyamide thin-film composite membranes were tested for effective CO_2/CH_4 separation. It was found that the water wettability has a key role for the separation of binary mixture representing a raw biogas, i.e. containing CO_2 and CH_4. The transport properties were analyzed by means of a mathematical model simulating gas permeation. A new modification of the mass transport coefficient model provided the concentration profiles of individual components on both sides of the membrane (inaccessible in experiments). Furthermore, the model enabled the evaluation of the mass transport coefficients of the gases in the mixture under varying stream flow rates and arrangements with respect to the membrane separation cell size. Therefore, the possibility of scale-up was discussed for both membranes and flow cell arrangement. Although the mathematical model was developed for a flat sheet membrane configuration, the results can be applied for a real spiral wound module with a wider surface.

The mass transfer coefficients of CO_2 and CH_4 in the Koch membrane as well as in the Sterlitech membrane were evaluated directly from experiments and by the developed permeation model for each individual experimental case. The final values of the mass transfer coefficients for CO_2 and CH_4 were evaluated as the averages and were used for the prediction model. The differences in CO_2 permeate molar fractions between the experimental data and prediction model were in the range from -19 to +10 % for Koch membrane and from -5 to +4 % for Sterlitech membrane, respectively.

The Sterlitech membrane showed CO_2 mass transfer coefficients of approximately one order of magnitude higher than the Koch membrane. On the other hand, the ratio of the CO_2 and CH_4 mass transfer coefficients was only fivefold in the case of the Sterlitech membrane, where as, in the Koch membrane, the difference was more than twentyfold. Based on our results, a real biogas upgrading plant can be designed. It should consist of the dual module arrangement with the first, highly permeable Sterlitech membrane module and the second selective Koch membrane module.

Acknowledgements
The financial support from the MŠMT by the project LH-14006 is gratefully acknowledged.

References

[1] Q. Xin, T. Liu, Z. Li, S. Wang, Y. Li, Z. Li, J. Ouyang, Z. Jiang, H. Wu, Mixed matrix membranes composed of sulfonated poly(ether ether ketone) and a sulfonated metal-organic framework for gas separation, Journal of Membrane Science, 488 (2015) 67-78.
[2] M. Kárászová, Z. Sedláková, K. Friess, P. Izák, Effective permeability of binary mixture of carbon dioxide and methane and pre-dried raw biogas in supported ionic liquid membranes, Separation and Purification Technology, 153 (2015) 14-18.
[3] Q. Xin, Y. Gao, X. Wu, C. Li, T. Liu, Y. Shi, Y. Li, Z. Jiang, H. Wu, X. Cao, Incorporating one-dimensional aminated titania nanotubes into sulfonated poly(ether ether ketone) membrane to construct CO2-facilitated transport pathways for enhanced CO2 separation, Journal of Membrane Science, 488 (2015) 13-29.
[4] M. Poloncarzova, J. Vejrazka, V. Vesely, P. Izak, Effective Purification of Biogas by a Condensing-Liquid Membrane, Angew. Chem.-Int. Edit., 50 (2011) 669-671.
[5] M. Kárászová, J. Vejražka, V. Veselý, K. Friess, A. Randová, V. Hejtmánek, L. Brabec, P. Izák, A water-swollen thin film composite membrane for effective upgrading of raw biogas by methane, Separation and Purification Technology, 89 (2012) 212-216.
[6] P. Dolejš, V. Poštulka, Z. Sedláková, V. Jandová, J. Vejražka, E. Esposito, J.C. Jansen, P. Izák, Simultaneous hydrogen sulphide and carbon dioxide removal from biogas by water-swollen reverse osmosis membrane, Separation and Purification Technology, 131 (2014) 108-116.

Modeling gas sorption in thermally rearranged polymers:

a thermodynamic approach

Michele Galizia, Kevin A. Stevens, Zachary P. Smith, Donald R. Paul, Benny D. Freeman

J.McKetta Department of Chemical Engineering, 200 E. Dean Keeton Street, 78712 Austin, and Center for Energy and Environmental Resources, 10100 Burnet Road, 78758 Austin, The University of Texas at Austin, Texas (USA)

Thermally rearranged (TR) polymers have attracted increasing attention in natural gas sweetening [1]. These materials, first reported by Park et al. [2], offer an unusual combination of high gas permeability and selectivity, which, in some cases surpass the upper bound defined by Robeson in 2008 [3]. TR polymers can be prepared via thermal rearrangement of solvent-soluble polyimides to form insoluble poly-benzoxazoles.

In this work, hydrogen, nitrogen, methane and carbon dioxide sorption isotherms in three TR samples, obtained via thermal conversion of HAB-6FDA polyimide, were theoretically interpreted using the Non-Equilibrium Lattice Fluid (NELF) model [4]. Since PVT data in the rubbery region are not accessible for HAB-6FDA and its thermally rearranged analogs, the polymer characteristic lattice fluid parameters, T^*, p^* and ρ^*, were estimated using a collection of mixture data at multiple temperatures [5]. The model outcomes show good agreement with the experimental sorption isotherms for both the pristine polyimide and the corresponding thermally rearranged samples. Interestingly, the model confirms by a completely independent perspective that the increase in gas solubility in TR samples is due mainly to entropic factors, i.e. the increase of excess free volume upon thermal rearrangement. This result is fully consistent with the experimental evidence [1,2]. Finally, the model results indicate that TR samples have, in the presence of carbon dioxide, greater dimensional stability relative to HAB-6FDA polyimide.

[1] D.F. Sanders et al., J. Membr. Sci. 2012 409-410 232-241

[2] H.B. Park et al., Science 2007 318 254-258

[3] L.M. Robeson, J. Membr. Sci. 2008 320 390-400

[4] F. Doghieri et al., Macromolecules 1996 29 7885-7896

[5] M. Galizia et al., J. Membr. Sci. 2012 405-406 201-211

Synthesis and Characterization of Tetraamino Diphenyl Sulfone (TADPS) Based Polybenzimidazoles and Post-Modification for Gas Separation Membranes.

A series of tetraamine diphenyl sulfone (TADPS) based polybenzimidazoles (PBIs) were synthesized. The monomer, TADPS, was synthesized via a novel and economical route. The monomer has a sulfonyl linking moiety, which results in a kinked morphology in the polymer backbone (**Figure 1**). Compared to the commercial PBI (Celazole®), these novel PBIs exhibit better solubilities in common solvents without sacrificing thermal or chemical integrity. Additionally, these PBIs have excellent gas transport properties at elevated temperatures.[1]

Our approaches to improve the gas transport properties of TADPS-based PBIs are: 1. blending the PBI with polyethylene oxide (PEO) or polypropylene oxide (PPO) that decompose and produce volatile components on controlled heat treatment, and 2. grafting propylene oxide onto the polybenzimidazole backbone by ring opening polymerization from the imidazole amines by anionic polymerization. In both methods, by degrading volatile components, blended PEO, PPO or grafted PPO blocks, through controlled heat treatment, nanoscale voids are expected to be introduced into the film, enhancing gas permeability.

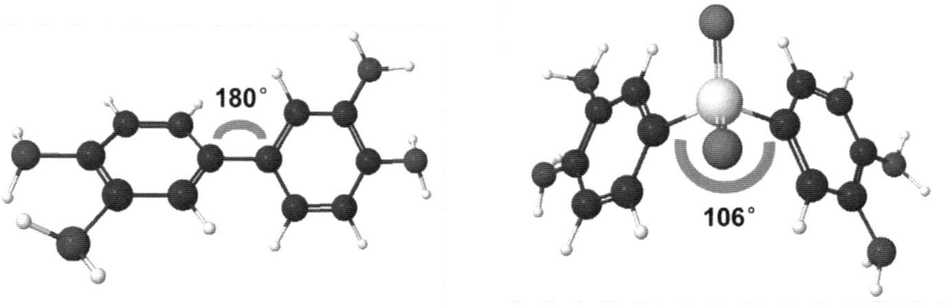

Figure 1. Difference in chemical structure between Celazole® monomer (left) and TADPS (right) illustrating introduction of kinked morphology in the polymer backbone.

References:

1. H. Borjigin, K. A. Stevens, R. Liu, J. D. Moon, A. Shaver, S. Swinnea, B. D. Freeman, J. S. Riffle, J. E. Mcgrath. *Polymer.*, **2015**, *71*,135-142

CO_2-Selective Membranes Containing Sterically Hindered Polyvinylamine

Zi Tong, The Ohio State University, tong.87@osu.edu
W.S. Winston Ho, The Ohio State University, ho.192@osu.edu

This abstract consists of a brief description of CO_2 Facilitated Transport Membranes, Sterically Hindered Amine, and New Membranes based on Sterically Hindered Polyvinylamines in the following paragraphs for CO_2 separation.

CO_2 Facilitated Transport Membranes: CO_2 capture from flue gas (main component: N_2) is essential for eliminating greenhouse gas emission, and CO_2 separation from synthesis gas is effective for hydrogen purification. Due to the inherent advantages of membrane separation, e.g., high energy efficiency, operational simplicity, small footprint, and the elimination of thermodynamic equilibrium capacity limitations, membrane technology is receiving more and more attention for CO_2 separation. For most conventional polymer membranes, CO_2/N_2 separation follows the solution-diffusion mechanism. The key challenge of solution-diffusion membranes is the trade-off between permeability and selectivity, i.e., the Robeson's Upper Bound. However, for facilitated transport membranes, the non-reactive gas component, e.g., N_2, diffuses through the membrane via the solution-diffusion mechanism, whereas CO_2 follows the facilitated transport mechanism via its reversible chemical reaction with carriers (often amines). Therefore, CO_2 transport is facilitated, and both high CO_2 permeability and high CO_2/N_2 selectivity can be achieved.

Sterically Hindered Amine: Amine carriers are the key for the CO_2 separation performance of a facilitated transport membrane. Amine carriers exhibiting a high CO_2 loading capacity and reaction rate are always desirable. A sterically unhindered amine forms a stable carbamate ion with CO_2, as shown in Eq. (1), and the stoichiometric CO_2 loading is one mole of CO_2 per two moles of the unhindered amine. However, for a sterically hindered amine, its carbamate ion is not stable due to the incorporation of a bulky substituent group to the nitrogen atom and thus is converted to the bicarbonate ion as the sink for CO_2 with the resultant reaction shown in Eq. (2). Therefore, the stoichiometric CO_2 loading is doubled, with one mole of CO_2 per mole of the sterically hindered amine.

$$CO_2 + 2\,R\text{-}NH_2 \rightleftharpoons R\text{-}NH\text{-}COO^- + R\text{-}NH_3^+ \quad (1)$$
$$CO_2 + R\text{-}NH_2 + H_2O \rightleftharpoons HCO_3^- + R\text{-}NH_3^+ \quad (2)$$

A sterically hindered amine is defined as either a primary amine in which the amino group is attached to a tertiary carbon or a secondary amine in which the amino group is attached to at least one secondary or tertiary carbon. The effects of sterically hindered amines have been reported in a lot of literature on aqueous amine solutions for CO_2 absorption. Those works focus on the small molecules of sterically hindered amines in the aqueous phase. For solid membranes, amine carriers anchored to the polymer matrix are more desired, since it can eliminate the carrier loss. Zhao *et al.* in our group were the first to demonstrate the steric hindrance effect in solid polyamine-containing membranes. In that work, the polyallylamine was modified into sterically hindered amine, and the steric hindrance effect on separation performance was investigated for syngas purification applications.

New Membranes based on Sterically Hindered Polyvinylamines: The goal of this work on sterically hindered polyamines is to extend the research of steric hindrance effect demonstrated in polyallylamine-based membranes to other polyamines for the synthesis of new, improved membranes for carbon capture applications. In this work, polyvinylamine was chosen as the amine carrier and modified into sterically hindered polyamines since it possesses the highest content of amino groups among all the common polyamines for CO_2 facilitated transport. The membranes synthesized in this work have been investigated mainly for CO_2 capture from flue gas in coal-fired power plants.

The amine modification reactions were conducted as follows: to synthesize poly(N-methyl-N-vinylamine), one equivalent amount of iodomethane as well as potassium hydroxide was added into the aqueous solution of polyvinylamine (PVAm) and reacted with the PVAm at 45°C for 48 h. After the reaction, one methyl group should be attached onto the nitrogen atom of the amino group in each repeating unit of polyvinylamine to form poly(N-methyl-N-vinylamine), with one equivalent amount of potassium iodide salt and water generated as byproducts. The salt was removed from the polyamine solution by dialysis. Similarly, poly(N-isopropyl-N-vinylamine) and poly(N-tert-butyl-N-vinylamine) were synthesized by reacting PVAm with 2-bromopropane and 2-bromo-2-methylpropane, respectively, under the same reaction conditions. The steric hindrance degree can be quantitatively described by Taft's steric hindrance parameter, $\sum E_s$, which is determined by the group contribution method. The $\sum E_s$ values of polyvinylamine, poly(N-methyl-N-vinylamine), poly(N-isopropyl-N-vinylamine) and poly(N-tert-butyl-N-vinylamine) were calculated to be 0.93, 1.00, 1.86 and 2.67, respectively. The obtained sterically hindered polyamine solutions were concentrated by N_2 purging to have an appropriate viscosity for coating onto nanoporous polyethersulfone (PES) substrates and dried in the fumed hood at ambient conditions.

The membranes were all tested at 57°C, the typical flue gas temperature. The dry feed gas consisted of 20% CO_2 and 80% N_2, and the sweep gas was purified argon. Both feed gas and sweep gas were humidified and saturated by water vapor at the testing temperature (the saturated water vapor content in each of the feed and sweep sides was 17 mol% at 57°C). The pressure of the feed side was controlled at 1.5 psig while the sweep side pressure was maintained at 1.4 psig. Under the testing conditions, polyvinylamine membranes demonstrated an average CO_2 permeability of 213.9 Barrers and CO_2/N_2 selectivity of 48.5. Although the content of amino groups reduced by incorporation of alkyl groups, CO_2 permeability was improved by 24% and CO_2/N_2 selectivity was increased by 14% by modifying polyvinylamine into poly(N-methyl-N-vinylamine). These results have demonstrated the effect of steric hindrance on CO_2 separation performance with the incorporation of the methyl group for enhancing CO_2 loading capacity. Poly(N-isopropyl-N-vinylamine) and poly(N-tert-butyl-N-vinylamine) membranes showed 15% and 11% improvements on CO_2 permeability, respectively, with no significant changes in CO_2/N_2 selectivity. The improvement on selectivity was not as significant as CO_2 permeability, which could be attributed to the reduced polarity caused by the incorporation of bulky alkyl groups, thus increasing the solubility of N_2 in the membranes. Comparing the CO_2 separation performances of the above polyamine membranes and their corresponding steric hindrance degrees, we found that the CO_2 permeability increased first significantly with increasing the steric hindrance degree and then less pronouncedly with further increasing the steric hindrance degree. This could be explained by the reduced amino group content in the membranes and the lower reaction rate due to the incorporation of bulky groups onto amino groups. It should be noted that the new poly(N-methyl-N-vinylamine) membrane can give higher CO_2 permeability than the best sterically hindered polyamine membrane, i.e., poly(N-isopropyl-N-allylamine), reported in the literature.

Title: CFD Numerical Simulations for Gas Separation Membrane Module Designs

Authors: Jiefu Ma[1], David Hasse, Ed Sanders

Hollow fiber membrane modules are widely used in gas separation applications, such as airplane on-board oxygen and fuel tank inert gas generation, biogas sweetening, separation of CO_2 from natural gas and flue gases etc. Air Liquide is an industry leader in the gas separation membrane research and manufacturing. To further develop the technology and enhance the gas separation efficiencies, it is critical to understand the gas flows inside the membrane modules and improve the fiber utilization through proper module designs. A computational fluid dynamics (CFD) model has been developed at Air Liquide to simulate the gas separation process in a membrane module that can visualize the pressure and different gas distributions, retentate and permeate flow fields, driving force distributions, etc. Information obtained are being used to analyse the separation efficiencies and guide the module design modifications.

The simulation model is developed based on general purpose CFD software FLUENT. The bore-side and the shell-side of a membrane module are modeled separately using different porous media domains. Both domains are meshed with exactly the same grid cells. The porosity of each domain is derived from the fiber dimensions and the local packing density. Gas flow resistances are calculated based on the appropriate pressure drop formulations. Gas mass transfer rates are calculated based on the local driving force and the gas permeance at each mesh cell of the module. The CFD model has been validated against experimental data on Air Liquide membrane modules and can be applied to both bore-feed and shell-feed configurations.

To illustrate the applications of the CFD simulations, a baseline module design is simulated together with several modifications to the baseline design. Results reveal the mechanisms of the performance improvements and show that the shell-side flow patterns are critical to the separation performance. Digital experiments using CFD provide a detailed understanding of fluid dynamics within the bundle that could be difficult to obtain experimentally. Results have proven to be very helpful in understanding the membrane module separation process for different configurations and to provide guidance for module designs. Module design becomes a more critical parameter as the module size increases and/or the intrinsic fiber performance improves.

Conference Details:

North American Membrane Society Annual Meeting, May 21-25, 2016, Bellevue, WA USA
http://www.membranes.org/2016/index.html

[1] Corresponding author.
Email: jiefu.ma@airliquide.com

METALLIC MEMBRANES AND N_2 SEPARATION

Simona Liguori, Kyoungjin Lee, Jennifer Wilcox,
Dept. of Energy Resources Engineering, Stanford University, CA

High permeability, high selectivity and good mechanical/chemical resistance are the four important characteristics required for a commercial gas separating membranes. Nevertheless, the development of membranes able to show both high permeability and selectivity is still in progress. On this route, metallic membranes seem to offer improved performance compared to polymeric membranes. The aim of this work is to explore the potential of metallic membranes for N_2 separation. In particular, the effect of different temperature and pressure conditions as well as the effect of different gas exposure on membrane performance, in terms of N_2 permeating flux and ideal selectivity of N_2 with respect to another gas is studied and analyzed. Moreover, scanning electron microscopy (SEM), and an electron microprobe analyzer (EMPA) are used for investigating the effect of the different operating conditions on the membrane surface.

Based on a preliminary theoretical investigation using density functional theory, the Group V transition metals (e.g., vanadium (V), niobium (Nb) and tantalum (Ta)) show strong affinity toward N_2. Moreover, from solubility and diffusivity values taken from the literature, iron (Fe) seems to be a good fit for this application; therefore, these metals are chosen as membrane materials in this study and in particular, pure V, Nb, Ta and Fe foils with a thickness of 40 μm have been investigated.

Permeation tests with pure gases (He, N_2, CO_2 and CH_4) are performed to characterize the membranes in terms of the N_2 permeating flux and ideal selectivity at different temperature and pressure conditions, which vary from 400 to 600 °C and from 2.0 to 60 bar, respectively. From the experimental results, all of the membranes show infinite selectivity towards N_2 separation at each operating condition tested. V shows better performance in term of the N_2 permeating flux compared to Nb, Ta and Fe.

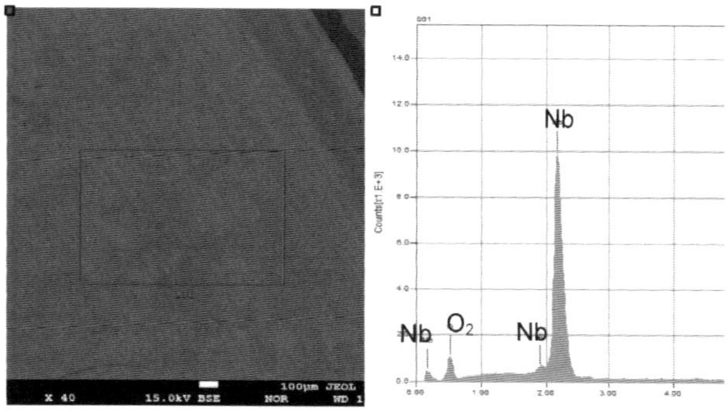

Figure 1 – EMPA analysis after tests – Nb membrane

Nb and Ta are completely destroyed when the temperature is raised from 400 to 500 °C. EMPA analysis on these membranes has shown the presence of oxygen on the surface, as illustrated in *Figure 1* where EMPA is shown for the Nb membrane. Moreover, although Fe shows lower N_2 permeating flux with respect to V, it seems less brittle than V after having worked up to 600 °C and 60 bar. The results of this work suggest that V and Fe membranes can be used for N_2 separation leading to a near-pure stream of N_2, regardless of their low permeability; therefore, future work will be focused on the synthesis and analysis of V and Fe alloys in order to enhance the N_2 permeability and improve membrane resistance toward oxidation.

ZIF-67 filled PDMS Mixed Matrix Membranes for comprising of recovery of bio ethanol via Pervaporation

Amin Khan.[1], Asim Laeeq Khan[1]
[1]Chemical Engineering Department. COMSATS Institute of Information Technology, Defense road, off Raiwind road, Lahore Pakistan
*aminkhan@ciitlahore.edu.pk

ABSTRACT

Due to the drastic effects of global warming and concerns on energy scarcity, it has become necessary to explore and develop both energy-saving and environment friendly processes to meet the energy demands. It has been reported that separation alone currently contains about 60 to 80% of the process cost in most of the chemical processes. Several techniques (e.g. Distillation, Gas stripping and Molecular sieve adsorption) have been proposed for the effective and sustainable separation of alcohol/water mixtures but they all are very costly and high energy intensive due to the formation of azeotropic mixtures. Among these techniques, membranes provide a promising opportunity due to their several advantages such as low ecological footprint, high energy efficiency and low capital cost[1].

Zeolitic Imidazole Frameworks (ZIFs) are composed of metal ions linked by organic ligands and offer exceptionally high surface area and relatively high thermal stability[2]. In this study, Mixed Matrix Membranes (MMM) were prepared by using PDMS as a hydrophobic polymer and ZIF-67 as a filler material. According to the best of our knowledge, this is the first study using this type of ZIF in MMM for any application. The synthesized membranes were tested in custom built pervaporation testing set-up for the measurement of separation factor and flux. The results showed that incorporation of ZIF-67 not only enhanced the flux but also helped in further improving the separation properties. The data of performance parameters and SEM images of the synthesized membranes suggest that the fillers adhered well to the polymer matrix. The ease of synthesis of these filler materials and improved performance make them attractive choice as filler in the MMMs for pervaporation applications. **a)**
b)

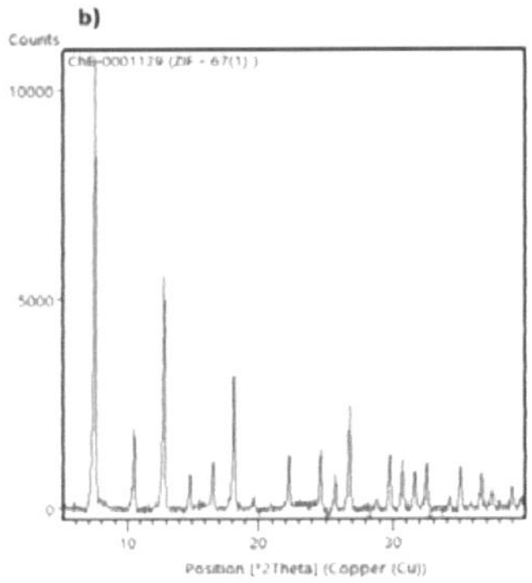

Fig. 1. a)SEM images of ZIF-67 particles b) XRD Pattern of ZIF-67 particles

Keywords: Pervaporation; Zeolitic Imidazole Framework; Mixed Matrix Membranes

BIOSKETCH

0Amin Khan earned his Masters of Science in Chemical Engineering from COMSATS Lahore with majoring in membrane based separation. He obtained his BS Chemical Engineering from COMSASTS Lahore with majoring in petroleum refinery and polymer engineering. He is currently working as a faculty member at the department of Chemical Engineering, COMSATS Lahore. He is also the head of Membrane based separation Lab. His research interests include, membrane based separation, synthesis of green membranes, designing of testing setups of membranes.

Sorption Enhanced Mixed Matrix Membranes for H_2/CO_2 Separation at Elevated Temperature

Lingxiang Zhu, Deqiang Yin, Shailesh Ganesh Konda, Mark T. Swihart and Haiqing Lin*

Department of Chemical and Biological Engineering, University at Buffalo, The State University of New York, Buffalo, NY 14260, USA

* Corresponding author. Tel: +1-716-645-1856, Email: haiqingl@buffalo.edu (H. Lin)

Membrane technology has been widely explored for high temperature H_2 purification and CO_2 capture in the integrated gasification combined cycle (IGCC) processes. Polymeric membranes permeating H_2 and rejecting CO_2 at 100 - 200 °C have demonstrated promise for economical CO_2 capture. Conventional approaches are focused on rigid polymers with strong size sieving ability for H_2/CO_2 separation. In contrast, we investigate mixed matrix membranes (MMMs) containing palladium (Pd) nanoparticles, which have strong affinity towards H_2 and thus exhibit H_2/CO_2 solubility selectivity as high as 840 at 50 - 200 °C. Poly[2,2'-(m-phenylene)-5,5'-bisbenzimidazole] (m-PBI) is selected as a model polymer matrix. The m-PBI can also be crosslinked to improve the size-sieving ability of polymer phase. Uniform Pd nanoparticles (~10 nm) are prepared via hot-injection approaches, and effect of Pd loading (0 – 30 wt%) on the H_2/CO_2 separation properties is thoroughly evaluated. The loading of Pd nanoparticles increases H_2 permeability and H_2/CO_2 selectivity at temperature of 50 – 200 °C. For example, addition of 10 wt% Pd in the m-PBI increases the H_2 permeability from 2.3 Barrers to 3.5 Barrers, and H_2/CO_2 selectivity from 12 to 18 at 50 °C. This presentation will discuss pure-gas (H_2, N_2 and CO_2) permeation properties of the MMMs in details, and H_2 and CO_2 solubility in the MMMs to reveal the influence of H_2-Pd interaction on H_2/CO_2 separation performance.

Keywords: Mixed matrix membranes; Palladium nanoparticles; Polybenzimidazole; H_2/CO_2 separation; CO_2 capture

The investigation of perfluorocompound gases separation with hollow fiber membrane.

Joo Hwan Lim, Su Jung Jeong, Hyung Chul Koh, Seong Yong HA*

Airrane Co., Ltd.,151-5 Gwahaksaneop 4-ro, Oksan-myeon, Heungdeok-gu, Cheongju-si, Korea

* haven@airrane.com

At the present time, there is a growing interest in the development of gas separation membranes based on materials providing better selectivity, thermal stability and chemical stability than those already existing. Carbon molecular sieve (CMS) membranes are promising materials for membrane-based gas separation because of their excellent permeation and separation properties and thermal and chemical stability under harsh conditions, such as at high pressure and elevated high temperatures [1]. CMS membranes fabricated by pyrolysis of different polymeric materials are characterized by a very narrow microporosity (3–5Å), which allows very sharp discrimination between gas molecules having different sizes and shape. In particular, thermally stable polyimides acting as organic molecular sieves are excellent precursors for obtaining CMS membranes [2]. Polyimides are rigid, high melting, high Tg, thermally stable polymers formed by the condensation reactions dianhydrides with diamines. Because of their high thermal stabilities, polyimides are used in machined devices and other high temperature applications [3]. The separation of gas molecules with a molecular size under 4Å by CMS membranes can be effectively achieved via a molecular sieve mechanism. The important factors determining the microstructure and gas separation properties of the CMS membranes are the choice of polymeric precursor and the pyrolysis conditions [4-5].

In this study, the permeation characteristics of fluorinated gases N_2 and SF_6 used in the semiconductor processes were examined by study through the processes that occur during the hollow fiber membrane separation technology. The precursor hollow fiber membranes were prepared using PI. It (synthesized polyimides) was received from Aekyung Petrochemical Company Limited (Korea). The precursor was spun into asymmetric hollow fiber form by a dry-wet spinning process. The polymeric precursors were pyrolyzed in a quartz tube furnace. To investigate the gas permeation properties of CMS, N_2, SF_6 and perfluorocompound Mixed gases were measured for the gas permeation.

References

1. C. W. Jones, W. J. Koros, *Carbon*, **32**, 1419 (1994).

2. H. Hatori, Y. Yamada, M. Shiraishi, H. Nakata, S. Yoshitomi, *Carbon*, **32**, 305 (1992).

3. M. B. Rao, S. Sircar, *J. Membr. Sci.*, **3**, 253 (1993).

4. H. Suda, K. Haraya, *J. Chem. Soc. Chem. Commun.*, **11**, 1179 (1995).

5. C. G. Vincent, W. J. Koros, *Ind. Eng. Chem. Res.*, **35**, 299 (1996).

Acknowledgment

This project is supported by the "R&D Center for reduction of Non-CO_2 Greenhouse gases" funded by Korea Ministry of Environment(MOE) as "Global Top Environment R&D Program".

The study of solid polymer electrolyte composite hollow fiber membrane for olefin gas separation

Joo Hwan Lim, Su Jung Jeong, Hyung Chul Koh, Seong Yong HA*

Airrane Co., Ltd.,120, Sinseongnam-ro, Yusung-gu, Daejeon 305-805, Korea

* haven@airrane.com

Key Words: olefin/paraffin, gas separation, facilitated transport membranes, silver, π-complexation

Development of olefin/paraffin separation has been performed in various industries and institution. Especially, huge amount of light olefins such as ethylene and propylene are produced in chemical and petroleum industry due to their diverse utilization as raw materials for many products[1]. However, it is very difficult to separate olefin from paraffin because their physical and chemical properties are very similar like molecular sizes and boiling point. Distillation process used to separate mixture gas so far, but it needs large amount of capital cost and energy. For this reason, many technologies are being investigated. Facilitated transport membrane is in the limelight for the separation of olefin/paraffin because of its simplicity and low energy consumption. Olefins and transition metal ions act electron donor/acceptor each other so that synthesis π-complexation[2]. Silver ( ) ions is commonly used as a carrier for facilitated olefin transport. In this study, $AgBF_4$ solution was produced as electrolyte for the olefin separation composite membrane[3]. Polyetherimide (PEI) hollow fiber membrane was fabricated through the dry-wet spinning process. The skin layer of the membrane is in the lumen side, so $AgBF_4$ solution was coated inside. The composite hollow fiber membrane was evaluated using pure gas and mixed gas.

References

1. J.L. Pellegrino, Chemical Industry of The Future: Energy and Environmental Profile of the U. S. Chemical Industry, Energetics, Inc., Columbia, MD220.

2. Safarik, D.J., Eldridge, R.B., 1998. Olefin/paraffin separations by reactive absorption: A review. Industrial and Engineering Chemistry Research 37, 2571.

3. Y. S. Kang, *Journal of membrane science* 2014, 463, 11

Investigation of high permeable polyimide hollow fiber membrane for gas separation.

Joo Hwan Lim, Su Jung Jeong, Hyung Chul Koh, Seong Yong Ha*

Airrane Co., Ltd.,120, Sinseongnam-ro, Yusung-gu, Daejeon 305-805, Korea

* haven@airrane.com

Gas separation using polymeric membranes has achieved important commercial success in some industrial processes since the first commercial-scale membrane gas separation system was produced in the late of 1970s. In order to extend its application and compete successfully with traditional gas separation processes such as cryogenics, pressure swing adsorption and absorption, great attention has been made in fabricating high separation performance membranes in both academia and industry. Development of ultra-thin skinned asymmetric hollow fiber membranes from highly selective polymers has been a challenge in gas separation membrane processes. Although thousands of polymers exhibit perm-selective properties for gas mixtures, only a few glassy polymers are useful in making asymmetric gas separation membranes, particularly, polyimide (PI), polysulfone (PSf), polyethersulfone (PESf), and polyetherimide (PEI). [1]

Polyimides have been attractive as membrane materials for gas separation because of their excellent thermal, chemical and mechanical properties, as well as excellent film-forming properties. Relationship between the chemical structure of polyimides and their gas permeation and separation properties has been investigated extensively. For a given polymer, how to make a high separation performance membrane by studying the membrane formation protocol, is an important topic in the membrane research. Formation of asymmetric hollow fiber contains many variables such as the nature of the solvent, the additive and the coagulant, the spinning conditions etc. all of which may affect the structure of the hollow fiber membranes and separation performance. Reviewing advances in preparing asymmetric gas separation membranes, the selection of a suitable solvent, additive and coagulant is the most important. From the view point of economic and environmental basis, the solvents, additives and coagulants used must be commercially available and low toxicity. In addition, the manufacture method must be easy, economic and reliable. However, in practice, the selection of solvent and coagulant is quite limited. Water is the best choice as the coagulant to remove solvent from hollow fiber membrane. For water soluble solvents used, the most common solvents are N-methyl-2-pyrrolidone (NMP) and dimethylacetamide (DMAc). Therefore, the use of a suitable additive became a most efficient tool to prepare desired membranes. In fact, all high separation performance membranes have been prepared by the use of right additives [2].

In this study, PI hollow fiber membranes have been prepared NMP/IPA/THF systems. It (synthesized polyimides) was received from Aekyung Petrochemical Company Limited (Korea). The effect of polymer, solvent, non-solvent, additives concentration, synthesized polyimide molecular weight and air-gap on the membrane structure and gas permeation were investigated.

References

1. D. Wang, K. Li, W. K. Teo, *J. Membr. Sci.*, **115**, 85 (1996).

2. D. Wang, K. Li, W. K. Teo, *J. Membr. Sci.*, **138**, 193 (1998).

Tuning PIM-PI/Z-MOF-Based Mixed-Matrix Membranes for Highly Efficient Propylene/Propane Separation

Ramy J. Swaidan, Xiaohua Ma, Ingo Pinnau*

King Abdullah University of Science and Technology (KAUST), Advanced Membranes and Porous Materials Center (AMPMC), Division of Physical Sciences and Engineering, Chemical and Biological Engineering Program, Thuwal 23955-6900, Saudi Arabia

*ingo.pinnau@kaust.edu.sa

Energy-efficient separation of propylene (C_3H_6) from propane (C_3H_8) is vital to the chemical process industry where C_3H_6 is used as feedstock for the production of a variety of important chemicals and plastics. A potentially significant marketplace exists for membrane technology in this separation if it can demonstrate the necessary performance. Currently, the required C_3H_6 purity is achieved by extremely energy-intensive cryogenic distillation. Recently, we reported the use of hydroxyl-functionalized polyimides of intrinsic microporosity (PIM-PIs) for C_3H_6/C_3H_8 separation. The best PIM-PI to date (PIM-6FDA-OH) had a mixed-gas C_3H_6/C_3H_8 selectivity of 15 combined with a C_3H_6 permeability of 2.5 Barrer at a feed pressure of 4 bar. One potential route to improving the permeability *and* selectivity is development of high-performance polymer-nanoparticle hybrid materials, also known as mixed-matrix membranes (MMM). For the first time, we report the fabrication and exceptional pure- and mixed-gas C_3H_6/C_3H_8 gas permeation properties of a PIM-PI-based MMM. Using PIM-6FDA-OH as the polymer matrix, we integrated a zeolitic metal-organic framework (ZMOF), ZIF-8, at various loadings ranging from 20% to 65% by weight. The membranes were tested with a 50:50 C_3H_6/C_3H_8 feed mixture at pressures up to 7 bar (p/p_{sat} ~0.8). Not only did the membranes demonstrate exceptional pure-gas properties (α >35, $P(C_3H_6)$ >10 Barrer), but the high performance was also strongly preserved in mixed-gas conditions. Exceptional C_3H_6/C_3H_8 mixed-gas selectivities ranging from 20 at 30% loading to 30 at 65% loading were achieved. We also observed minimal onset of plasticization, highlighting the strong potential of PIM-6FDA-OH/ZIF-8 mixed-matrix membranes for C_3H_6/C_3H_8 separations.

Performance prediction of a tubular membrane contactor for CO_2 absorption in reactive AAIL-based solvent

Muhammad Usman[1], Magne Hillestad[1], Liyuan Deng[1,]*

[1]*Norwegian University of Science and Technology (NTNU), Trondheim 7491, Norway*

* liyuan.deng@nt.ntnu.no

Key words: CO_2 capture, Ionic liquid, Mass transfer, Membrane contactor

Membrane contactor has been incorporated to capture pre-combustion CO_2 at elevated pressure and temperature in this work. Pre-combustion process is considered to be more efficient to capture CO_2 due to high concentrations and pressure of CO_2. Membrane contactor gives the advantages of high surface area, non-dispersive contact between the phases without causing flooding and entrainment of absorbent. The challenging part in the application of membrane contactor for pre-combustion conditions is to find highly thermal and chemical stable membranes and absorbents. In this study, a glass membrane was chosen as membrane material and amino acid based ionic liquid (AAIL) [P_{4444}][Gly] with co-solvent PEG400 as absorbent. This blend solvent has negligible vapor pressure and superior thermal stability, while AAILs have the ability of reacting with CO_2 in the similar fashion as primary amines as it comprises of primary amine group and thus enhances the absorption capacity of the absorbent.

In this study, 1D-modelling of chemical absorption of CO_2 in the tubular membrane contactor using the blend absorbent is developed at elevated temperature and pressure conditions, which describes the mass transfer of CO_2 through gas, liquid and membrane phases. Earlier works [1-4] have been reported for post-combustion conditions and using aqueous amine solutions, and there are only a few publications describing the modeling of membrane contactor using ionic liquids [5, 6]. The model presented is for physical absorbent in hollow fiber membrane contactor and the liquid film is stagnant [5]. Chemical absorption enhances the solubility and hence increases the absorption capacity of solvent.

In this work the effect of chemical absorption on mass transfer flux in tubular membrane contactor has been studied and the results compared with physical absorbent. The liquid flows laminar through the lumen side while the gas flows in the shell side of the tubular membrane contactor in a counter current parallel flow configuration. Three resistances due to each phase occur for mass transfer of CO_2 in membrane contactor and these resistances are modeled in series using heat analogy. Mass transfer coefficients for gas and liquid phases are estimated using correlations based on Reynold and Schmidt numbers. Physical properties of ionic liquid-PEG400 are measured in the lab at elevated temperatures and empirical correlations are developed for these experimental data. Performance of membrane contactor has been evaluated on the basis of mass transfer flux and pressure drop. Operational parameters such as gas and liquid flow rates are also optimized to get maximum CO_2 mass transfer flux and reduced pressure drop.

References

1. Hoff, K.A., et al., *Modeling and experimental study of carbon dioxide absorption in aqueous alkanolamine solutions using a membrane contactor.* Industrial and Engineering Chemistry Research, 2004. **43**(16): p. 4908-4921.
2. Wang, R., D.F. Li, and D.T. Liang, *Modeling of CO2 capture by three typical amine solutions in hollow fiber membrane contactors.* Chemical Engineering and Processing: Process Intensification, 2004. **43**(7): p. 849-856.
3. Zhang, H.Y., et al., *Modeling and experimental study of CO2 absorption in a hollow fiber membrane contactor.* Journal of Membrane Science, 2006. **279**(1-2): p. 301-310.
4. Keshavarz, P., J. Fathikalajahi, and S. Ayatollahi, *Analysis of CO2 separation and simulation of a partially wetted hollow fiber membrane contactor.* Journal of Hazardous Materials, 2008. **152**(3): p. 1237-1247.
5. Chau, J., et al., *Pressure swing membrane absorption process for shifted syngas separation: Modeling vs. experiments for pure ionic liquid.* Journal of Membrane Science, 2014. **453**: p. 61-70.
6. Albo, J., P. Luis, and A. Irabien, *Carbon Dioxide Capture from Flue Gases Using a Cross-Flow Membrane Contactor and the Ionic Liquid 1-Ethyl-3-methylimidazolium Ethylsulfate.* Industrial & Engineering Chemistry Research, 2010. **49**(21): p. 11045-11051.

Dual CO_2- and N_2-selective intrinsically microporous hydroxyl-functionalized triptycene-based polyimide for natural gas upgrading

Fahd Alghunaimi, Bader Ghanem, Nasser Alaslai, Mohsin Mukaddam, Ingo Pinnau*

King Abdullah University of Science and Technology (KAUST), Advanced Membranes and Porous Materials Center (AMPMC), Division of Physical Sciences and Engineering, Chemical and Biological Engineering Program, Thuwal 23955-6900, Saudi Arabia

*ingo.pinnau@kaust.edu.sa

The synthesis and gas permeation properties of two novel PIM-polyimides (TDA1-DMN) and (TDA1-APAF) prepared from 9,10-dimethyl-2,3,6,7-triptycene tetracarboxylic dianhydride (TDA1) with commercially available 3,3 -dimethylnaphthidine (DMN) and 2,2-bis(3-amino-4-hydroxyphenyl)-hexafluoropropane (APAF) diamine monomers respectively are reported. The triptycene building block with bridgehead methyl side groups offers TDA1-DMN with high BET surface area and very high gas permeabilities (CO_2 = 3700 Barrer) but with only moderate selectivities (CO_2/CH_4 selectivity = 17). Thus, an *ortho*-hydroxyl-functionalized polyimide (TDA1-APAF) was designed to enhance selectivities, while maintaining high gas permeability. TDA1-DMN and TDA1-APAF had BET surface areas based on nitrogen adsorption of 760 m^2 g^{-1} and 260 m^2 g^{-1} respectively. The freshly prepared sample of TDA1-APAF exhibited excellent gas permeation properties: a H_2 permeability of 94 Barrer with a H_2/CH_4 selectivity of 129 and CO_2 permeability of 40 Barrer coupled with a CO_2/CH_4 selectivity of 55. Physical aging over 250 days resulted in enhanced H_2/CH_4 selectivity of 183 with only ~ 25% loss in H_2 permeability. Aged TDA1-APAF exhibited 5-fold higher pure-gas CO_2 permeability (30 Barrer) and 2-fold increase in CO_2/CH_4 permselectivity (~75) over conventional dense cellulose triacetate membranes at 2 bar. Gas mixture permeation experiments with a 1:1 CO_2/CH_4 feed mixture demonstrated higher mixed- than pure-gas selectivity (~88) and plasticization resistance up to 30 bar. Moreover, aged membrane exhibited N_2/CH_4 selectivity of 2.3 which offer a simultaneous removal of N_2 and CO_2 for natural gas upgrading. These results suggest that intrinsically microporous hydroxyl-functionalized triptycene-based polyimides are promising candidate membrane materials for *simultaneous* removal of CO_2 and N_2 on natural gas sweetening and hydrogen purification on petrochemical refinery applications.

Adsorption Behavior and Self-Aggregation of 5,10,15,20-tetrakis-(4-sulfonatophenyl)-porphyrin on Quaternized Polysulfone Membrane

Manman Liu[ab], Lizhi Zhao[ab], Yuzhong Zhang[ab*]

[a] *School of Materials Science and Engineering, Tianjin Polytechnic University, Tianjin 300387, PR China*

[b] *State Key Laboratory of Separation Membranes and Membrane Processes, Tianjin Polytechnic University, Tianjin 300387, PR China*

* Corresponding author: e-mail: zhangyz2004cn@vip.163.com

Highlights

- TPPS species in solution showed different adsorption behavior on QAPSF membrane.
- TPPS adsorbed on membrane performed different pH-dependence compared with in solution.
- The self-assembly and spectroscopic features of TPPS on the porous membrane may have potential applications in fabrication of smart membranes and optical devices.

Graphical Abstract

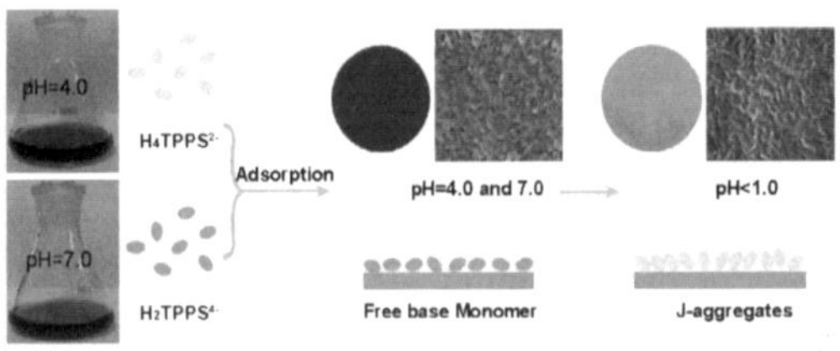

Abstract

5,10,15,20-Tetrakis-(4-sulfonatophenyl)-porphyrin (TPPS) can be used in many different fields according to its forms. We select quaternized polysulfone QAPSF) as a type of carrier, and QAPSF/TPPS membrane is prepared by adsorption based on electrostatic interaction. TPPS species in solution showed different adsorption behavior. The diacid (H_4TPPS^{2-}) monomer and the free base (H_2TPPS^{4-}) monomer of the porphyrin showed almost equal adsorption capacity on QAPSF membrane, which is due to the identical existence form of TPPS on membrane proved by the identical color and UV-vis spectra. H_4TPPS^{2-} showed slower adsorption rate than H_2TPPS^{4-}. Because the electrostatic repulsion between the positively-charged center of H_4TPPS^{2-} and quaternary ammonium groups on QAPSF membrane influences the migration of H_4TPPS^{2-} from solution to the membrane surface. J-aggregates of TPPS in aqueous solution affected the adsorption and led to a smaller adsorption capacity. Compared with in aqueous solution, the apparent pK_a for the equilibrium between H_4TPPS^{2-} and H_2TPPS^{4-} of the adsorbed porphyrin was significantly decreased by the interaction between TPPS and QAPSF. The adsorbed TPPS can form J-aggregates with nanorod morphology by equilibrating the QAPSF/TPPS membrane with buffer at much lower pH. Higher adsorption capacity promoted the J-aggregation.

The present systems provide valuable insights into adsorption behavior and pH-controlled species of TPPS on porous membrane; self-assembly and spectroscopic features of TPPS may have potential applications in fabrication of smart membranes and optical devices.

***Keywords*:** TPPS; adsorption; membrane-TPPS interaction; aggregates

Free Diffusivity of Viruses: Experimental Measurements and Model Predictions

Ruth E. Baltus
Department of Chemical and Biomolecular Engineering
Clarkson University
Potsdam, NY 13699-5705

Appala Raju Badireddy
Department of Civil and Environmental Engineering
University of Vermont
Burlington, VT 05405

Armin Delavari
Department of Chemical and Biomolecular Engineering
Clarkson University
Potsdam, NY 13699-5705

Shankar Chellam
Department of Civil Engineering
Texas A&M University
College Station, TX 77843-3136

Porous membranes are extensively used in the pharmaceutical industry to remove microorganisms. Biological products produced from cell lines require tight orthogonal virus filtration steps in order to remove retroviruses and other undesirable species. Virus contamination is a significant safety threat in the production of plasma-derived pharmaceuticals, requiring tight virus filtration. Virus diffusion coefficients and their effective hydrodynamic dimensions are crucial in quantitatively understanding their behavior in microfiltration and ultrafiltration systems employed for such bioseparations. To date, investigators have largely employed dimensions obtained from crystallography or microscopy where viruses had been dehydrated prior to observations; diffusivities were then calculated using the Stokes-Einstein relationship. The applicability of such approximations even for near-spherical viruses has not yet been rigorously tested. Additionally, several viruses of public health and environmental interest are non-spherical for which models relating size and shape and diffusivity are not available.

In this work, the aqueous bulk diffusivities of several near-spherical and non-spherical bacterial viruses were experimentally measured using two techniques. The diffusive flux of various near-spherical (Qβ, MS2, ΦX174, and PRD1), and non-spherical ('tailed') (Mu1, P1, P22 and T4) bacteriophages of varying dimensions through large pore (8 micron diameter) track-etched polycarbonate membranes was measured. Experiments were conducted in a specially constructed diffusion cell using phosphate buffered saline solutions to reduce electrostatic interactions and to maintain virus viability during the course of the measurements. Virus diffusion coefficients were obtained from measured concentrations using Fick's first law and a simple model of diffusive mass transfer across the membrane.

Virus self-diffusivities were also measured using dynamic light scattering or photon correlation spectroscopy (Zetasizer Nano ZS, Malvern Instrument Ltd., Worcestershire, UK). For our measurements, a He-Ne laser (633 nm) was employed, the sample cell temperature was maintained at 22 °C, and the decay of the scattered light intensity arising from Brownian motion was monitored at a fixed 90° angle.

Diffusivity values determined from the membrane flux measurements were in good agreement with values determined from dynamic light scattering, with values ranging from 3.8 x 10^{-12} m^2/s to 16.6 x 10^{-12} m^2/s. Values are qualitatively consistent with expectations based on bacteriophage dimensions obtained from electron microscopy images for the near-spherical viruses.

Two modeling approaches were used to predict diffusivity values for the bacteriophages, models that incorporate the unique shape of each organism. Both approaches involved calculation of the hydrodynamic properties of model particles with structure (size and shape) consistent with electron microscope images. In one approach, each virus was modeled as an array of spheres assembled to match the electron micrograph. The model particle was constructed by specifying the size and relative position of each sphere in the structure. The program HYDRO++ (freely available at http://leonardo.inf.um.es/macromol/programs/hydro++/hydro++.htm) was then used to calculate the hydrodynamic coefficients (including the translational diffusivity) of the model structure. This program was developed by Garcia de la Torre; the theoretical foundation of the hydrodynamic calculations in the program is the bead model originally developed by Kirkwood and Riseman with a number of modifications and updates over the years by Garcia de la Torre and coworkers. As an example, Figure 1 shows an electron micrograph of the tailed P1 virus and the bead model assembled to represent this organism.

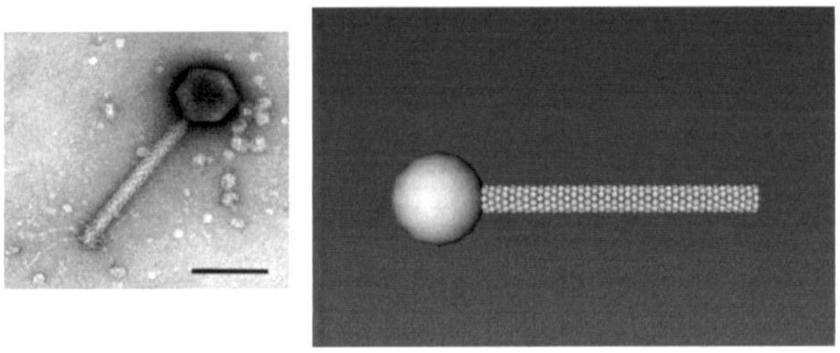

Figure 1 P1 virus from scanning electron microscopy (left) and bead model used with HYDRO++ to calculate its diffusivity (right). Measured diffusivity is 4.1 $\pm$0.1 x 10^{-12} m^2/s from membrane flux measurements and 4.5 $\pm$0.3 x 10^{-12} m^2/s from light scattering. Diffusivity predicted for the bead model structure with head and tail is 4.0 x 10^{-12} m^2/s. Stokes Einstein prediction for a sphere with diameter equal to the diameter of the P1 head (65 nm) is 6.65 x 10^{-12} m^2/s.

In the second approach, the computational fluid dynamics program COMSOL was used to determine the hydrodynamic forces on the surface of a model particle as it moves through an unbounded fluid at a fixed velocity. The size and shape of each model particle were defined by matching the structure obtained from SEM imaging. The particle diffusivity (D) is related to the force (F) and velocity by $D = k_B\, T/(F/velocity)$ where k_B is the Boltzmann constant and T is the absolute temperature.

In this presentation, experimental diffusivity values will be compared to model predictions made using both approaches. We will address questions such as 'can sizes determined from SEM imaging be used to characterize the diffusive properties of viruses?', 'what impact does the 'tail' have on the diffusive flux of viruses such as Mu1, P1, P22 and T4?' and 'can these comparisons be used to assess the flexibility of these organisms?'

Functionalization of silica membranes with lipid bilayers for sugar recovery

Shanshan Zhou, Daniel Schlipf, Stephen E. Rankin, Barbara L. Knutson

Department of Chemical and Materials Engineering

University of Kentucky Lexington, KY 40506

Corresponding Author: bknutson@engr.uky.edu

The ability of biological membranes to facilitate the selective transport of ions or hydrophilic molecules through carrier protein inspired the design of facilitated transport membrane for highly selective separations in dilute aqueous solute systems. For constructing an artificial membrane, thin film mesoporous silica, which has tunable pore morphology, chemical, thermal stability and ease of organic functionalization, are an ideal support for lipid bilayers. Manipulating the pore size of the silica thin films, an artificial lipid bilayer can be prepared in the form of pore enveloping or pore filling. Combining selective permeable lipid bilayers with high surface area silica materials suggests many potential applications for efficient, selective dilute aqueous solute separations. The recovery of aqueous glucose, which can be bound and released by boronic acid carriers, has been chosen as model separation, representing the recovery of dilute carbohydrates from the depolymerization of biomass. The separation of glucose using lipid bilayer supported within the pores of mesoporous silica thin film composite membranes is examined, in which the carbohydrate carrier (boronic acid) is immobilized in the bilayer. The selective separation of carbohydrates from biomass hydrolysate mixtures will address significant challenges to the biochemical production of biofuels and chemicals from lignocellulose.

In this work, vertical oriented hexagonally close packed (HCP) silica thin films on macroporous support were used as the lipid supports, which was demonstrated to have ordered pore structure and accessible porosity by TEM, solvent flux of ethanol and permeability of fluorescent molecules. Lipid filled membranes (in the absence of boronic acid) prevent the transport of glucose across the membrane, while bare silica and lipid-enveloped supports did not act as effective barriers to aqueous sugars. The ability of lipid filled pores to act as a barrier to sugar transport provides basis for selective separation with the incorporation of biomimetic carriers in the bilayer. For example, the glucose permeability was significantly improved after immobilizing the boronic acid into the lipid filling system, which confirmed the ability of boronic acid to facilitate the transport of glucose through the lipid bilayer. Based on the different affinity of boronic acid to the sugars (glucose, fructose, xylose and sucrose), factors like pH gradient, temperature (related to the lipid bilayer diffusivity) and boronic acid composition are explored to improve the transport and selectivity for sugars across lipid filled mesoporous thin film silica membranes.

Keywords: sugar recovery, boronic acid, lipid bilayer, silica membrane

Living Biofouling-Resistant Membranes

Manish Kumar*, Thammajun L. Wood, Rajarshi Guha, Li Tang, Michael Geitner, Thomas K. Wood
Department of Chemical Engineering, Pennsylvania State University, University Park, PA, 16802, USA
*Corresponding author: 155 Fenske Laboratory, Pennsylvania State University, University Park, PA, 16802, USA, (814) 865-7519, manish.kumar@psu.edu

Membrane systems are being increasingly used for water treatment, recycling water from wastewater, during food processing, and energy production. They are thus a key technology to ensure water, energy and food sustainability. However, biofouling, the build-up of microbes and their polymeric matrix, clogs these systems and reduces their efficiency. Realizing that a microbial film is inevitable, we engineered a beneficial biofilm that prevents membrane biofouling by limiting its own thickness by sensing the number of its cells that are present via a quorum-sensing circuit (**Figure 1**). The beneficial biofilm also prevents biofilm formation by deleterious bacteria by secreting nitric oxide, a general biofilm dispersal agent as demonstrated by both short-term dead-end filtration and long-term crossflow filtration tests. In addition, the beneficial biofilm was engineered to produce an epoxide hydrolase so that it efficiently removes the environmental pollutant epichlorohydrin. Thus, we have created a living biofouling-resistant membrane system that simultaneously reduces biofouling while providing a platform for biodegradation of persistent organic pollutants.

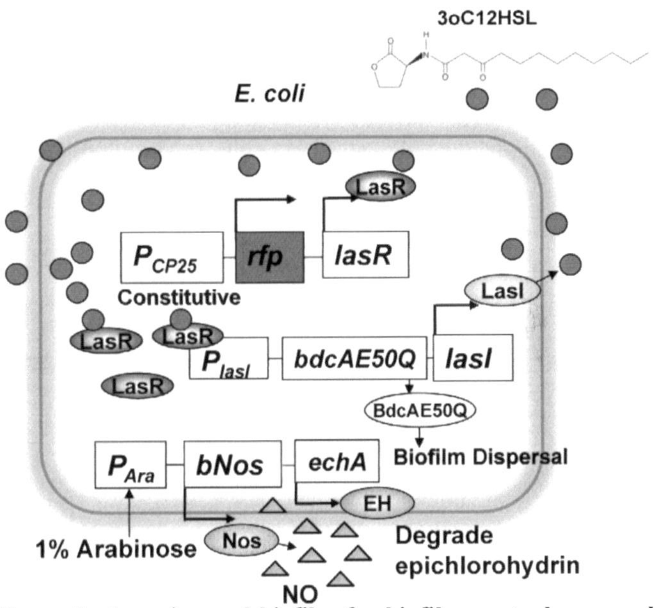

Figure 1: A engineered biofilm for biofilm control on membranes. The engineered *E. coli* biofilm can control its own thickness by using a novel "self-control" circuit while releasing quorum sensing molecules (nitric oxide using the enzyme nitric oxide synthase) that prevent colonization of membranes by other species. It was also engineered to release an enzyme (epoxide hydrolase) that can degrade the micropollutant epichlorohydrin. The self-control circuit was based on the LasI/LasR QS module of *P. aeruginosa*. The LasI protein and the engineered biofilm dispersal protein BdcA E50Q are controlled by the lasI promoter. When LasI is produced, it synthesizes the QS signal 3oC12HSL, and upon reaching a threshold value based on increasing cell density, the QS signal binds to LasR (. The 3oC12HSL+LasR complex activates the lasI promoter which leads to increasing production of dispersal protein BdcA E50Q as cell density increases.

A Weak Cation Exchange Macroporous Hydrogel Membrane for Monoclonal Antibody Capture

Huayu Niu and Christine Moresoli

Department of Chemical Engineering, University of Waterloo
200 University Avenue West, Waterloo, ON, N2L 3G1, Canada

Abstract

The recent development of ion exchange membrane materials with increased binding capacity represents a promising purification avenue for the capture of monoclonal antibodies (mAbs). Compared to conventional Protein A resins, the macropores of ion exchange membranes reduce pore diffusion and display higher throughput due to the convective flow characteristics. Weak cation exchange membranes capture protein by electrostatic interactions. These materials possess variable ionization according to pH, which can be used to tailor protein selectivity. This charge modulation will also affect charge repulsion and alter the 3-dimensional porous structure of the material. While ion exchange materials represent a promising alternative to Protein A resins for mAbs capture, their binding mechanisms and performance are not well understood.

The objective of our work is to better understand protein binding and the effect of charge and buffer environment for macroporous weak cation exchange membrane materials. Specifically we have used the Langmuir and the steric mass action binding models to outline key features of protein binding and comment on the contributions of the buffer co-ions and counter-ions in protein binding for weak cation exchange membrane materials. The benefits and limitations of these binding models will also be discussed. The characterization of the porous structure for macroporous hydrogel materials will be presented and related to protein binding.

Concentrating membrane proteins without concentrating detergent using ultrafiltration

Membrane proteins (MPs) are of rapidly growing interest in the design of pharmaceutical products, novel sensors and synthetic membranes (Kim et al. 2012; Kowalczyk et al. ; Shen et al. 2014). The use of MPs in engineered systems such as biomimetic membranes will require purification scales well beyond what is currently available. The hydrophobic exterior of MPs necessitates their stabilization in detergent micelles. Ultrafiltration using commercially available centrifugal concentrators is typically employed for concentration of low-yield MPs, but its use is accompanied by a concomitant increase in concentration of detergent micelles. The resulting high detergent concentrations can destabilize MPs, interfere with subsequent structural and biophysical characterization of MPs, and present challenges in applications. We will present a detailed analysis of the hydrodynamic processes that control detergent passage during ultrafiltration of MPs and propose methods to optimize detergent passage during protein concentration. Experiments were conducted using nonionic detergents, octyl–β-D glucoside (OG) and decyl-β-D maltoside (DM) with the bacterial water channel protein, Aquaporin Z (AqpZ) and the light driven chloride pump, Halorhodopsin (HR), respectively. The observed sieving coefficient (S_o), a measure of detergent passage, was evaluated in the stirred cell and centrifugal systems. The sieving coefficient for detergent was much greater in the stirred cell systems than in the commercial filters (**Figure 1**). Similar effects

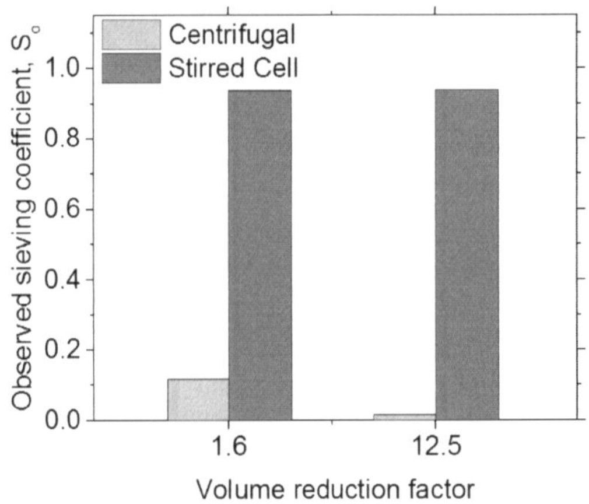

Figure 1. Concentration polarization promotes passage of detergent, DM during ultrafiltration. DM shows higher transmission in stirred cell versus centrifugal filter under equivalent conditions of transmembrane pressure of 11 psi (75.9 kPa) and 14 psi (96.6 kPa) and concentration factors of 1.6 and 12.5 respectively.

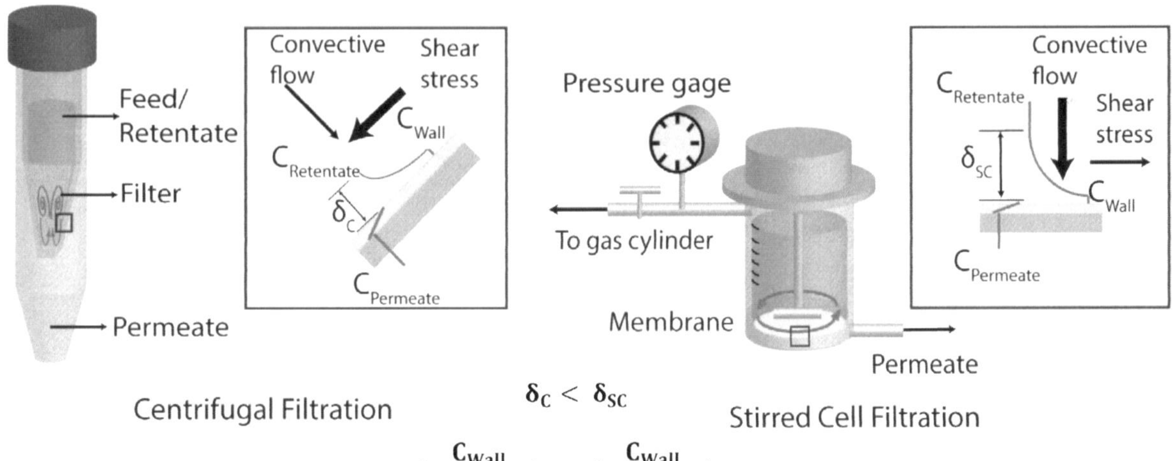

$$\delta_C < \delta_{SC}$$

$$\left(\frac{C_{Wall}}{C_{Retemtate}}\right)_C < \left(\frac{C_{Wall}}{C_{Retemtate}}\right)_{SC}$$

Figure 2. Protein concentration using centrifugal filtration is designed to minimize concentration polarization by maximizing shear stress. This is achieved by inducing circulatory flow by using a specific geometry of the filter. In pressure- driven filtration using a stirred cell, the shear rates can be controlled and thus the extent of concentration polarization can be tuned to study the effect of concentration polarization on detergent passage.

were observed during filtration of protein-detergent micelles. Thus, almost complete transmission of detergent can be achieved by operating under conditions that encourage concentration polarization. This is in complete contrast to current approaches in soluble protein concentration systems, including centrifugal filters, which are designed to maximize flux by minimizing concentration polarization (**Figure 2**). The results in this study clearly indicate the importance of CP in achieving greater detergent passage during ultrafiltration of MP-detergent systems, which has important implications for downstream processing of MPs.

References

Kim Y-R, Jung S, Ryu H, Yoo Y-E, Kim SM, Jeon T-J. 2012. Synthetic biomimetic membranes and their sensor applications. Sensors 12(7):9530-9550.

Kowalczyk SW, Blosser TR, Dekker C. Biomimetic nanopores: learning from and about nature. Trends in Biotechnology 29(12):607-614.

Shen Y, Saboe PO, Sines IT, Erbakan M, Kumar M. 2014. Biomimetic membranes: a review. Journal of Membrane Science 454:359-381.

Protein adsorption and desorption of pH-responsive membrane based on ethylene vinyl alcohol copolymer

Hui Ye, Zhan Xu, Qi Liu, YuZhong Zhang*, Hong Li

State Key Laboratory of Separation Membranes and Membrane Processes, Tianjin Polytechnic University, Tianjin 300387, China

Abstract

The protein adsorption and desorption was investigated for pH-responsive ethylene vinyl alcohol copolymer (EVAL) membrane with the interconnected porous structure. The poly(dimethylaminoethyl methacrylate) (poly(DMAEMA)) chain acted as a pH sensor and the valve was used to regulate the protein adsorption properties of the membrane. The maximum static protein adsorption capacity was obtained at pH 6.4. The dynamic adsorption capacities of membranes EVAL5, EVAL10 and EVAL15 were 58.8, 102.1 and 140.3 mg BSA/g membrane (equivalent to 14.2, 24.6 and 33.8 mg BSA/mL membrane), which was 90-99% of the static adsorption capacity. The dynamic absorption capability of the membrane was not negatively influenced by increasing the flow rate to 3-fold. The protein was more quickly desorbed from the membrane in the elution process at pH 9.0 than that at pH 3.6. Full recovery of protein was observed at pH 9.0. The transition of electrostatic behavior and conformation change of the poly(DMAEMA) chain contributed a pH-responsive protein adsorption and desorption of the membrane. Such pH-responsive EVAL membrane shows good prospects for protein purification.

* Corresponding author's Email: Zhangyz2004cn@163.com

Measurement of Aquaporin Z Compatibility with Biomimetic Membranes Using A Developed Membrane Protein Compatibility Evaluation Method

Tingwei Ren[a], Mustafa Erbakan[a,b], Mariusz Grzelakowski[c], Peter Butler[d], Manish Kumar[a]

a. Department of Chemical Engineering, The Pennsylvania State University, University Park, PA 16802; b. Current address: Department of Biosystem Engineering, Bozok University, Yozgat 66000, Turkey; c. Applied Biomimetic, Cincinnati, OH 45237, USA; d. Department of Biomedical Engineering, The Pennsylvania State University, University Park, PA 16802

Cell membrane matrices are mainly composed of lipids and membrane proteins. Membrane proteins, present in the membrane, performing a vast majority of functions including signal reception[1], material transport and biochemical reactions[2,3], that occur at membrane-liquid interfaces. High compatibility between cell membranes and membrane proteins are necessary for living organisms to finish complicated tasks and thrive in the nature[4]. Inspired by living cell systems, researchers are using lipids and lipid-like membrane matrices as models for functional biomimetic membranes development and the number of these membrane matrices are increasing. However, only compatible biomimetic membranes specific to different membrane proteins will provide desired membrane performance[5]. Highly compatible biomimetic membranes have high membrane protein reconstitution density without compromising membrane protein single molecule properties. However, there are few accurate methods for compatibility evaluation[4].

Here, we propose and test a method to evaluate membrane protein compatibility with biomimetic membranes (shown as **Figure**). We use aquaporin (AQP) reconstituted biomimetic membranes (ABMs) as examples to verify the accuracy of this evaluation method. ABM is a widely accepted model system for novel biomimetic membrane research. Previous researchers found that ABMs have higher water transport rate and better salt rejection performance than traditional reverse osmosis membranes[6,7]. However, AQPs compatibility with different biomimetic membranes hasn't been studied systematically. We combined Stopped Flow Light Scattering technology and Fluorescence Correlation Spectroscopy technology for compatibility evaluation[8]. AQPs reconstitution density and single channel water permeability were accurately measured at the single molecule level. By reconstituting AQPs into different ABMs (both block copolymers and lipids), we show that AQP single channel water permeability stays similar in different membranes. However, reconstitution density varies in different materials, and AQPs have higher reconstitution density in block copolymers. We have also implemented this evaluation method on other membrane proteins and artificial water channels[9] and these results will be summarized as well.

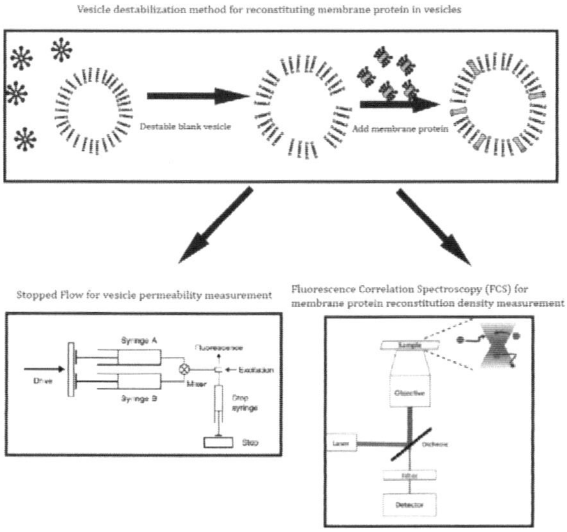

Figure: Illustration of developed method for measuring membrane protein compatibility.

References:

1. Broselid, S. et al. G protein-coupled Receptor 30 (GPR30) Forms a Plasma Membrane Complex with Membrane-associated Guanylate Kinases (MAGUKs) and Protein Kinase A-anchoring Protein 5 (AKAP5) That Constitutively Inhibits cAMP Production. Journal of Biological Chemistry 289, 22117-22127,(2014).
2. Parker, M. D., Myers, E. J. & Schelling, J. R. Na+-H+ exchanger-1 (NHE1) regulation in kidney proximal tubule. Cell. Mol. Life Sci. 72, 2061-2074,(2015)
3. Krall, J. F. & Korenman, S. G. Soluble Precursor of Membrane-Associated Protein-Kinase in Uterine Smooth-Muscle Cells. Arch. Biochem. Biophys. 199, 479-487,(1980).

4. Shen, Y.-x., Saboe, P. O., Sines, I. T., Erbakan, M. & Kumar, M. Biomimetic membranes: A review. Journal of Membrane Science 454, 359-381,(2014)
5. Nichols, S. P., Koh, A., Storm, W. L., Shin, J. H. & Schoenfisch, M. H. Biocompatible Materials for Continuous Glucose Monitoring Devices. Chem. Rev. 113, 2528-2549,(2013)
6. Kumar, M., Grzelakowski, M., Zilles, J., Clark, M. & Meier, W. Highly permeable polymeric membranes based on the incorporation of the functional water channel protein Aquaporin Z. Proceedings of the National Academy of Sciences of the United States of America 104, 20719-20724,(2007)
7. Kowal, J., Zhang, X., Dinu, I. A., Palivan, C. G. & Meier, W. Planar Biomimetic Membranes Based on Amphiphilic Block Copolymers. ACS Macro Letters 3, 59-63,(2014)
8. Erbakan, M., Shen, Y.-x., Butler, P. J., Kumar, M. & Curtis, W. R. Molecular cloning, overexpression and characterization of a novel water channel protein from Rhodobacter sphaeroides. PLoS ONE 9, e86830,(2014)
9. Shen, Y. X. et al. Highly permeable artificial water channels that can self-assemble into two-dimensional arrays. Proceedings of the National Academy of Sciences of the United States of America 112, 9810-9815, (2015).

Functionalization of Silver Nanoparticles On Membranes and Its Influence On Biofouling

Authors: Conor Sprick[1,*], Sneha Chede[1], Vinka Craver[2], Tequila Harris[3], and Isabel C. Escobar[1]
[1]University of Kentucky, USA,
[2]University of Rhode Island, USA
[3]Georgia Institute of Technology, USA

Presenter: Conor Sprick, Conor.Sprick@uky.edu

Abstract:

The aim of this project is modifying the attachment of silver nanoparticles (Ag-NPs) on water treatment membranes, such as cellulose acetate (CA), and to observe its effect towards biofouling. Biofouling results from the accumulation of live/dead microorganisms present in water on the membrane surface and pores, and it creates several performance problems such as clogging of pores, higher operating cost, higher pressure drop, etc. Minimizing this would be ideal to lower operating cost and save expensive materials. In this project, *Pseudomonas Fluorescens Migula* are used because this species generates extracellular polymeric substances (EPS). EPS produced from bacteria helps create a viable structural foundation for biofilm accumulation with densely packed matrices, which in turn fouls the polymeric membranes. In this study, Ag-NPs are added to CA membranes for biofouling minimization. A concern with combining Ag-NP with membranes is the leaching of nanoparticles. If the Ag-NPs are chemically attached to the membranes, it is believed that leaching can be prevented. Different methods of introducing the Ag-NPs to the membranes studied here will include physical blending, chemically-activated blending, and chemically-crosslinked blending.

Nicotine delivery via programmable carbon nanotube membrane transdermal delivery devices

Gaurav Kumar Gulati, Lily Berger, Bruce Jackson Hinds*

Materials Science and Engineering, University of Washington, Seattle, Washington, USA

* Address correspondence to bjhinds@uw.edu

Abstract:

Transdermal nicotine patches currently have low success rates in smoking cessation therapy, but coupling psychological counseling with replacement therapy is expected to dramatically improve cessation rates. This could be achieved with a low-power switchable nicotine delivery device that can automatically couple psychological counseling with dosing. Carbon nanotube (CNT) membranes are energy efficient electroosmotic pumps that can be used to control drug delivery rates through the skin for a variable-rate transdermal dosing system. In particular, this would be applied to nicotine addiction treatment via nicotine replacement therapy (NRT) where dosing rate (between therapeutically used rates) is set through counseling on a smartphone using Bluetooth communications with the membrane device. This system is highly favorable for compact devices due to its low power consumption and ability to be switched by small applied biases.

We evaluated the performance of switchable CNT devices for transdermal nicotine delivery by measuring variable fluxes of nicotine through CNT membranes. CNT membranes were placed over a flow-cell and nicotine levels collected in the cross-flow increased ~6-8 times when the CNT membrane was turned from OFF to ON state by application of -1.0V bias, and ~12-13 times with a -1.5V bias. To evaluate the transdermal performance of the CNT membranes, we developed an *in-vitro* microdialysis method that allows us to detect variable transdermal fluxes of nicotine through the CNT devices and can be applied directly to *in-vivo* studies. Fluxes in the ON state were approximately 3 times that of commercial Nicoderm patches and switching times were less than 2 hours, thus showing improved therapeutic potential of our device.

Title

Dynamic Electrochemical Membranes for Cyclic Protein Transport

Authors

Daniel E Shea, Susrut Akkineni, Bruce J. Hinds

Abstract Text

Recently, a mesoporous anodized alumina oxide membrane has been integrated with electrochemically functionalized nanoscale electrodes for selective and continuous protein separations [1]. This electrochemically functional membrane selectively binds genetically-modified, hexahistidine-tagged proteins using chemistry borrowed from immobilized metal ion affinity chromatography and subsequently pumps the protein across the membrane with alternating, sequential voltage pulses. The electrodes are templated across the mesoporous membrane surface and approximately 20 nm into the pore entrances, providing a convenient geometry for specific target capture while blocking transmembrane transport of non-specific proteins during the binding cycle. During the release cycle, imidazole is electrophoretically pumped towards the bound protein to facilitate release of the protein. Imidazole concentration is carefully controlled to keep the top surface of the membrane pores blocked while facilitating protein release at the bottom edge of the electrode inside the membrane pores. This strategy balanced protein release efficiency to improve selectivity, resulting in a separation factor for GFP:BSA of 16:1 (mass/mass). The existing, unoptimized platform demonstrates throughput of the 0.75 cm^2 area membrane has comparable throughput as 1 mL of commercially available chromatography columns. Our current work reported here focuses on improving selectivity and throughput to improve platform viability by measuring the kinetics of NTA-mediated protein binding and imidazole-based protein release on the modified electrode surface.

[1] "Dynamic Electrochemical Membranes for Continuous affinity protein separation", Z. Chen, T. Chen, X. Sun and B.J. Hinds. Advanced Functional Materials **2014** 24(27) 4317-23

Utrafiltration to Concentrate Membrane Proteins without Concentrating Detergents

Membrane proteins (MPs) are of rapidly growing interest in the design of pharmaceutical products, novel sensors and synthetic membranes (Kim et al. 2012; Kowalczyk et al. ; Shen et al. 2014). The use of MPs in engineered systems such as biomimetic membranes will require purification scales well beyond what is currently available. The hydrophobic exterior of MPs necessitates their stabilization in detergent micelles. Ultrafiltration using commercially available centrifugal concentrators is typically employed for concentration of low-yield MPs, but its use is accompanied by a concomitant increase in concentration of detergent micelles. The resulting high detergent concentrations can destabilize MPs, interfere with subsequent structural and biophysical characterization of MPs, and present challenges in applications. We will present a detailed analysis of the hydrodynamic processes that control detergent passage during ultrafiltration of MPs and propose methods to optimize detergent passage during protein concentration. Experiments were conducted using nonionic detergents, octyl–β-D glucoside (OG) and decyl-β-D maltoside (DM) with the bacterial water channel protein, Aquaporin Z (AqpZ) and the light driven chloride pump, Halorhodopsin (HR), respectively. The observed sieving coefficient (S_o), a measure of detergent passage, was evaluated in the stirred cell and centrifugal systems. The sieving coefficient for detergent was much greater in the stirred cell systems than in the commercial filters (**Figure 1**). Similar effects

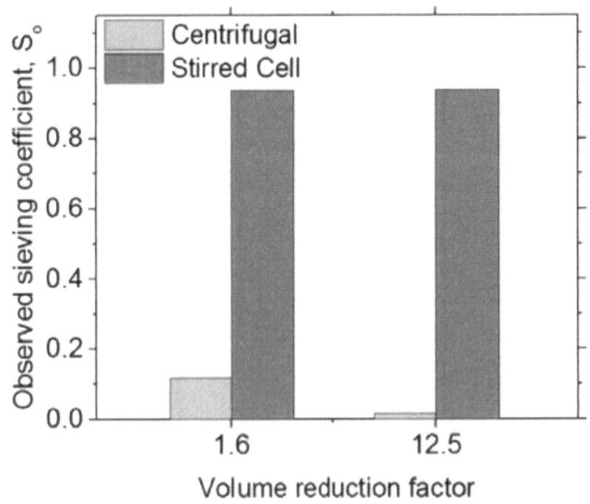

Figure 1. Concentration polarization promotes passage of detergent, DM during ultrafiltration. DM shows higher transmission in stirred cell versus centrifugal filter under equivalent conditions of transmembrane pressure of 11 psi (75.9 kPa) and 14 psi (96.6 kPa) and concentration factors of 1.6 and 12.5 respectively.

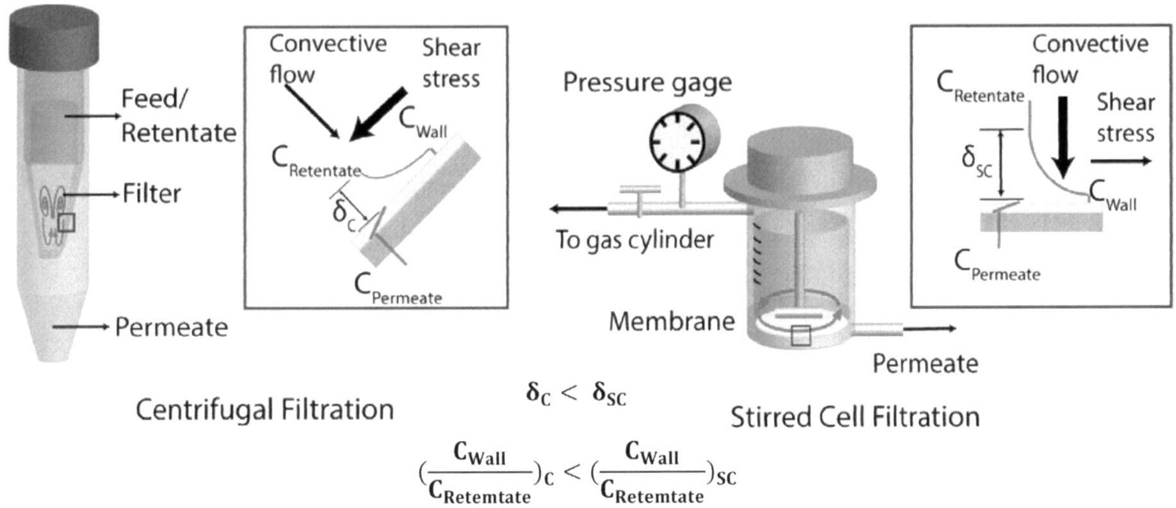

Figure 2. Protein concentration using centrifugal filtration is designed to minimize concentration polarization by maximizing shear stress. This is achieved by inducing circulatory flow by using a specific geometry of the filter. In pressure- driven filtration using a stirred cell, the shear rates can be controlled and thus the extent of concentration polarization can be tuned to study the effect of concentration polarization on detergent passage.

were observed during filtration of protein-detergent micelles. Thus, almost complete transmission of detergent can be achieved by operating under conditions that encourage concentration polarization. This is in complete contrast to current approaches in soluble protein concentration systems, including centrifugal filters, which are designed to maximize flux by minimizing concentration polarization (**Figure 2**). The results in this study clearly indicate the importance of CP in achieving greater detergent passage during ultrafiltration of MP-detergent systems, which has important implications for downstream processing of MPs.

References

Kim Y-R, Jung S, Ryu H, Yoo Y-E, Kim SM, Jeon T-J. 2012. Synthetic biomimetic membranes and their sensor applications. Sensors 12(7):9530-9550.

Kowalczyk SW, Blosser TR, Dekker C. Biomimetic nanopores: learning from and about nature. Trends in Biotechnology 29(12):607-614.

Shen Y, Saboe PO, Sines IT, Erbakan M, Kumar M. 2014. Biomimetic membranes: a review. Journal of Membrane Science 454:359-381.

Fabrication and characterization of Polybenzimidazole-multiwall carbon nanotubes composite membrane for gas separation

Moon Ki Jeong, Sang Yong Nam*

Department of Materials Engineering and Convergence Technology, Engineering Research Institute, Gyeongsang National University, Jinju 660-701 Korea

*walden@gnu.ac.kr

High temperature gas separation techniques are of great interest for reduction in green-house gas emissions from hydrocarbon fuels such as natural gas, coal or biomass used in power and chemical industry. The material used in this study is polybenzimidazole well known commercially as PBI. It has exceptional thermal and chemical stability and does not readily ignite. PBI's excellent chemical, thermal and physical properties demonstrate that it can be a promising material as a flue gas filter fabric for coal fired boilers. for this reason, we selected this polymer to reduction in green-house gas emissions. Meanwhile, carbon nano tube has enormous mechanical and electrical property. So, it can be applied various field such as nanotechnology, electrical engineering, optical science and material engineering. PBI-CNT composit material is expected to have excellent physical properties and good proton conductivity. In this respect, PBI-CNT is anticipated being able to promising material in high temperature gas separation process. Membrane was obtained through well dispersed PBI-CNT solution. morphology of PBI-CNT membrane was observed by scanning electron microscope(SEM). Permeablity and selelctivity of this membrane about H_2, CO_2 was measured through pure gas.

Fabrication of a superhydrophilic PVDF based membrane using a facile prefabricated TiO_2 nanoparticles surface

Hassan Younas[a,1], **Jiahui Shao**[a,2*], **and Yiliang He**[a]

[a] *School of Environmental Science and Engineering, Shanghai Jiao Tong University, 800 Dongchuan road, Shanghai, 200240, PR China.*
[1] *Presenting author: younasseven@yahoo.com*
[2*] *Corresponding author: jhshao@sjtu.edu.cn*

Abstract

The hydrophilicity of PVDF membrane is playing an enormously important role in its widespread water treatment fields considering the excellent intrinsic properties of PVDF raw materials. Rather than the conventional surface modification or hybridization, herein, we report a novel approach to prepare highly hydrophilic PVDF UF membrane by creating a prefabricated super-hydrophilic surface of inorganic TiO_2 nanoparticles (NPs). The resultant membrane [prepared by prefabrication surface adhesion of TiO_2 NPs on PVDF-PEG-TiO_2 hybrid membrane (SaT-PPT)] has a uniform distribution of TiO_2 NPs not only on the membrane surface but also within membrane matrix, this will maximize the super-hydrophilic feature throughout the membrane: from inner porous structures to outer surfaces, and will avoid "dead-corner" to block fast water pass through. The performances of SaT-PPT membrane as compared with the control membranes (PVDF-PEG, and PVDF-PEG with TiO_2 anchored only at surface) were investigated in terms of humic acid (HA) rejection, flux and flux decline in lab-made cross flow UF experiments with and without UV irradiation. It indicated that SaT-PPT membrane exhibited the highest hydrophilicity and flux, lowest flux decline and total resistance, but still with the highest HA rejection rate. Moreover, it was found that the presence of TiO2 NPs on the membrane surface increased the surface energy of the membrane (i.e. 51 mJ m^{-2}) that increased the wettability potential of the membrane. In addition, SaT-PPT membrane showed the highest flux recovery after simple physical cleaning to extend the longer life span of membrane. It is reasonable to believe that our developed SaT-PPT membranes will provide insightful engineering practices to benefit the broad water treatment applications.

Acknowledgement

This research was supported by National Natural Science Foundation of China (21277090) and National Key Science & Technology Project of Water Body Pollution Control and Reclamation (2014ZX07206001).

The Potential to Enhance Membrane Module Design with 3D Printing Technology

Jian-Yuan Lee[a,b,c,*], **Wen See Tan**[a,b,c], **Jia An**[c], **Chee Kai Chua**[c], **Chuyang Y. Tang**[d], **Anthony G. Fane**[b,e], **Tzyy Haur Chong**[b,e]

[a] *Nanyang Environment & Water Research Institute, Interdisciplinary Graduate School, Nanyang Technological University, Singapore 639798*

[b] *Singapore Membrane Technology Centre, Nanyang Environment & Water Research Institute, Nanyang Technological University, Singapore 637141*

[c] *Singapore Centre for 3D Printing, School of Mechanical and Aerospace Engineering, Nanyang Technological University, 50 Nanyang Avenue, Singapore 639798*

[d] *Department of Civil Engineering, the University of Hong Kong, Pokfulam Road, Hong Kong*

[e] *School of Civil and Environmental Engineering, Nanyang Technological University, Singapore 639798*

Highlights

1. The design of membrane modules via 3D printing technology will be discussed.
2. Resolution and materials play important roles during the selection of a 3D printer.
3. Considerations in designing 3D-printed membrane modules will be presented.

Abstract

3D printing is an emerging technology and has attracted massive attention in recent years. This article focuses on the recent developments on enhancing the membrane module design with 3D printing technology. With the recent advancement of 3D printing technology, breakthroughs in fabricating novel membrane module components are expected in the near future. Improvement of 3D printing technologies in terms of resolution, materials and speed should assure the production of various membrane module components with high efficiency.

References

1) Lee, J.-Y.; Tan, W. S.; An, J.; Chua, C. K.; Tang, C. Y.; Fane, A.G.; Chong, T. H. *The Potential to Enhance Membrane Module Design with 3D Printing Technology J. Membr. Sci.* **2016**, *499*, 480.

* Presenting author's e-mail address: leej0127@e.ntu.edu.sg, phone number: (+65) 9715 2529.

Theoretical Study of Fluid Flow through Screened Channel in a Tangential Flow Filtration Process

Armin Delavari, Yu Zou

ABSTRACT. Design and optimization of feed screens in filtration devices assembled of layers of membranes and screens as turbulence promoters is of great importance in filtration process. In this study, performance of different screens were modeled using computational fluid dynamics in order to understand the impact of screen characteristic parameters. The screens are all made of polypropylene fibers weaved in form of warp and weft in which the feed solution passes through the free spaces between the fibers. The effect of membrane as a permeable wall was also considered in the computational models using the Brinkman equation. The radius of the fibers, the distance between fibers and different weave pattern of fibers were varied to examine their impact on the average magnitude of the flow velocity, pressure drop through the channel length and the shear rate on the membrane surface and on the fibers surface. Also, the impact of compression of layers on the performance of the feed screens was investigated using microscopic analysis along with the computational calculations. The results enable us to optimize the feed screens according to our purpose of manipulating the shear rate and the pressure drop.

Keywords: Filtration feed screens, CFD, Ultrafiltration, Optimization

Session topic:
Micro, ultra, and nano-filtration applications

Title of contribution:
Weak polyelectrolyte multilayers as tunable membranes for solvent resistant nanofiltration (SRNF)

Shazia Ilyas [a,b], Nithya Joseph [b], Kitty Nijmeijer [a], Ivo F.J. Vankelecom [b], Wiebe M. de Vos [a]

[a] Membrane Science and Technology, MESA+ Institute for Nanotechnology, University of Twente, Faculty of Science and Technology, P.O. Box 217, 7500 AE Enschede, The Netherlands
[b] Centre for Surface Chemistry and Catalysis, Faculty of Bioengineering Sciences, Katholieke Universiteit Leuven, Celestijnenlaan 200F - box 2461, 3001 Leuven, Belgium

Abstract
One of the key current challenges for nanofiltration (NF) is to broaden the range of applications from aqueous to organic feeds in the form of solvent-resistant nanofiltration (SRNF) membranes. As SRNF does not involve an energy demanding phase transition, it has drawn considerable attention as an alternative separation technique to alleviate the energy costs and related environmental problems caused by conventional separation methods such as distillation. SRNF has potential applications in strategic fields such as the petrochemical industry, pharmaceutical industry and food industry, to separate desired molecules from solvents and/or to recover solvents and solutes from waste streams._ENREF_4 Unfortunately, there is lack of membranes that are thermally and chemically stable, that have a high permeability and that provide effective separation of molecules in the range of 200–1000 g·mol^{-1} in various organic solvents over long time-scales. Polymers are an interesting material for SRNF related applications but very few polymeric membranes are stable: swelling and/or dissolution of the polymeric matrix often results in a loss of membrane selectivity. This work encompasses the investigation into the SRNF performance of polyelectrolyte multilayer (PEM) based SRNF membranes prepared from weak polyelectrolytes. These weak polyelectrolytes are unique in that the charge density is not fixed and depends on the coating pH, adding an extra variable as tuning parameter for SNRF performance. The weak polyelectrolyte multilayers were prepared on a charged support from poly(allylamine hydrochloride) (PAH) as polycation and poly(acrylic acid) (PAA) as polyanion. Detailed investigations into the role of the pH of the coating solution on the performance of the prepared SRNF-membranes were carried out with organic dyes of different size (~300 – 1000 Dalton) and charge. Variation in pH of the coating solutions was found to lead to a large degree of control over the separation performance of the prepared SRNF-membranes for the different dyes. The solvent permeabilities and the dye retentions were measured and correlated to variations in the PEM membrane structures, with more dye adsorption being found for membranes with more free acid and amine groups. The membranes were also found to be stable for long term-filtrations in solvents such as isopropyl alcohol (IPA), acetonitrile (ACN), tetrahydrofuran (THF) and in the challenging polar aprotic solvent, N,N-dimethylformamide (DMF). Results of this study demonstrate the potential of using pH as tuning parameter for weak PEMs to prepare SNRF-membranes optimized for specific industrial applications.

Keywords: *Nanofiltration, Weak polyelectrolyte multilayers, SRNF, Layer by layer assembly, Polyelectrolyte multilayer membranes, Organic solvents*

Study on Insoluble Polyimide Solvent-resistant Nanofiltration Membranes Fabricated by a Two-step Method

Pei Li*, Yuan Li, Jingjing Xue, Bing Cao

Beijing University of Chemical Technology, Beijing, 100029, China.

Abstract

Two insoluble polyimides, pyromellitic dianhydride oxydianiline-4,4'-oxydianiline (PMDA-ODA) and pyromellitic dianhydride oxydianiline-4,4'-methylenedianiline (PMDA-MDA), have been utilized to form nanofiltration membranes using a two-step method. Specifically, the polyimides' precursor, polyamic acid(PAA), was used to form integrated asymmetric membranes via the non-solvent induced phase separation method, and then the PAA membrane was converted to PI membrane via thermal or chemical induced imidization. We systematically studied the effect of molecular weight of the polyamic acid (PAA) precursors to the skin layer formation and separation performance. It is believed that the polyimide membranes with better performances on chemical and thermal stability are achieved by higher a molecular weight of PAA. We also investigated the method of heating induced imidization, and found that by heating the membrane from 2 to 2.5h at a heat temperature ranged from 200 to 300℃ in vacuum resulted in the rejections to Crystal Violet and Eriochrome Blue Black R up to 99.8% and 97.9% with a relatively high level fluxes of some common organic solvents. Additionally, a gelation process is taken to prepare PAA precursors before phase inversion, while the chemical imidization method is used for later process instead. The PAA membranes after gelation process using NMP as solvent show the largest flux of 98 L • m^{-2}• h^{-1} and the highest rejection of 98% to Bengal Rose aqueous solution (10mg • L^{-1}). However, the chemical imidization may result in the reduction on performances of PI membrane. Subsequently, the non-waven fabrics are served as substrates for casting process to stabilize the performances, it is found that the flux of pure water is steady at 50.3-53.4 L • m^{-2} • h^{-1}, while the flux and the rejection maintains at 37-42 L • m^{-2} • h^{-1} and 99% in Bengal Rose aqueous solution (100ppm), respectively.

***Keywords*:** solvent-resistant nanofiltration; insoluble polyimide; a two-step method

Permeation-deformation relationships for porous polymer membranes

Yifu Ding*, Sajjad H. Maruf, Masoud Aghajani, and Alan R. Greenberg

Membrane Science, Engineering and Technology Center
Department of Mechanical Engineering
University of Colorado Boulder, Boulder, CO 80309-0427, USA

Abstract

Most polymeric membranes are necessarily fabricated with high overall porosity to enable large flux values for desired penetrants. However, high porosity inevitably reduces both the stiffness and yield strength of the membranes, which causes meaningful mechanical deformation during manufacturing operations such as pleating as well as during separation applications that involve elevated pressures (compressive creep). The deformation resulting from mechanical loading, whether intended or not, has a significant impact on membrane permeability. To date, permeation-deformation relationships for porous membranes are not well understood, despite their obvious importance. Recently, we utilized thermal embossing-based nanoimprint lithography (NIL) to impart regular sub-micron patterns on the surface of membranes to improve their antifouling characteristics. This presentation describes experimental studies of the permeation-deformation relationships for these patterned polymer membranes as a function of polymer chemistry and pore structure. Uniaxial compression of the membranes produced decidedly different levels of strain by careful control of the NIL processing parameters. The permeability of these compressed membranes was characterized using DI water to avoid complications due to fouling. The results indicate that DI water permeation values for ultrafiltration (UF) and microfiltration (MF) membranes decrease with increasing compressive strain. The upper limit of the dependency appears to well match the relationship observed for open-cell polymer foams. Furthermore, the potential effects of pore-preserving agents as well as the asymmetry of the pore structure on the permeation-deformation relationships are quantified. These findings, together with those from our previous work on membrane mechanical behavior, provide both a more comprehensive understanding of important mechanical aspects and practical guidance for applications such as surface patterning of membranes with NIL.

* Corresponding author at :Department of Mechanical Engineering, 427 UCB, University of Colorado, Boulder, CO 80309-0427; Tel.: +1 303 492 2036; E-mail address: Yifu.Ding@Colorado.Edu.

Phase Inversion Kinetics of Polymer-Ionic Liquid Solutions in Relation with Membrane Morphology and Macrovoid Formation

Elif Nur Durmaz, P. Zeynep Çulfaz-Emecen

Middle East Technical University, Chemical Engineering Department, Ankara, Turkey

Cellulose is an abundant bio-polymer and it is stable against dissolution in many solvents due to its inter- and intra- molecular hydrogen bonds. Ionic liquids can dissolve cellulose and allow membrane fabrication via phase inversion. On the other hand, it is known that morphology of membranes made by using ionic liquid solvents can be quite different than those made using conventional solvents such as polar aprotic solvents [1].

The aim of this study is to understand the relationship between phase inversion kinetics of three different polymers in their ionic liquid solutions and the morphology of resultant membranes. Eight different solutions with different polymers and solvents were chosen to be studied and these solutions were coagulated either in water or in ethanol which were chosen as two readily available solvents of different nonsolvent strength. 1-ethyl-3-methylimidazolium acetate ([EMIM]OAc) is the primary solvent for all solutions and dimethyl sulfoxide (DMSO) is the co-solvent (Table 1). To estimate solvent or nonsolvent strength Hansen solubility parameters were used and relative energy difference (RED) values were calculated in order to get better understanding of affinity between two components [2]. RED values of less and more than 1.0 represent high and low affinity, respectively. Therefore, for a good solvent RED value is expected to be less than 1.0 and similarly for a good nonsolvent, it is expected to be much higher. RED value is found by dividing the solubility parameter distance value, Ra to radius of interaction, Ro of the solute; which is the limit of solubility parameter distance, above which solvent does not dissolve the polymer [2]. In Table-1 RED values of polymer-solvent, polymer-water and polymer-ethanol is given. Since Ro values are not available for solvent-nonsolvent pairs, to assess the interaction between solvents and nonsolvent solubility parameter distance (Ra) values are used.

Table 1: Composition and relevant RED values for polymer solutions used [2,3]

Solution Code	Polymer	Solvent	$RED_{P/S}$	$RED_{P/W}$	$RED_{P/E}$	Ra_{S-W}	Ra_{S-E}
A	8.0% Cellulose	92.0% [EMIM]OAc	0.922	1.027	1.024	9.44	7.52
B	8.0% Cellulose	46.0% DMSO 46.0% [EMIM]OAc	0.921	1.027	1.024	9.29	10.16
C	8.0% Cellulose	61.3 % DMSO 30.7% [EMIM]OAc	0.924	1.027	1.024	9.42	11.10
D	10.6% Cellulose	59.6% DMSO 29.8% [EMIM]OAc	0.924	1.027	1.024	9.42	11.10
E	8.0% Cellulose Acetate	92.0% [EMIM]OAc	0.523	0.999	1.347	9.44	7.52
F	8.0% Cellulose Acetate	46.0% DMSO 46.0% [EMIM]OAc	0.244	0.999	1.347	9.29	10.16
G	8.0% Polyether Sulfone (PES)	92.0% [EMIM]OAc	1.066	2.427	2.077	9.44	7.52
H	8.0% Polyether Sulfone (PES)	46.0% DMSO 46.0% [EMIM]OAc	0.930	2.427	2.077	9.29	10.16

All membranes were prepared via immersion precipitation and to measure phase inversion rate, light transmission through the solution during coagulation was measured. The light transmission data were normalized in the form of I/I_0 where I_0 is the initial value for illuminance before the polymer solution is immersed and the I is the value at any time after the immersion. The magnitude of the slope of normalized light transmission (I/I_0) against time graphs were expressed as the phase inversion rate (Figure 1).

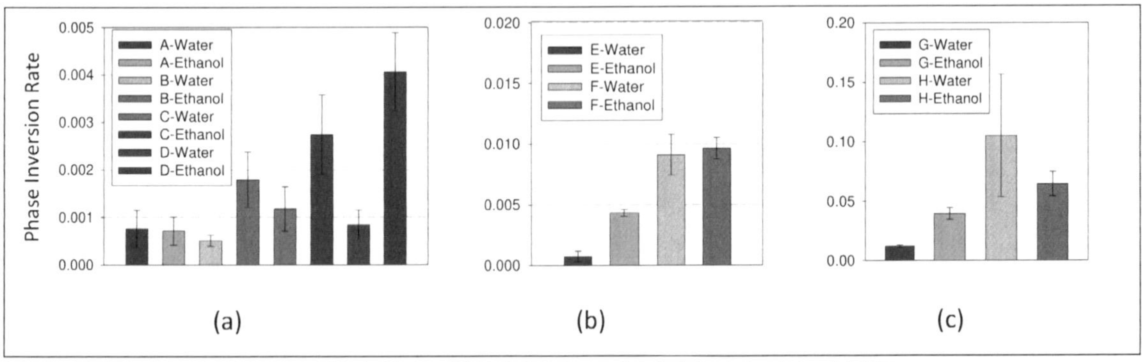

Figure 1: Phase inversion rates of solutions with (a) Cellulose, (b) Cellulose Acetate, (c) PES

Phase inversion rates for cellulose solutions coagulated in water are very close to each other (Figure 1-a). SEM images of these membranes indicate that morphologies are nodular and quite dense. Bovine Serum Albumin (BSA) rejections of these membranes are all measured to be above 90%. Although there is a small increase in phase separation rate for the solutions coagulated in ethanol when DMSO is used as cosolvent, SEM images show that the morphologies are still similar to each other and to solutions coagulated in water (Figure 2). The nodular morphologies are typically attributed to low diffusivities of nonsolvent in solvent due to the high viscosity of ionic liquids compared to conventional polymer solvents [4].

On the other hand, there is a considerable increase in phase inversion rate of cellulose acetate solutions when compared to cellulose solutions in both nonsolvents (Figure 1-b). For *Solution E* which is 8% cellulose acetate in [EMIM]OAc, SEM images show a nodular morphology with large circular macrovoids near the bottom of the membrane facing the substrate during coagulation in water. Although the phase inversion rate is very similar to *Solution A* coagulated in water, the presence of macrovoids in *Solution E*, may be attributed to difference in non-solvent quality of water for cellulose and cellulose acetate. When DMSO was added as cosolvent, and water was the coagulant, a typical asymmetric membrane morphology with a selective layer and macrovoids was observed. For cellulose acetate solutions coagulated in ethanol, phase inversion begins after a time delay of approximately 1 minute. When *Solution E* is considered there is a clear increase in phase inversion rate when compared to the solution coagulated in water. This may be attributed to the nonsolvent strength of ethanol being higher than water as RED values indicate, assuming diffusivities of ethanol and water in [EMIM]OAc are similar. On the other hand, phase inversion rates are similar for *Solution F* for both nonsolvents. Resultant membranes of cellulose acetate solutions coagulated in ethanol could not be broken with liquid nitrogen disabling to examine cross-section morphology via SEM. However, surface images indicate that those membranes are macroporous with lacy and open structure.

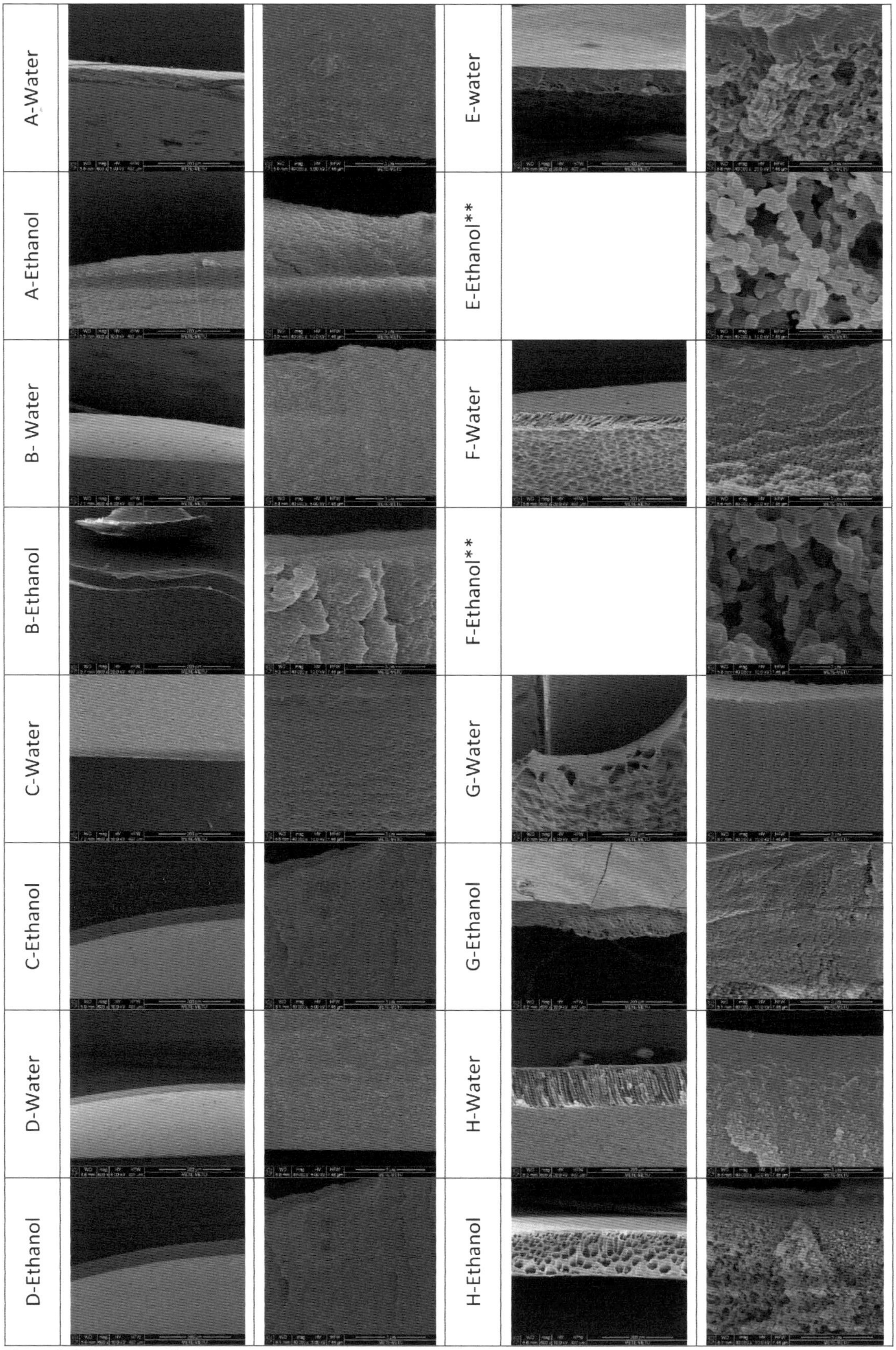

Figure 2: SEM images of membranes

*** SEM image of top surface, not the cross section; because those membranes could not be broken with liquid nitrogen.*

Phase inversion rate of PES solutions was the fastest when compared to other polymer solutions. Coagulation in ethanol was faster than in water in PES-[EMIM]OAc solutions as with Cellulose Acetate-[EMIM]OAc solutions. The rate is increased when DMSO is added to the solution; and, water and ethanol result in similar phase inversion rates and morphologies with selective layer and macrovoids. (Figure 1-c). SEM images of the top of the membranes show that coagulation in ethanol results in a looser skin layer for PES solutions, in both solvents.

In general, it was concluded that the phase inversion rate of solutions was faster in ethanol, with the exception of solution A. Phase inversion rates of Solution A in water and ethanol is very close to each other and RED values of these nonsolvents with cellulose indicate that the nonsolvent strengths are also very similar. Among the three polymers, the rate increased in the order of Cellulose < Cellulose Acetate < PES, which in general again agrees with increasing nonsolvent strength for the polymer. All cellulose solutions with or without DMSO as cosolvent resulted in membranes with nodular morphology with no macrovoids. For cellulose acetate and PES, on the other hand, addition of DMSO resulted in typical asymmetric membrane morphologies, whereas the solutions in [EMIM]OAc only resulted in nodular morphologies with large circular macrovoids only near the bottom of the membranes.

REFERENCES

[1] Xing, D., Peng, N. and Chung, T. (2010). Formation of Cellulose Acetate Membranes via Phase Inversion Using Ionic Liquid, [BMIM]SCN, As the Solvent. *Industrial & Engineering Chemistry Research*, 49(18), pp.8761-8769.

[2] Hansen, C. (2007). *Hansen solubility parameters*. 2nd ed. Boca Raton: CRC Press.

[3] Weerachanchai, P., Wong, Y., Lim, K., Tan, T. and Lee, J. (2014). Determination of Solubility Parameters of Ionic Liquids and Ionic Liquid/Solvent Mixtures from Intrinsic Viscosity. *ChemPhysChem*, 15(16), pp.3580-3591.

[4] Xing, D., Chan, S. and Chung, T. (2013). Fabrication of porous and interconnected PBI/P84 ultrafiltration membranes using [EMIM]OAc as the green solvent. *Chemical Engineering Science*, 87, pp.194-203.

Cellulose membranes for Organic Solvent Nanofiltration

Faqih Muhamad Sukma, P. Zeynep Culfaz-Emecen
Middle East Technical University, Chemical Engineering Department, Ankara, Turkey

Being a young yet promising technology, Organic Solvent Nanofiltration (OSN) has many potentials for different applications such as homogeneous catalyst recycling, recovery of organic solvents in food, painting, and pharmaceutical industries by allowing separations of organic mixtures by simply applying a pressure gradient over a membrane. However, limited number of suitable polymers that can withstand a wide range of organic solvents are available. Polyimide, among those is the polymer mostly studied [1]. Cellulose offers an alternative for this field of study.

Although abundant in nature, use of cellulose as a membrane polymer is hindered by the lack of suitable nonderivatizing solvent. Common solvent systems like NaOH/CS_2 or NMMO are either toxic or unstable. Recent studies showed that there is an alternative way of dissolving cellulose using ionic liquids [2].

The aim of this study is to process cellulose via phase inversion using ionic liquids as solvents to fabricate membranes for Organic Solvent Nanofiltration. 1-ethyl-3-methylimidazolium ([EMIM]OAc) is used as the ionic liquid to dissolve cellulose, while acetone is used as cosolvent and water as nonsolvent. Membrane percursor solutions and membrane types used in this study are listed in Table 1. For solution containing acetone, a cosolvent evaporation step was used to increase the cellulose concentration in solution before precipitation in water.

Table 1. Composition of cellulose solutions used to fabricate membranes

Membrane	Membrane percursor solution (%wt)	Cosolvent evaporation before coagulation
A	8% Cellulose – 92% [EMIM]OAc	No
B	8% Cellulose – 20% Acetone – 72% [EMIM]OAc	No
C	8% Cellulose – 20% Acetone – 72% [EMIM]OAc	Yes
D	12% Cellulose – 25% Acetone – 63% [EMIM]OAc	No
E	12% Cellulose – 25% Acetone – 63% [EMIM]OAc	Yes
F	20% Cellulose – 80% [EMIM]OAc	No

To study the retention of membranes, Bromothymol Blue (624 Da) and Brilliant Blue R (826 Da) solutions in ethanol are used. Besides, different solute-solvent system based on water are studied using Brilliant Blue R (826 Da), Blue Dextran (5 kDa), and Blue Dextran (20 kDa) as the probes.

It is observed that increasing cellulose concentration in the membrane precursor solution, either by having more cellulose in the starting solution or by evaporating the acetone prior to coagulation, decreases membrane permeance and increases rejection of Bromothymol Blue as shown in Figure 1. Drying post-treatment of membranes further increased the retention (Figure 1).

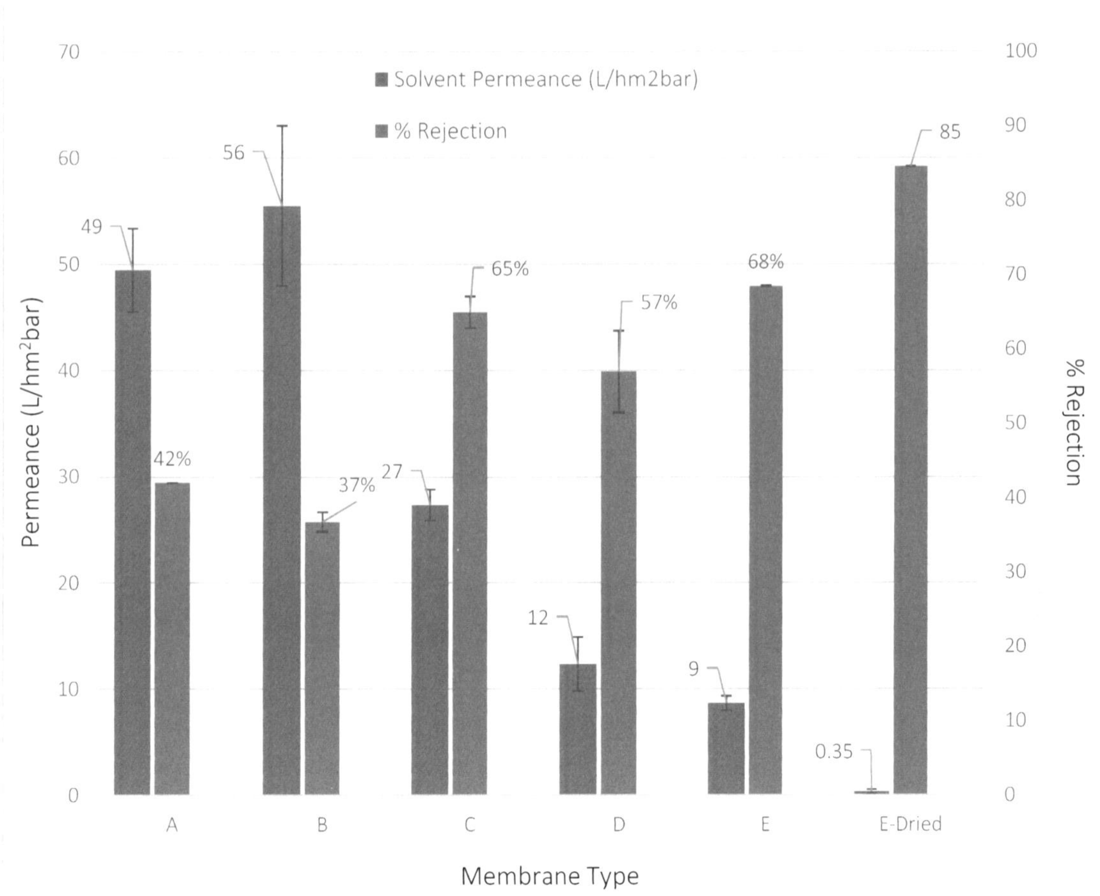

Figure 1. Bromothymol Blue rejection performance in five different non-dried membranes

It is noticed that solute-solvent-membrane interaction may hold an important role in determining the retention performance as a higher molecular weight dye (Brilliant Blue R, 826 Da) is retained less than Bromothymol Blue (624 Da) dissolved in ethanol, which may be related to the interaction of the ionizable groups in Brilliant Blue R with the membrane as shown in Figure 2. Change of solvent also affects the rejection as shown by higher value of Brilliant Blue R rejection in water compared to in ethanol, probably due to the presence of a hydration shell around Brilliant Blue R in aqueous solution and its ionized groups (Figure 2). Based on Blue Dextran retention, membrane E has MWCO between 5 and 20 kDa and membrane F lower than 5 kDa. Brilliant Blue R rejection in ethanol and water show differences due to solute-solvent, membrane-solvent, and membrane-solute interactions (Figure 2 & 3). Further increase of cellulose concentration to 20% combined with drying post-treatment increased Bromothymol Blue (in ethanol) rejection to 94%, Brilliant Blue R (in ethanol) rejection to 40%, and Blue Dextran 5 kDa (in water) rejection to 94% (Figure 3).

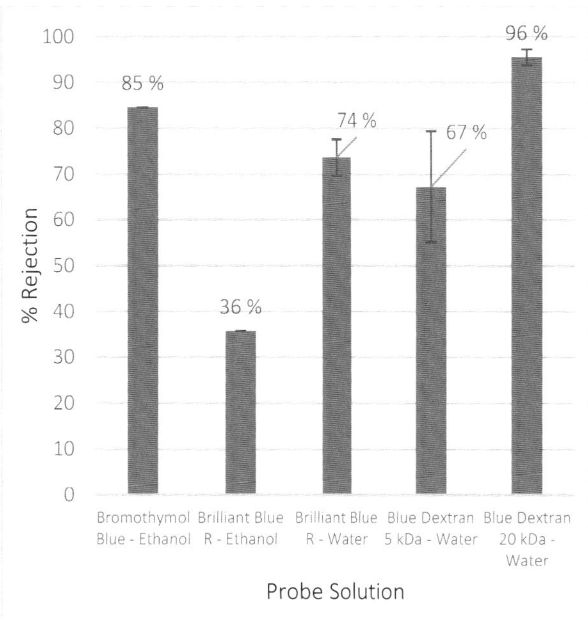

Figure 2. Rejection value on membrane E by using five different solute-solvent systems

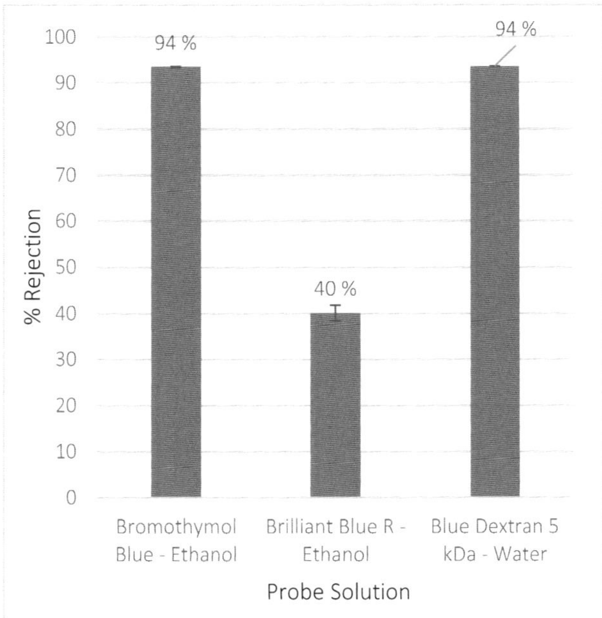

Figure 3. Rejection value on membrane F by using three different solute-solvent systems

References

[1] S. Hermans, E. Dom, H. Mariën, G. Koeckelberghs, and I. F. J. Vankelecom, "Efficient synthesis of interfacially polymerized membranes for solvent resistant nanofiltration," vol. 476, pp. 356–363, 2015.

[2] J. Vitz, T. Erdmenger, C. Haensch, and U. S. Schubert, "Extended dissolution studies of cellulose in imidazolium based ionic liquids," *Green Chem.*, vol. 11, no. 3, p. 417, 2009.

Carbon nanotubes/polyethersulfone (CNT/PES) nanocomposite membranes fabrication and characterization

Vildan Cansiz Selcuk[1], Derya Karsi[2], Vedat Uyak[1*], Ismail Koyuncu[2]

[1]Department of Environ. Engineering, Istanbul Technical University, 34469 Maslak, Istanbul, Turkey
[2]Department of Environmental Engineering, Pamukkale University, 20020 Kinikli, Denizli, Turkey
*corresponding author: vuyak@pau.edu.tr

Multi-walled carbon nanotubes/polyethersulfone (CNT/PES) blend membranes were prepared by a phase inversion process, using N-methyl-2-pyrrolidinone (NMP) as a solvent and water as a coagulant. The prepared CNT/PES blend membranes were then characterized using the several analytical methods such as a Fourier transform infrared (FTIR) spectroscopy, a contact angle goniometer, a scanning electron microscopy (SEM) and permeation tests. Because of the hydrophic CNTs, the surface of the CNT/PES nanocomposite membranes appeared to be more hydrophilic than a just PES membrane. The carboxylic acid functional groups on the surface of CNTs seemed to act to increase hydrophilicity of the CNT/PES membranes. The morphology and permeation properties of the CNT/PES nanocomposite membranes were also found to be dependent on the amounts of CNTs used. The pore size of the nanocomposite membranes increased along with the contents of CNTs up to 1.0%, then decreased, and at 4.0% of CNTs, it showed smaller increase. The CNT/PES membrane with 1.0% of CNT content showed higher flux (434 $L/m^{2}*h*bar$) and rejection than the pristine PES membrane.

Key words: Carbon nanotubes, Polyethersulfone, Nanocomposite membrane, FTIR

Charged nanopore membranes by amphiphilic random copolymer micelle assembly

Membranes that can separate solutes based not only on their size but also their chemical features (e.g. charge, hydrophobicity, specific functional groups) can potentially broaden the applications where membranes are used, replacing less energy-efficient methods such as chromatography and extraction. However, finding an easy and scalable approach which confers the membrane with high permeability, selectivity, and systematic tunability is still a great challenge. Integration of chemical functionalities into membrane nanopores requires engineering the pores and functionalizing them with specific groups to induce desired chemical interactions with the target molecule. Biological pores are the best representative of this specific interaction. For example, ion channels feature pores slightly larger than the target solute, mostly lined with hydrophobic residues but also featuring selectivity regions with charged functional groups. These characteristics bring about superior selectivity and permeability for target solute in these pores.

Inspired by these biological channels, we designed a new and easy approach for charge-based separation of sub-nanometer size molecules, with potential for further functionalization. We prepared an amphiphilic random copolymer that combines highly hydrophobic, fluorinated repeat units of 2,2,2-trifluroethyl methacrylate (TFEMA) with charged, hydrophilic repeat units of methacrylic acid (MAA), PTFEMA-r-PMAA, by free radical polymerization. The copolymer consists of 45% MAA and is insoluble in water. We found that this copolymer forms micelles in methanol. When a solution of this copolymer containing micelles are coated onto the surface of a porous support membrane a thin selective layer is formed. Using this method, we achieved membranes with pure water permeabilities up to ~6 L/m^2.h.bar by tuning the copolymer composition. The resultant membranes exhibited rejection properties expected from negatively charged membranes with ~2 nm pore size, as indicated by the relative rejections of monovalent and divalent salts as well as polyethylene glycol (PEG) with different molecular weights. The membrane showed charge-based selectivity between organic dye molecules, with a negatively charged dye being retained by >85% whereas a positively charged dye with a similar molecular size as retained by <30%. This is the first demonstration of a new approach to forming membrane selective layers by the self-assembly of easily synthesized random copolymers with highly incompatible repeat units. We believe this approach can be adapted to address various pore sizes and pore surface chemistries, leading to novel membranes with promising selectivity.

The Effect of Pore Penetration on Composite Membranes

Pingjiao Hao*, J.G. Wijmans, Timothy C. Merkel, Jenny He, Lloyd S. White

Membrane Technology and Research, Inc.,
39630 Eureka Drive, Newark, CA 94560, U.S.A.

Composite membranes typically consist of at least one nonporous layer on top of a porous structure. The porous support restricts diffusion in the top layer because molecules can exit the layer only where a pore is present. Computational fluid dynamics (CFD) has been used to simulate the geometric restriction and an empirical model was developed to quantify the effect of the porous support on the permeance of composite membranes.

In this work, modeling results are compared to experimental data for composite membranes consisting of a dense polydimethylsiloxane (PDMS) layer on a polyetherimide (PEI) ultrafiltration support. Significant discrepancies are observed between the experimental and modeling results. Pore penetration is believed to be a likely reason for the difference between the theoretical and measured permeances. A resistance-in-series model and CFD were employed to calculate the effect of pore penetration on the permeance of composite membranes. Results from these two approaches are in good agreement, and provide a good fit to experimental composite membrane data if pore penetration depth is allowed to vary. An analysis of the PDMS/PEI composite membrane data, suggests that the depth of PDMS penetration into the PEI pores increases as the thickness of the PDMS selective layer increases.

A Study of the Effect of Pore Geometry on the Rejection of Nano Particles from Porous Membranes

Armin Delavari, Ruth Baltus

Department of Chemical and Biomolecular Engineering, Clarkson University, Potsdam, NY

ABSTRACT. The study of particle motion in porous membranes requires an understanding of the impact of fluid characteristics and membrane geometrical properties on motion. The complexity of such media demands a detailed examination of the effect of membrane structure which has not been previously investigated. In this study, the motion of spherical and non-spherical particles in pore channels were modeled and the impact of pore entrance and pore bends were analyzed. A finite element method was employed to evaluate the hydrodynamic forces on particle boundaries using the Navier-Stokes equation. Steric exclusion and the rotational forces for non-spherical particle were included in the models.

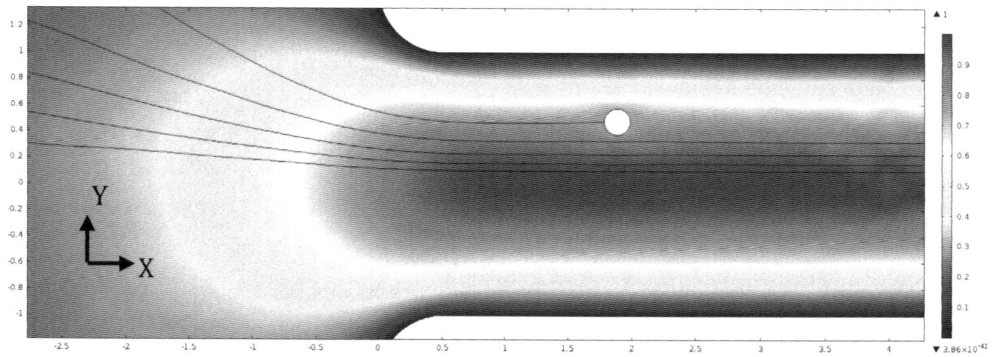

Figure 1: Particle trajectories for different initial positions in the reservoir at $X = -5$ and $Y = 5$ to 1

Several representative results are presented here. The position of a particle was tracked as it traveled from a reservoir into a pore (Figure 1). The results including the 2-D components of velocity, drag force and position were collected and used to statistically analyze the hydrodynamic effect of the pore (Figure 2). The results indicated that the particles are concentrated near the pore centerline and do sample all the available positions inside the pore unlike the assumption used in previous studies where the particles were expected to equally sample all the sterically allowed positions based on the Boltzmann expression.

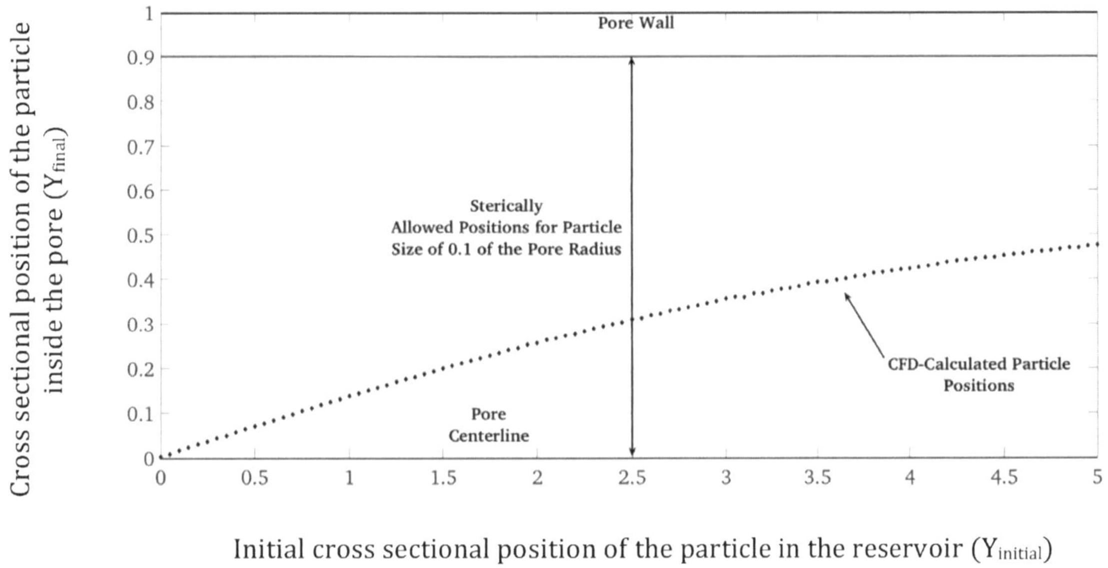

Figure 2: Cross sectional position of a particle in a pore as a function of its initial position in the Reservoir. $Y_{final}=0$* represents the centerline of the pore. The radius of the pore was assumed to be at $Y_{final}=1$*.

*** A non-dimensional scale was used for any position on the graph based on pore radius.**

The effective entrance length is defined as the distance from the pore entrance where the particle accelerates as it reaches its steady velocity. Understanding the impact of different parameters such as flow velocity, geometry of the pore entrance, etc. on this length is of great importance in determining the steady position of a particle inside a pore. The particle cross sectional position is required when evaluating the rejection coefficient of a particle in a certain pore structure.

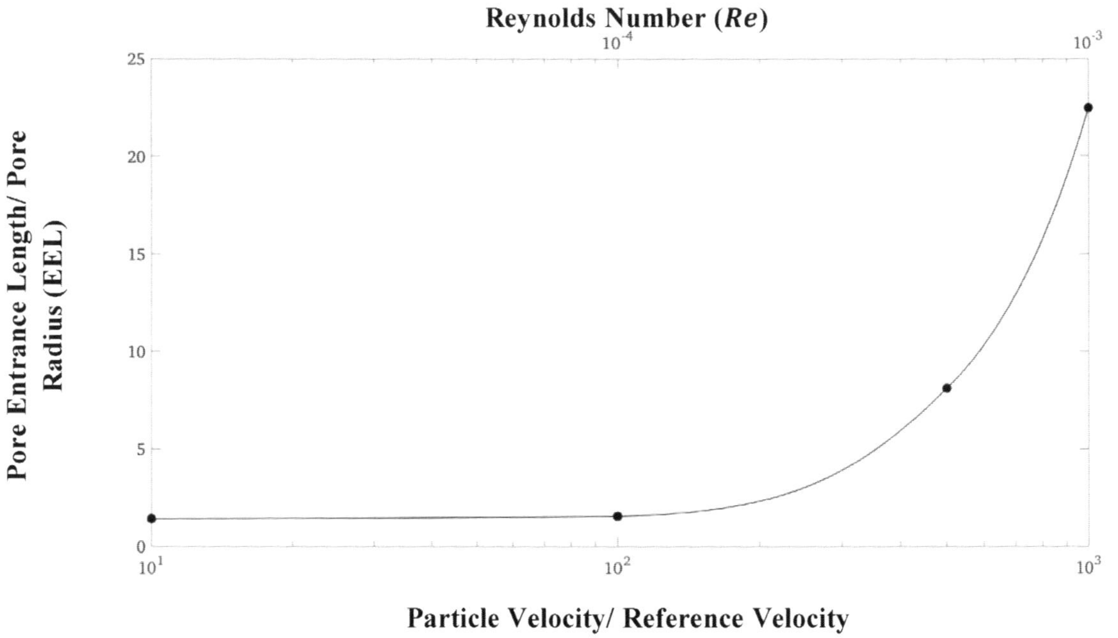

Figure 3: Effect of particle velocity on the dimensionless pore entrance length. These results span from Re = 10^{-6} to 10^{-3}. Calculations were performed for a particle with size 10% of the pore size ($\lambda = 0.1$).

The results also showed that the flow velocity has the strongest effect among all the geometrical and flow parameters on the effective entrance length where it increased exponentially with increasing the velocity (Figure 3). In Addition, Particles may not achieve a steady velocity for pores with length of less than calculated EEL for a specific Reynolds number (e.g. particles in a pore with length of 10× pore radius reach steady velocity only in Reynolds number of less than 5×10^{-4}).

This on-going project will provide a better perspective of the rejection characteristics of complex porous structures compared to previous calculations on the simple geometries (straight cylinder or channel) using the mentioned assumption for position of a particle inside a pore. This work will continue by investigating the effect of bends in pores which can greatly impact the position of the particle.

Nanoparticle Deposition with Controlled Density and Placement.

Hydraulic fracturing utilizes water containing chemical additives and propping agents to increase oil and gas production in low-permeability reservoirs. Recent increase in the use of hydraulic fracturing to develop oil shales has resulted in a rise of water usage in well completion processes. Stimulating a single well can use up to 19,000 cubic meters, or 5 million gallons, of water. Between ten and thirty percent of this water usage is returned to the surface in the form of flowback water. Flowback water is contaminated with hydraulic fracturing additives and naturally occurring organic compounds, making it unsafe for immediate reuse, resulting in fiscal and environmental costs. Flowback water is often trucked to an offsite location for disposal in injection wells. Trucking costs and disposal fees make this process expensive, especially from rural wells. Environmental impacts from vehicular traffic and exposure to contaminated water present other drawbacks to this disposal method. Another method, using ultrafiltration membrane modules, shows promise in treating flowback water for reuse. Recycling flowback water through treatment reduces current handling costs of contaminated water, enabling hydraulic fracturing in acrid areas while protecting the environment through reuse of a vital resource. However, membrane treatment processes have been held back by fouling attributed to dissolved organic compounds and high total dissolved solids (TDS).

In this poster presentation, we focus on addressing membrane fouling resulting from dissolved organics by using catalytic nanoparticles to modify commercial membrane surfaces. Catalytic nanoparticles such as hydrophilic TiO_2 has been utilized in surface modification to convert fouling organic molecules to smaller molecules such as CO_2 and H_2O. Existing deposition methods, such as dip coating, result in poor nanoparticle dispersion and, consequently, particle aggregation, which significantly reduces water flux over time. We demonstrate that this problem can be mitigated by understanding and optimizing nanoparticle deposition using polymeric membranes that have different surface properties and pore sizes. We have studied in-situ deposition in liquid phase via covalent bonding and a well-controlled, self-limiting deposition approach in vapor phase. Both methods improved nanoparticle dispersion and performance of the resulting membranes. The effect of pore size using microfiltration and ultrafiltration membranes will be discussed in the poster, in addition to other process parameters such as illumination time when photocatalytic nanoparticles were used. It is worth pointing out that the optimized nanoparticle deposition and treatment platform can be easily extended to treat water with natural organic matters (NOM), in addition to flowback water management.

Alkali Metal Chloride Sorption in Cross-linked Polymer Membranes

Eui-Soung Jang, Donald R. Paul**, Benny D. Freeman*

McKetta Department of Chemical Engineering, The University of Texas at Austin, Austin, TX, 78758, U.S.A.

**freeman@che.utexas.edu, **drp@che.utexas.edu*

Various alkali metal chloride sorption isotherms in cross-linked polymer membranes were experimentally determined to investigate the effect of fixed charge groups on ion sorption in polymers. Cross-linked poly(ethylene glycol)diacrylate hydrogels (XLPEGDA) were chosen as control uncharged polymers and negatively charged cross-linked poly(2-acrylamido-2-methylpropane sulfonic acid-r-diethylene glycol dimethacrylate) (XL(AMPS-DEGDMA)) was chosen as a charged counterpart. Three alkali metal chlorides (LiCl, NaCl, and KCl) were chosen to study the significance of ion size and hydration state on ion sorption in hydrated polymers.

The equilibrium membrane ion concentrations were measured using a desorption technique in the range of 0.01-1.0 mol L^{-1} external aqueous alkali metal chloride solutions. In the uncharged polymer, the concentrations of cations and anions in the membrane were equal to one another due to electro-neutrality. The order of salt sorption coefficients in XLPEGDA was: KCl > LiCl > NaCl, which does not correspond with the order of ion hydration size, i.e., Li^+ > Na^+ > K^+. These results suggest that ion sorption in uncharged XLPEGDA is influenced by both ion hydration and polymer-ion specific interactions.

The charged polymer sorbed a significantly greater amount of cations (i.e., counter-ions) than anions (i.e., co-ions or mobile ions) due to the presence of fixed charge group attached to the polymer backbone (i.e., sulfonate groups). The order of mobile salt sorption coefficients in XL(AMPS-DEGDMA) was: KCl > NaCl > LiCl, in contrast to the order observed in uncharged polymers. These results suggest that ion sorption in XL(AMPS-DEGDMA) is influenced mainly by the hydration tendency of counter-ions. Furthermore, based on the ion sorption measurements, electrolyte activity coefficients in the polymers were calculated to quantify the thermodynamic non-ideality of the ion-polymer-water mixing system. All of these findings assist in characterizing and understanding ion sorption properties of polymers as part of a long-term effort to improve fundamental understanding regarding ion uptake and ion transport in both charged and uncharged polymers.

Antifouling Modification of Ultrafiltration Membranes by Grafting Polyzwitterionic Brushes

Douglas Davenport, Jongho Lee, and Menachem Elimelech

Department of Chemical and Environmental Engineering, Yale University, USA

Fouling control is critical in membrane-based water treatment processes due to its detrimental effects on system performance and overall energy efficiency. In this work, we develop an antifouling modification for ultrafiltration (UF) membranes by grafting polyzwitterionic brushes via surface-initiated atom transfer radical polymerization (ATRP). Bio-inspired polydopamine (PDA) was first used to immobilize 2-bromoisobutyryl bromide (BiBB) on the membrane surface, which initiated polymerization of zwitterionic sulfobetaine methacrylate (SBMA). Various characterization techniques, including electron microscopy, contact angle measurement, and FTIR spectra proved the successful modification of the membranes with zwitterionic polymer brushes.

Dead-end filtration experiments for the modified membranes showed the effects of various modification conditions, including duration of PDA deposition and thickness of the polymer brush, on water permeability and molecular weight cutoff characteristics. An optimal duration of PDA deposition was found at which a dense polymer brush was obtained with minimal impact on the water permeability. Antifouling properties were assessed by a bovine serum albumin (BSA) adsorption test. The modified membrane exhibited a significantly lower adsorption of BSA than the pristine membrane, indicating a high fouling resistance against proteins. This simple antifouling modification technique has the potential to significantly mitigate membrane fouling, without compromising transport properties.

Mechanical enhancement of thin-film composite membranes using helically-coiled carbon nanotubes for engineered osmosis

Jaime Idarraga-Mora[a], Jinxiang Zhou[a], Apparao Rao[b], Scott Husson[a]

[a] Department of Chemical & Biomolecular Engineering, Clemson University, Clemson, SC 29634

[b] Department of Physics and Astronomy, Clemson University, Clemson, SC, 29634

The objective of this work is to enhance the mechanical robustness of thin-film composite (TFC) membranes and thereby improve their performance. Reverse osmosis (RO) and osmotically driven membrane processes (ODMP), namely forward osmosis and pressure-retarded osmosis often require a high pressure gradient across the membrane. Therefore, mechanical robustness is a required characteristic for a RO and ODMP membranes. Unfortunately, mechanical robustness commonly is inversely proportional to the water permeability of the membrane, which is a critical performance indicator of the membrane.

Previous works have used nano-engineered materials to cast membranes with improved properties, including mechanical strength. This contribution shows the effect on the mechanical strength of TFC membranes by the addition of never-before-tested helically-coiled carbon nanotubes (HCNT) in the membrane porous support. These membranes were fabricated by casting a support via phase inversion of a Matrimid 5218® solution with dispersed HCNT, followed by interfacial polymerization using m-phenylenediamine and trimesoyl chloride. Characterization and testing included compressive stress measurements using an INSTRON machine to evaluate the effect of HCNT content on the membrane mechanical strength. Water permeation and salt rejection tests using a direct flow cell in RO mode were done to compare the performance between the HCNT-TFC membranes and standard TFC membranes. Additionally, cross-sectional SEM images were taken before and after testing the membranes to evaluate the reversibility of the support compaction after testing due to the addition of the resistant HCNT. Finally, data also were compared to commercial RO membranes.

Results show a significant improvement on the mechanical strength compared to the standard TFC membranes and commercial RO membranes. Additionally, water permeation and salt rejection values were not affected by the addition of the HCNT relative to standard TFC membranes. The inclusion of HCNTs in the membrane support has potential for reducing support layer thickness, which is expected to yield higher water flux and generally better ODMP performance.

Keywords: thin-film composite membrane, carbon nanotubes, membrane mechanical resistance

Modification of UF Membranes with a Polyampholyte Hydrogel for Fouling Mitigation.

A major challenge in the desalination and wastewater treatment using membrane technology is to develop membranes having low fouling and biofouling propensity. Prior researches have shown that zwitterion polymers, polymers which contain a homologous arrangement of mixed charges, exhibit excellent antifouling properties.

The goal of this research was to study the non-fouling properties of a new polyampholyte (a co-polymer of monomers with anionic and cationic functional groups) hydrogel, using a mixtures of negatively charged vinyl sulfonic acid (VSA) and positively charged [2-(Methacryloyloxy)ethy]ltrimethylammonim chloride solution (METMAC) monomers, with N,N′-Methylenebis (acrylamide) (BIS) as a cross-linker, which was grafted to a polyethersulfone UF membrane by a UV-photoinitiation method. The modified membranes were characterized using: ATR-FTIR for chemical composition; zeta potential for surface charge; and contact angle (captive bubble) for hydrophilicity. The swelling was evaluated by the gravimetric method. The nonfouling properties were estimated based on adsorption measurements using bovine serum albumin (BSA) as a model foulant.

The ATR-FTIR showed that the modification was successful. The contact angel decreased from 70° to ca. 30° following modification demonstrating that the modified membranes are much more hydrophilic. The zeta potential was close to zero (pH 3-9) at a monomer ratio of 3:1 mole VSA:METMAC and generally positive at other monomers ratios, indicating that at a ratio of 3:1 a zwitterion was obtained. This was also supported be the swelling measurement - at a ratio of 1:1 and 1:3 mole VSA:METMAC, the swelling with DDW was very high as expected for charged polyelectrolyte (polyelectrolyte behavior), whereas at a ratio of 3:1 VSA:METMAC the swelling was very low showing an anti-polyelectrolyte behavior as usually found for zwitterion polymers. Furthermore, the results clearly demonstrated that at this ratio the membrane had the lowest protein adsorption compared to the pristine membrane and membranes following modification at other monomers ratios.

The nonfouling proprieties of this polyampholyte will be further studied using other model foulants (e.g., humic acid, extracellular polymeric substances), and later in a short and long term experiments with real wastewater.

One-step synthesis of hybrid zeolite membranes containing catalytic metal nanoclusters

*Seok-Jhin Kim, Shuai Tan, Micaela Taborga, Laura Briones Gil, David S. Sholl, Christopher W. Jones, and Sankar Nair**
School of Chemical & Biomolecular Engineering, Georgia Institute of Technology, Atlanta, GA 30332

Yujun Liu, Jason S. Moore, Ravindra S. Dixit, and John G. Pendergast Jr.
Engineering & Process Sciences, Core R&D, The Dow Chemical Company, Freeport, TX 77541

Karren L. More
Center for Nanophase Materials Sciences, Oak Ridge National Laboratory, Oak Ridge, TN 37831

ABSTRACT

Metal-loaded zeolitic membranes have promising applications as catalytic membranes that function simultaneously as reactors and separation devices. Post-treatment methods such as impregnation and ion exchange of metals need at least two or three steps, which increase the membrane production cost. Recent research has shown that a one-step synthesis method to encapsulate metal nanoparticles in zeolite powders offers a promising alternative to greatly improve the synthesis efficiency. However, to the best of our knowledge, the use of a one-step synthesis of zeolite membranes containing metal clusters has not yet been described in the literature.

We will discuss a one-step synthesis method to incorporate metal nanoparticles in zeolite membranes. The hybrid metal-zeolite membranes were prepared by one-step colloidal hydrothermal synthesis with addition of 3-mercaptopropyl-trimethoxysilane (MPS) and a metal precursor to the initial synthesis mixture. We show that Pt ions/clusters are uniformly distributed along the membrane cross sectional layer. High magnification scanning transmission electron microscopy (STEM) shows that the Pt metal clusters in the hybrid zeolite membrane have a narrow size distribution of 0.5-2.0 nm. In contrast, a pure-silica MFI membrane obtained from MPS-free precursor, shows negligible loading of Pt metal clusters throughout the zeolite membrane layer. We then show that hybrid metal-zeolite membranes can be used as catalytic membrane reactors (CMR) to investigate a high-temperature propane dehydrogenation (PDH) at 600 °C and 1 atm. The results indicate that Pt metal clusters formed within hybrid MFI zeolite membrane provide effective catalysts for high-temperature PDH reaction along with successful H_2 removal through the membrane permeation. The hybrid zeolite MR was further tested for high-temperature PDH reaction at extended operating times, and showed excellent stability in performance.

*Corresponding author: E-mail: sankar.nair@chbe.gatech.edu

Preparation of PVDF UF Membrane using Lithium Chloride and Amphiphilic Block Copolymer *via* Non-solvent Induced Phase Separation

Sang-Hee Park,[1,2] , Jinhan Cho[2], Soon Man Hong[1] and Kyung-Youl Baek[1*]

[1]Materials Architecturing Research Center, Korea Institute of Science and Technology, Seoul, 02792, Korea

[2]Department of Chemical & Biological Engineering, Korea University, Seoul, 02841, Korea

[3] KIST-UNIST Ulsan Center for Convergent Materials, Ulsan 689-798, Korea

A series of amphiphilic block copolymers with different molecular weights and weight fractions were systematically examined as a pore forming agent to control the pore structure of poly(vinylidene fluoride) (PVDF) ultrafiltration membranes. For this, amphiphilic diblock copolymers consisting of poly(methyl methacrylate) (PMMA) and poly(polyethylene glycol methyl ether methacrylate) (PPEGMEMA) (PMMA-*b*-PPEGMEMA) were synthesized by Cu-based atom transfer radical polymerization (ATRP). These amphiphilic block copolymers with lithium chloride were employed as an additive in the polymer dope solution to prepare PVDF UF membranes *via* non-solvent induced phase separation (NIPS). Pore size and pore size distribution of the obtained PVDF membranes were determined by capillary flow porometer (CFP). Structure morphologies and roughness of the obtained PVDF membranes were investigated by scanning electron microscope (SEM) and atomic force microscope (AFM). Small and uniform pores on PVDF membrane were achieved in the combination system of lithium chloride and PMMA-*b*-PPEGMEMA in comparison to those with separately, which showed the highest water flux (140 LMH/bar) with 100 k poly(ethylene oxide) (PEO) rejection (90%).

Keywords: PVDF UF membrane, non-solvent induced phase separation, amphiphilic block copolymer, lithium chloride

Preperation and characterization of novel polyetheretherketone based polymer cation exchange materials

Moon Ki Jeong, Deuk Ju Kim, Sang Yong Nam*
Department of Materials Engineering and Convergence Technology, Engineering Research Institute, Gyeongsang National University,
Jinju 660-701 Korea
*walden@gnu.ac.kr

In this study, polyetheretherketone based novel cation exchange membranes that are used for energy applications are synthesized and characterized. The sulfonated PEEK(sPEEK) is synthesized via nucleophilic aromatic substitution polycondensation between bisphenol A, 4,4-difluorobenzophenone, and sulfonated monomer due to their good mechanical and chemical stabilities, thermal stability and conductivity. The sulfonation degree of the polymer is calculated using NMR spectroscopy. The sulfonated PEEK membranes are dissolved in solvent. The solution is poured into a petri dish at room temperature until finish evaporation of the solvent. The prepared sulfonated PEEK membranes are characterized for potential fuel cell applications through determining their water uptake, proton conductivity and thermal stability. The properties are compared with a commercial Nafion membrane. Furthermore, the effect of functional groups in the polymer chains of the ion exchange membranes and their potential use in future technology are discussed.

Pulsed Low Energy Positron System (PLEPS):
A promising tool to investigate dispersion in mixed matrix membrane

Ahmad Choudhary*[%], [#], Tönjes Koschine*, Werner Egger[#], Peter Van Kessel[%], Yujiro Itami[%], Ivo F.J. Vankelecom*

*Centre for Surface Chemistry and Catalysis (COK), Faculty of Bioscience Engineering, KU Leuven, Kasteelpark Arenberg 23 PO Box 2461, 3001 Leuven, Belgium

[#] Institut für Angewandte Physik und Messtechnik, Universität der Bundeswehr München, Werner-Heisenberg-Weg 39, 85577 Neubiberg, Germany

[%] Tilburg Research Laboratory, Fujifilm Manufacturing Europe B.V., P.O. Box 90156, 5000 LJ Tilburg, The Netherlands

ahmad.choudhary@biw.kuleuven.be, ivo.vankelecom@biw.kuleuven.be

Abstract

The methodology of mixed matrix membranes (MMMs) was introduced years ago to boost the performance of polymeric membranes by passing the so called upper bound of Robeson's plot [1-3]. MMMs are composite in nature, comprising one continuous phase (i.e. polymers) and a discontinuous phase (i.e. fillers) which is dispersed in continuous matrix. Many fillers have been studied such as zeolites, silica, carbon molecular sieves, carbon nanotubes (CNT), nano particles, metal organic frame work (MOFs), etc [4-5]. The dispersion of the fillers plays a vital role in performance and properties of MMMs. An ideal MMM is one in which fillers are homogenously distributed over the polymer matrix with good adhesion. In gas separation MMMs, the defects are very critical because gas can easily pass through even very small voids without any selectivity [6-8]. Mostly in literature, researchers investigate these defects by using x-ray diffraction, scanning electron microscopy and other conventional techniques. But these characterization techniques don't tell anything about size and distribution of defects and free volume in MMMs [4-5].

Positron lifetime spectroscopy (PALS) is a valuable non-destructive technique which gives information about free volume in membranes [9]. To detect, characterize, and depth-profile free volume in thin membrane layers with PALS, monochromatic positron beams of variable energy are necessary. By varying the positron implantation energy, the mean implantation depth of the positrons can be controlled. The pulsed low energy positron system (PLEPS) at the high energy positron source NEPOMUC at the MLZ in Garching is a unique tool for detecting, characterizing and depth profiling of free volume in membranes and thin membrane layers with PALS from the surface down to some microns below the surface with a depth resolution in the nm-range[10-13].

In this work, polyimide based (Matrimid®, 6FDA-DABA) MMMs were fabricated using self-synthesized and functionalized UiO-66 based MOFs particles. These MMMs are characterized by FTIR/ATR, WAXD, and SEM. Effects of MOF functionality and loading on gas separation performance are studied for CO2/CH4 gas mixture. In addition, correlation of MOF dispersion and distribution in the polymer matrix with gas separation behavior is studied by depth profiling with different positron implantation energy (1KeV to 16KeV) using PLEPS. Increase or decrease in the gas selectivities and permeabilities of different MMMs could easily be understood using PLEPS. Results suggested that PLEPS is a promising characterization tool to get insight in free volume distributions starting from the membrane surface to deep inside the membrane matrix.

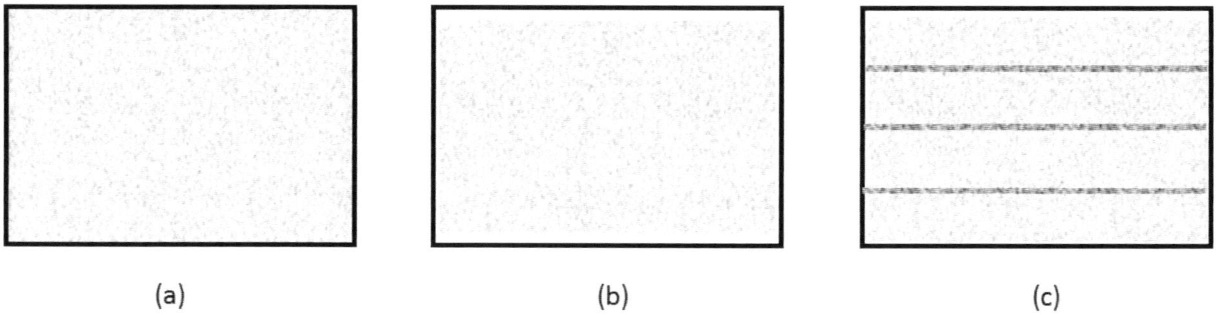

Figure: (a) Ideal dispersion of filler in polymer matrix,
(b) Polymer rich phase on surface and filler rich phase under the surface,
(c) Filler is not homogenously dispersed.

Key words: Mixed Matrix Membranes, Pulsed low energy positron system, Gas separation

References:

[1] L.M. Robeson, J. Membr. Sci. 62 (1991) 165–185.
[2] I.F.J. Vankelecom, E. Merckx, M. Luts, J.B. Uytterhoeven, J. Phys. Chem. 99 (1995) 13187–13192
[3] C.M. Zimmerman, A. Singh, W.J. Koros, J. Membr. Sci. 137 (1997) 145–154
[4] R.D. Noble, J. Membr. Sci. 378 (2011) 393– 397
[5] J. Mater. Chem. A, 2013, 1, 10058-10077
[6] M. Moaddeb, W.J. Koros, Journal of Membrane Science 125 (1997) 143
[7] T. S. Chung, L. Y. Jiang, Y. Li, S. Kulprathipanja, *Prog. Polym. Sci.* **32** (2007) 483–507
[8] M.J.C. Ordoñez, K.J. Balkus, J.P. Ferraris, I.H. Musselman, J. Membr. Sci. 361 (2010) 28–37
[9] G. Consolati, I. Genco, M. Pegoraro, L. Zanderighi, J. Polym. Sci. B Polym. Phys. 34 (1996) 357–367
[10] W. Egger, Positron sources and positron beams. In Physics with many positrons: Proceedings of the International School of Physics "Enrico Fermi" Course CLXXIV, R. Brusa, A. Dupasquier, A. P. Mills Jr, Eds.; IOS Press: Amsterdam (2010) p 419.
[11] J. Algers,P. Sperr, W. Egger, G. Kogel, FHJ. Maurer, Phys. Rev. B 67 (2003) 125404
[12] C. Hugenschmidt, H. Ceeh, T. Gigl, F. Lippert, C. Piochacz, M. Reiner, K. Schreckenbach, S. Vohburger, J. Weber, and S. Zimnik, J. Phys. Conf. Ser. 505 (2014) 012029
[13] T. Koschine, K. Rätzke, F. Faupel, M. M. Khan,T. Emmler,V. Filiz, V. Abetz, L. Ravelli, W. Egger, (2015), J. Polym. Sci. B Polym. Phys. 53 (2015) 213–217

Composite Self-assembled Block Polymer Hollow Fiber Nanofiltration Membranes with Chemically-tailored Pore Walls

Yizhou Zhang, Jessica L. Sargent, Bryan W. Boudouris, William A. Phillip

Block polymer-based membranes fabricated using the self-assembly and non-solvent induced phase separation (SNIPS) methodology show great promise for a wide range of advanced membrane applications. In this study, we demonstrate a technique to produce dual-layer block polymer hollow fiber membranes by combining dip-coating and the SNIPS technique. In addition to exploiting the high density of well-defined pores that result from block polymer self-assembly, the composite membranes have the additional advantage of providing a higher surface area to volume ratio than flat sheet membranes. The influence that different coating parameters (e.g., polymer solution concentration and solvent evaporation time) have on the membrane nano- and microstructure and transport properties when a poly(isoprene-*b*-styrene-*b*-*N,N*-dimethylacrylamide) (PI-PS-PDMA) triblock polymer is coated on a poly(vinylidene fluoride) (PVDF) support are investigated. The performance of the composite hollow fiber membrane as a size-selective filter is compared with that of a flat sheet PI-PS-PDMA membrane fabricated using the same SNIPS condition. The hydraulic permeability evaluation of a PI-PS-PDMA hollow fiber membrane demonstrates that the membrane has permeability of 1.4 L m^{-2} h^{-1} bar^{-1} with an averaged pore diameter of 4.2 nm, while the flat-sheet membrane has permeability of 34 L m^{-2} h^{-1} bar^{-1} and an average pore diameter of 7.6 nm.

The poly(*N,N*-dimethylacrylamide) (PDMA) moiety is readily hydrolyzed using heterogeneous (i.e., after membrane fabrication) reactions into a poly(acrylic acid) (PAA) moiety. The PI-PS-PAA membrane is pH-responsive. Solvent permeability tests demonstrate that the PI-PS-PAA hollow fiber membrane has the pH-dependent permeability ranging from 0.58 to 1.6 L m^{-2} h^{-1} bar^{-1} with the smallest pore diameter achieved at a pH of 5.5 is 2.6 nm. Meanwhile, the PI-PS-PAA flat-sheet membrane has a pH-dependent permeability ranging from 2 to 55 L m^{-2} h^{-1} $bar^{-1}$with the smallest pore diameter of 3 nm at a pH of 5.5. Moreover, the differences between the performance of the PI-PS-PDMA and PI-PS-PAA membranes in the hollow fiber and flat geometries are correlated with the asymmetric microstructures of the two membrane types. Then, a hollow fiber membrane optimization strategy is proposed, in order to provide a more efficient separation process. These composite hollow fiber membranes are promising candidates for fabrication of novel membrane application devices for a variety of separation needs.

Rapid Uranium Isotopic Analysis using Ultrafiltration Membranes and Alpha Spectroscopy

Christine E. Duval[A], Timothy A. DeVol[B], Scott M. Husson[A]
[A]*Department of Chemical and Biomolecular Engineering, Clemson University, SC, USA*
[B]*Department of Environmental Engineering and Earth Sciences, Clemson University, SC USA*

Clandestine activities involving the separation, concentration or manipulation of special nuclear material for the express purpose of developing a weapon of mass destruction is likely to result in the contamination of environmental water sources. The capability to conduct isotopic analyses for waterborne special nuclear material (SNM) would be a powerful nuclear forensics tool. Despite widespread interest, there currently is no fieldable measurement system available for low-level quantification or isotopic determination of waterborne uranium.

This contribution describes our efforts to develop a high throughput analytical technique for waterborne isotopic analysis with the following procedure: (1) deposition of ligand-containing, water-soluble polymer film on an ultrafiltration membrane; (2) separation of the uranium from near-neutral pH water by filtration through the selective film; and (3) isotopic analysis of the uranium-containing polymer film with high resolution alpha spectroscopy. The uranium-selective, water soluble polymer was prepared by the copolymerization of 2-hydroxylethyl methacrylate and 2-ethylene glycol methacrylate phosphate. The polymer was characterized by Fourier-transform infrared spectroscopy and nuclear magnetic resonance spectroscopy to support the incorporation of the ligand into the polymer backbone. Film morphology and thickness were observed by atomic force microscopy and laser scanning profilometry after deposition of the polymer onto the ultrafiltration membrane.

The effect of membrane type, membrane pore size and polymer film thickness on peak resolution in the alpha spectrum was investigated for pure uranium containing solutions at pH = 6. To mimic more realistic conditions, the selectivity of the membrane was tested using natural uranium in simulated groundwater. Uranium-coated membranes prepared from distilled water and simulated groundwater both showed resolutions of 60 keV in the alpha spectrum and detection efficiencies of 12% for uranium-233. The high resolution is attributed in part to the concentration of uranium on the surface of the thin polymer coating. Using a thin coating also was important for maintaining high membrane permeability. The permeability of the polymer coated, polyethersulfone ultrafiltration membranes (MWCO 50 kDa) was determined to be 780 LMH/bar. The results of this research are laying the foundation for the development of a novel rapid, field portable method for isotopic identification of SNM in environmental media.

Thin Film Nano-composite (TFN) Membrane with Metal Organic Framework (MOF) for Antimicrobial Property

Sang-Hee Park,[1,2], Jinhan Cho[2], Soon Man Hong[1,3] and Kyung-Youl Baek[1,3*]

[1]Materials Architecturing Research Center, Korea Institute of Science and Technology, Seoul, 02792, Korea

[2]Department of Chemical & Biological Engineering, Korea University, Seoul, 02841, Korea

[3] KIST-UNIST Ulsan Center for Convergent Materials, Ulsan 689-798, Korea

Novel thin film nano-composite (TFN) membranes were prepared by interfacial polymerization of *m*-phenylenediamine and trimesoyl chloride with iron 1,3,5-benzenetricharboxylate (Fe-BTC) metal organic framework (MOF) on a porous polysulfone (PSF) support. Fe-BTC particles with less than 500 nm with 1300 ~ 1600 m^2/g of surface area were employed to improve both membrane performance and antimicrobial activity. The presence of Fe-BTC in the polyamide layer of TFN membrane was confirmed by the analysis of FT-IR and SEM with EDS. In addition, various surface properties of the Fe-BTC based TFN membranes were determined by surface (zeta) potential, water contact angle, and AFM. Reduced surface energy with the free volume between polyamide and Fe-BTC particles in the obtained TFN membrane improved the water performances. In particular, the TFN membrane with 1 wt.% of Fe-BTC showed 3.0 LMH/bar of water flux and 97.2% of NaCl rejection. In the antimicrobial test, the Fe-BTC based TFN membranes revealed a significant reduction in the number of live bacteria on the surfaces by 93.5% and 65.3% for *E. coli* and *S. aureus*, respectively, indicating its strong antimicrobial properties against both gram-negative (*E. coli*) and gram-positive (*S. aureus*) bacteria.

Keywords: TFN membrane, metal organic framework, interfacial polymerization, antimicrobial property

High fouling resistance thin-film composite membranes with copolyazole substrates for osmotically driven processes

P. H.H. Duong,[1] S. Chisca,[1] P. Y.Hong,[2] H. Cheng,[2,3] S. P. Nunes,[1,*] T. S. Chung[2,4,*]

[1]King Abdullah University of Science and Technology (KAUST), Biological and Environmental Science and Engineering Division (BESE), Thuwal, 23955-6900, Saudi Arabia

[2]King Abdullah University of Science and Technology (KAUST), Water Desalination & Reuse Center (WDRC), Thuwal, 23955-6900, Saudi Arabia

[3]Chinese Academy of Sciences, Chongqing Institute of Green and Intelligent Technology, Chongqing, 401122, China

[4]National University of Singapore, Department of Chemical and Biomolecular Engineering, Singapore, 117585, Singapore

* Corresponding Authors

Abstract:

New synthesized copolyazoles, hydroxyl functionalized polytriazole-co-polyoxadiazoles (PTA–POD), with various hydrophilicity have been synthesized and developed as ultrafiltration substrates by phase inversion for thin-film composite (TFC) forward osmosis (FO) membranes. Besides the suitable morphology for high performance TFC-FO membrane formation, PTA-POD substrates are highly thermal stable, chemical resistant and good anti-organic/biological fouling. The effects of PTA-POD substrate hydrophilicity on the TFC layer formation and TFC-FO membrane performance have been investigated by varying the roles PTA/POD ratios in the membrane substrates. Organic and biological fouling resistances of the PTA-POD TFC membrane have been compared to a polysulfone (PSU)-TFC membrane under the similar fabricated conditions. The results show that the TFC-FO membrane fabricated from the PTA-POD substrates surpasses the PSU-TFC membrane in FO performance and anti-fouling properties. The newly investigated copolymers may be potentially used as a new class of material for FO membrane substrates as well as for other applications.

Keywords: polytriazole, membrane, water treatment, forward osmosis.

~~Optimization of~~Optimized ~~UF~~ Ultrafiltration Backwash with Real-Time ~~Optimization~~ Control of Coagulant Dosing

Over the past two decades, reverse osmosis (RO) has emerged as the leading technology for seawater and brackish water desalination. However, membrane fouling remains a major challenge for robust RO operation. Hence, effective RO feed pre-treatment is vital to long-term operation of an RO desalination system. Among ~~st~~ the existing RO pre-treatment technologies, ~~U~~ultrafiltration (UF) membrane technology ~~filtration~~ has proven ~~itself~~ to be effective in ~~a~~ consistently producing high quality filtrate for RO feed. One of the major advantages of UF membrane pre-filtration is the ability to reverse flow through an UF module and remove the foulant layer in a process known as UF backwash. After a physical UF backwash, irreversible fouling (i.e., the foulant portion ~~of the foulant layer~~ which is not removed and remains on the membrane surface or pores) ~~surface is irreversible fouling.~~ must eventually be removed ~~Irreversible fouling is typically removed~~ through chemical cleaning. This latter step adds to the cost of water treatment ~~washing which is economically inefficient~~ due to chemical consumption and possible system stoppage. In order to make UF operation more robust (i.e., lengthen operation before chemical cleaning is needed), the efficiency of UF backwash must be improved. In this regard, t~~T~~he use of coagulation for pre-treatment has ~~proven to be effective~~improved both ~~at improving UF~~ efficiency of UF foulant removal and the quality of UF filtrate ~~quality and improving UF fouling removal~~. Studies have shown that UF backwash effectiveness improves with increasing ~~as~~ coagulant dose ~~increases~~ up to a certain threshold~~, beyond which UF backwash effectiveness no longer improves as coagulant dose increases~~. Excessive coagulant dosing can lead to fouling of the RO membrane (i.e., due to coagulant passage through the RO membrane), suboptimal UF performance and increased the cost of UF treatment. Therefore, it is imperative to utilize the optimal coagulant dose which in turn is dependent on feed water quality that is often temporally variable. Accordingly, in the present work the optimal coagulant dose to use for UF operation is this threshold, where maximum UF backwash effectiveness is achieved while coagulant is being used efficiently. This dose may be difficult to calculate since feed water quality is known to fluctuate both short term as well as seasonally, and having a static coagulant dose can lead to suboptimal UF operation. Therefore, a method which is able a new approach was developed for to adjust the real time optimization of coagulant dose, along with self-adaptive (self-triggering) of UF backwash, making use of the in real time is vital for robust, long term UF operation. A method is proposed where the change in the initial UF filtration resistance is used as a measure of UF backwash effectiveness. A novel controller was is designed where the value of change in initial UF filtration resistance is used to determine whether the coagulant is being underdosed or overdosed, and coagulant dose is changed accordingly. This The controller was implemented on a pilot seawater desalination plant (capable of 18,000 GPD production of fresh water) and deployed in a seawater desalination field study. The controller performance was evaluated with respect to its performance during a storm event (i.e., worsening feed water quality) and its ability to reduce coagulant consumption. Results of the study demonstrated that significant coagulant savings can be achieved through real-time coagulant dose optimization while allowing the UF system to cope even with the most severe episodes of poor water feed quality. UF product water was consistently of high quality and enabled fouling free operation of the RO system over a period of several years without the need for chemical cleaning of the RO elements.

Removal of organic micro-pollutants (phenol, aniline and nitrobenzene) via forward osmosis (FO) process: Evaluation of FO as an alternative method to reverse osmosis (RO)

Yue Cui[1,2], Xiang-Yang Liu[1,3], Tai-Shung Chung[2,4]
Martin Weber[5], Claudia Staudt[5], and Christian Maletzko[6]

[1] Department of Chemistry
[2]Department of Chemical and Biomolecular Engineering,
[3]Department of Physics
National University of Singapore, Singapore 117542
[4]Water Desalination & Reuse (WDR) Center, King Abdullah University of Science and Technology, 23955-6900 Saudi Arabia
[5]Advanced Materials & Systems Research, BASF SE, GM-B001, 67056 Ludwigshafen, Germany
[6]Performance Materials, BASF SE, G-PM/PU-F206, 67056 Ludwigshafen, Germany
*Corresponding author: Prof. Tai-Shung Chung
E-mail address: chencts@nus.edu.sg
Tel.: +65 65166645
Fax: +65 67791936

ABSTRACT
In this study, we have explored and compared the effectiveness of using (1) lab-fabricated forward osmosis (FO) membranes under both FO and reverse osmosis (RO) modes and (2) commercially available RO membranes under the RO mode for the removal of organic micropollutants. The lab-fabricated FO membranes are thin film composite (TFC) membranes consisting of a polyamide layer and a porous substrate cast from three different materials; namely, Matrimid, polyethersulfone (PESU) and sulfonated polyphenylene sulfone (sPPSU). The results show that the FO mode is superior to the RO mode in the removal of phenol, aniline and nitrobenzene from wastewater. The rejections of all three TFC membranes to all the three organic micro-pollutants under the FO processes are higher than 72% and can be even higher than 90% for aniline when a 1000 ppm aromatic aqueous solution and 1 M NaCl are employed as feeds. These performances outperform the results obtained from themselves and commercially available RO membranes under the RO mode. In addition, the rejection can be maintained even when treating a more concentrated feed solution (2000 ppm). The removal performance can be further enhanced by using a more concentrated draw solution (2 M). The water flux is almost doubled, and the rejection increment can reach up to 17%. Moreover, it was observed that annealing as a post-treatment would help compact the membrane selective layer and further enhance the separating efficiency. The obtained organic micro-pollutant rejections and water fluxes under various feasible operating conditions indicate that the FO process has potential to be a viable treatment for wastewater containing organic micropollutants.

Key words: removal of organic micro-pollutants, forward osmosis, wastewater treatment

North American Membrane Society (NAMS)
University of Arkansas
Ralph E. Martin Department of Chemical Engineering
3202 Bell Engineering Center
Fayetteville, AR 72701-1201
USA

ISBN 978-1-5108-5115-3

9 781510 851153

CURRAN ASSOCIATES INC.
proceedings.com